NEW PERMAFROST AND GLACIER RESEARCH

NEW PERMAFROST AND GLACIER RESEARCH

MAX I. KRUGGER
AND
HARRY P. STERN
EDITORS

Nova Science Publishers, Inc.
New York

For permission to use material from this book please contact us:
Telephone 631-231-7269; Fax 631-231-8175
Web Site: http://www.novapublishers.com

LIBRARY OF CONGRESS CATALOGING-IN-PUBLICATION DATA

Krugger, Max I.
New permafrost and glacier research / Max I. Krugger and Harry P. Stern.
p. cm.
Includes index.
ISBN 978-1-60692-616-1 (hardcover)
1. Permafrost. 2. Glaciers. I. Stern, Harry P. II. Title.
GB641.K78 2009
551.3'84--dc22

2008054767

Published by Nova Science Publishers, Inc. ✦ New York

CONTENTS

PREFACE

In geology, permafrost or permafrost soil is soil at or below the freezing point of water (0 °C or 32 °F) for two or more years. Ice is not always present, as may be in the case of nonporous bedrock, but it frequently occurs and it may be in amounts exceeding the potential hydraulic saturation of the ground material. Most permafrost is located in high latitudes (i.e. land in close proximity to the North and South poles), but alpine permafrost may exist at high altitudes in much lower latitudes. The extent of permafrost can vary as the climate changes. Today, approximately 20% of the Earth's land mass is covered by permafrost (including discontinuous permafrost) or glacial ice. A glacier is a large, slow-moving mass of ice, formed from compacted layers of snow that slowly deforms and flows in response to gravity and high pressure. The word glacier comes from French via the Vulgar Latin glacia, and ultimately from Latin glacies meaning ice.

Glacier ice is the largest reservoir of fresh water on Earth, and second only to oceans as the largest reservoir of total water. Glaciers cover vast areas of polar regions, are found in mountain ranges of every continent, and are restricted to the highest mountains in the tropics. The processes and landforms caused by glaciers and related to them are referred to as glacial. The process of glacier growth and establishment is called glaciation. Glaciers are sensitive monitors of climate conditions and are crucial to both world water resources and sea level variation.

This new book presents the latest research on both permafrost and glaciers.

Chapter 1 - Designing infrastructure and assessing hazard for risks mapping in mountainous environments is a challenging task for every engineer and geoscientist. Steep and sometimes unstable terrain, heterogeneous geological settings, harsh climatic conditions with strong winds, rain, snow (including drifts and significant snow loads) and large temperature variations between summer and winter play key roles in the design process. Not only is a structure directly influenced by these factors in terms of foundation conditions, for example, but also indirectly by rock fall, debris flows or snow avalanches. The difficulties related to foundations in permafrost are largely controlled by the fact that the ground is frozen and may contain ice in various forms, such as ice rich layers, pore ice in coarse soils, ice lenses in fine soils or ice-filled joints in fractured rocks. The ground ice is the main problem affecting mountain infrastructure due to its susceptibility to creep, accrete and melt, hence changing the soil structure. In addition, the top layer thaws during the summer months further changing the strength and deformation characteristics of the ground. Climate change adds even more uncertainties to the foundation and load conditions of any mountainous infrastructure in the long term and needs to be addressed early in the design process. The ground is therefore in a transient state that has to be considered and characterised adequately.

Unique geotechnical characteristics and important features of permafrost soils and rocks, which focus on mountain permafrost, are highlighted related to the design and the construction of mountain infrastructure. The main objective of this chapter is to help with the design process, prolong the service life of structures and to lower the risks and damage potential when dealing with infrastructure located in, or in the proximity of, mountain permafrost environments.

Chapter 2 - This chapter briefly reviews current state-of-the-art in modeling permafrost in numerical weather prediction models (NWPMs), chemistry transport models (CTMs) and in general circulation models (GCMs) and earth system models (ESMs) for projecting the global climate. Pros and cons of various methods are assessed. Deficits of GCM/ESMs permafrost modeling practice are discussed based on gridded observed soil-temperature data; deficits of the treatment of permafrost in NWPMs and CTMs are elucidated by examples of site-by-site evaluations. In addition, the uncertainty in simulated soil moisture and heat fluxes due to uncertainty in soil physical and plant-physiological parameters is illustrated. The consequences of incorrect simulation of or even neglecting of permafrost processes for simulated weather and climate are discussed. Extreme changes in permafrost distribution and active layer depths, as they are associated with wildfires/fires and their impact on the simulated atmospheric conditions, are addressed as well. Finally the great challenges for improving permafrost simulations (grid resolution, lack of horizontally and vertically high resolved soil data, uncertainty in soil parameters, organic soils) in GCMs, ESMs, CTMs and NWPMs and how to address these challenges is outlined.

Chapter 3 - Bacterial strains resistant to beta-lactams, aminoglycosides, tetracycline, chloramphenicol, sulphathiazole and trimethoprim were isolated from more than 60 samples of Arctic and Antarctic permafrost subsoil sediments dated from 5 thousand to 3 million years of age. About 30% of the isolated strains were cross-resistant to two and more antibiotics of different classes. The diversity of multidrug-resistant ancient bacteria, the genetic structure of resistance determinants and their association with different mobile elements were studied.

Principal attention was given to multidrug-resistant strains of Gram-negative bacteria belonging to genera *Acinetobacte*r, *Pseudomonas*, *Psychrobacter, Stenotrophomonas*, and *Xanthomonas*. It was shown that multidrug resistance of some strains of *Acinetobacter* sp. can be transferred by transformation of chromosomal genes and most probably results from expression of efflux pumps. The authors also revealed that many of the strains contained antibiotic resistance genes closely related to those of modern bacteria. In particular, among different strains resistant to streptomycin, they identified strains with *strA-strB* genes, strains with *aadA* genes and strains containing both types of genes. Genes closely related to *tetR-tet(H)* genes were detected in a strain of *Psychrobacter psyhrophilus* resistant to tetracycline and streptomycin. Finally, the authors demonstrated that many of the resistance determinants are associated with mobile elements such as plasmids and transposons.

The results of the study strengthen the hypothesis that antibiotic resistance genes were present in natural bacterial populations long before the 'antibiotic era'. Also, the association of the resistance determinants with different mobile elements confirm an important role of horizontal transfer in distribution of these genes among environmental bacteria.

Chapter 4 - Pleistocene periglacial features were studied in different areas of Hungary in comparison with Canadian recent cryogenic features. Pleistocene periglacial activity forms an important component of the landscape of the Carpathian Basin. There is a general consensus about the study and interpretation of cryogenic deformation structures (e.g., frost fissures, cryoturbations and involutions) being helpful in paleoenvironmental reconstructions. During the glacial periods of the Pleistocene, the Carpathian Basin was ice-free and subject to a cryogenic environment that produced various periglacial features. The reason for the cold climate during these glacial periods is the Basin's unique geographic setting. The Carpathians, which surrounds this large basin, creates an almost closed climatic situation, producing climatic conditions not found elsewhere in Europe. In effect, the climate in Hungary during the glacial periods of the Pleistocene was somewhat similar to the recent climate of the dry tundra regions of North Siberia. But according to other researchers, the Carpathian Basin was mostly devoid of permafrost during the Quaternary. Previous researches described so many relict forms in different geomorphic positions, but none of these have been revised according to the most recent research methods and permafrost nomenclature. Our review addresses past periglacial processes and their coupling to paleoclimate. Permafrost distribution in Canada shows a latitudinal zonation. The main annual temperature associated with these permafrost zones are 0° to −5.5°C (SPZ), −5.5° to −8.3°C (WPZ) and −8.3° to −17°C (CPZ). Sand wedges, frost cracks and ice wedges all develop in permafrost environments and are currently actively forming in the Continuous Permafrost Zone in Canada. The well-developed sand-wedge polygons, frost cracks, ice wedge casts and cryoturbated soils found in Hungary suggest that the Carpathian Basin was a permafrost-affected area during a cold, glacial part of the Weichselian (Late Pleniglacial, 22,000–18,000 years BP). At that time this area was most likely underlain by continuous permafrost.

Chapter 5 - This paper reviews our current knowledge on periglacial landforms and processes on Disko Island, central West Greenland. Disko Island is located on the southern limit of the continuous permafrost zone, and permafrost and periglacial processes are therefore sensitive to future climate warming. The landscape of Disko Island contains a wide variety of periglacial landforms such as rock glaciers, pingos, palsas, patterned ground and active layer phenomena. Also, weathering processes and fluvial and coastal landforms and processes related to cold-climate environments are widespread.

Until now, most research has concerned the incidence, morphology and palaeoclimatic significance of rock glaciers. Although this research has improved our understanding of the late-Holocene history, there are limited studies on other process-landform interrelations, making holistic geomorphological reconstructions of the landscape evolution difficult.

Disko Island contains 247 glaciers larger than 1 km^2, and 75 of these are classified as surge-type glaciers. The recession after surge events leaves proglacial areas prone to formation of periglacial landforms, providing good conditions for field studies on landform evolution on decadal and centennial timescales.

Chapter 6 - Recent studies have described mountain permafrost degradation due to global warming in many mountain regions, such as European mountains, the Tibetan Plateau, the Tien Shan and mountainous areas of Mongolia. In this chapter, the authors describe the recent mountain permafrost degradation in the Nepal Himalayas and the Russia Altai Mountains. The Nepal Himalayas is one of the largest mountainous areas of the world. In 1973, the permafrost lower limit was estimated to be 5200–5300 m above sea level (ASL) on southern-aspect slopes in the Khumbu Himal, the eastern part of the Nepal Himalayas. Using ground-temperature measurements, the mountain permafrost lower limit on slopes with the same aspect was estimated in 2004. The results indicate that the permafrost lower limit was

5400–5500 m ASL in 2004. The permafrost lower limit was estimated to be 5400 to 5500 m on slopes with a southern aspect in the Khumbu Himal in 1991 using seismic reflection soundings. Thus, it is possible that the permafrost lower limit has risen 100–300 m between 1973 and 1991, followed by a stable limit of 5400 to 5500 m over the last decade. The Russia Altai Mountains is located on the southern fringe of the Siberia Plain. The altitudinal range of sporadic/patchy permafrost zone and that of discontinuous/continuous permafrost zone are 1800–2000 m ASL and above 2000 m ASL, respectively. The mean annual air temperature at Russian meteorological stations in Russian Altai exhibited remarkable warming trends. The authors observed the phenomena relating to permafrost degradation, such as landslide influenced by antecedent permafrost degradation, and rapid degradation of pingos around the lower limit of discontinuous/continuous permafrost zone.

Chapter 7 - Since 1973 one drillhole (5-14 warm times / 6-15 glacials) and 39 expeditions in the Himalayas, Tibet, the Karakorum, the Kuen Lun, the Tienshan, the Sayan Mountains, the Altai and other parts of Central Asia contributed to detailed knowlegde about extension (2.4 Mio km^2) and thickness (800–1000 m) of High Asian inland-ice. The data are best for the Last Glacial Period (Würm, Marine Isotop Stage (MIS) 4-2). The author thinks that during the last 2.75 Ma conditions that are comparable to the LGP ocurred several times in Central Asia. Geometric boundary conditions that resulted from the low latitude caused a substantial albedo–induced impact on the energy budget of the Earth during glacial times. The vast extension of the ice–sheets and the high elevation (~6000 m asl) contributed to this. A substantial albedo–induced cooling of the atmosphere is inferred.

Chapter 8 - Glacial ice was for a long time considered only as an extremely stable, frigid and static environment. However, recent investigations showed that glacial ice and glaciers are much more dynamic at the microscale, as well as at the geomorphological level, than previously assumed. Particularly in polythermal glaciers, characterized by a warm core, quick seismic shifts can occur and cause displacements of cryokarst formations and glacial ice masses. Increased pressure at the base of the glaciers can generate subglacial ice melting. These waters, supplemented with supraglacial melt-waters and groundwater, soak rocks and sediments below the glacier and become enriched with local solutes and suspended sediments. When frozen together with the base of the glacier, they constitute the subglacial environment.

Until recently, subglacial environments were thought to be abiotic. However, recent studies revealed the existence of aerobic heterotrophic bacterial communities able to survive the dynamic processes of thawing and freezing. To our knowledge, until our investigations, there were no reports on the presence of fungi in subglacial ice. Given the known adaptive behaviour of many fungi to low water activity (a_w) and a wide range of temperatures, the authors assumed that various types of ice can represent potential natural habitats for diverse halotolerant fungi. To evaluate this hypothesis, media with lowered a_w and incubations at low and "normal" temperatures were chosen to provide a selective advantage for the recovery of culturable fungi from supra- and subglacial environments of four different polythermal Arctic glaciers (Svalbard, Norway).

The dominant taxons isolated were basidiomycetous and ascomycetous yeasts, melanized yeast-like fungi, mainly represented by the genera *Cladosporium* and *Aureobasidium* and different species of the genus *Penicillium*. The fungal counts detected in the subglacial samples were two orders of magnitude greater when compared with those recovered from supraglacial samples, mainly due to yeasts (with counts reaching 4×10^6 CFU L^{-1}). Five ascomycetous and twenty-two basidiomycetous yeast species were isolated, including three new species. According to species diversity and abundance, the majority of species were assigned to the hymenomycetous yeasts (*Filobasidium/Cryptococcus albidus* taxa of the Tremellales). The stable core of the subglacial yeast communities were represented by *Cr. liquefaciens*, *Rhodotorula mucilaginosa*, *Debaryomyces hansenii* and *Pichia guillermondii*.

Among the isolated filamentous fungi the prevailing genus was *Penicillium*, with twenty-four different species being identified and a new species, *P. svalbardense* being described. The dominant species was *P. crustosum*, representing on the average half of all isolated strains from the studied glaciers. In contrast to yeasts, primarily associated with the clear subglacial ice, the highest counts for penicillia were obtained for debris-rich subglacial ice.

Enriched fungal populations in subglacial environments may represent a significant reservoir of biological activity with the potential to influence glacial melt-water composition, release of nitrogen and carbon in the polar environment and seeding of oceans with microbial life.

Chapter 9 - Many supraglacial lakes have appeared and expanded in the Himalayas since the 1950s by glacial retreat, probably due to global warming after the Little Ice Age. Some of these lakes have produced glacial lake outburst flood (GLOF) events since the 1960s, which have occurred, on average, once every three years somewhere in the Himalayas. The three glacial lakes, Tsho Rolpa (water level 4,580m a.s.l.) and Imja (5,009m a.s.l.) in the Nepal Himalayas, and Lugge (4,539m a.s.l.) in the Bhutan Himalayas, were typical supraglacial lakes in the 1950s to 1960s, but, at present, are moraine-dammed by the subsequent horizontal expansion, enough to touch the side moraine. The lakes were investigated from the hydrological and hydrodynamic viewpoints by field observations from 1995 to 1997, and in 2001 and 2002. In particular, Tsho Rolpa has the highest potentiality of GLOF, since the lake water directly contacts the end moraine, in addition to sufficient water pressure (maximum depth, 131m). Hence, the hydrodynamics of the lake were explored in the pre-monsoon season of 1996 by mooring current meters, temperature loggers and turbidimeters under water, and by obtaining vertical profiles of water turbidity and temperature. As a result, a dynamic model of the lake basin expansion, related to calving at the glacier terminus, was proposed.

A comparison of the thermal structure between Tsho Rolpa, Imja, and Lugge shows a definite difference in the lake hydrodynamics associated with the lake expansion rate. The end-moraine and the dead-ice zone of Imja are 10–25m higher than the lake level. The topographic screening of the end moraine on valley winds, commonly blowing along the elongated lakes, tends to decrease wind velocity near the lake surface, which weakens vertical thermal circulations inducing the ice melt below the lake bottom. The characteristic thermal structure of Imja Lake was probably produced by such a screening effect. A three-dimensional numerical simulation of wind around Imja definitely shows the upwind topographic screening effect. The hydrodynamics of Tsho Rolpa was simulated by creating a three-dimensional lake basin of actual size in the calculation domain. In the simulation, a return current, activated by the inflow–outflow system of the lake, was observed. Spatial distributions of the concentration and size for suspended sediment were also simulated, which was reasonable in pattern to the observation.

Chapter 10 - This study identifies 75 surge-type glaciers on Disko Island (Qeqertarsuaq), central West Greenland, using old maps, aerial photographs, satellite imageries and field observations. The surge-type glaciers comprise about 30% of the glaciers larger than 1 km^2

and 59% of the glacierised area. The duration of the surge cycle is at least 100 years or more, which is relatively long compared to surge clusters in other parts of the world. The surge events are very dramatic and include some of the longest frontal advances ever recorded such as 10.5 km advance of Kuannersuit Glacier in 1995-98, where the glacier advance velocity reached at least 70 m d^{-1}. The surge-type glaciers are responsible for valley floor geomorphology such as complex moraine systems and extensive dead ice areas. This implies that glacier surging has to be considered when glacier recession is related to climatic fluctuations.

PERMAFROST

In: New Permafrost and Glacier Research
Editors: Max I. Krugger and Harry P.Stern

ISBN: 978-1-60692-616-1

Chapter 1

GEOTECHNICAL CONSIDERATIONS AND TECHNICAL SOLUTIONS FOR INFRASTRUCTURE IN MOUNTAIN PERMAFROST

Lukas U. Arenson[1], Marcia Phillips[2] and Sarah M. Springman[3]
[1] BGC Engineering Inc., Vancouver, BC, Canada
[2] WSL Swiss Federal Institute for Snow and Avalanche Research SLF, Davos, Switzerland, [3]
Institute for Geotechnical Engineering, ETH Zurich, Switzerland

ABSTRACT

Designing infrastructure and assessing hazard for risks mapping in mountainous environments is a challenging task for every engineer and geoscientist. Steep and sometimes unstable terrain, heterogeneous geological settings, harsh climatic conditions with strong winds, rain, snow (including drifts and significant snow loads) and large temperature variations between summer and winter play key roles in the design process. Not only is a structure directly influenced by these factors in terms of foundation conditions, for example, but also indirectly by rock fall, debris flows or snow avalanches. The difficulties related to foundations in permafrost are largely controlled by the fact that the ground is frozen and may contain ice in various forms, such as ice rich layers, pore ice in coarse soils, ice lenses in fine soils or ice-filled joints in fractured rocks. The ground ice is the main problem affecting mountain infrastructure due to its susceptibility to creep, accrete and melt, hence changing the soil structure. In addition, the top layer thaws during the summer months further changing the strength and deformation characteristics of the ground. Climate change adds even more uncertainties to the foundation and load conditions of any mountainous infrastructure in the long term and needs to be addressed early in the design process. The ground is therefore in a transient state that has to be considered and characterised adequately.

Unique geotechnical characteristics and important features of permafrost soils and rocks, which focus on mountain permafrost, are highlighted related to the design and the construction of mountain infrastructure. The main objective of this chapter is to help with the design process, prolong the service life of structures and to lower the risks and damage potential when dealing with infrastructure located in, or in the proximity of, mountain permafrost environments.

Introduction

Design, construction and maintenance of civil infrastructure in mountainous permafrost terrain is an engineering challenge. The difficulties encountered are mainly governed by the fact that the ground is frozen and contains ice, top soil layers thaw during the warm summer months, steep and potentially unstable slopes prevail, the ground is covered with snow for long periods of the year and access can be complicated. The tendency of ground ice to creep, accrete, segregate layers and thaw causes most uncertainty when designing infrastructure. Specially adapted site investigation, construction, maintenance and monitoring techniques are required in this environment to ensure the longevity and the sustainability of the infrastructure. In addition to the potential impact caused by the presence of infrastructure, climate change is influencing air temperatures, precipitation regimes, snow cover distributions, active layer thicknesses and lower limits of permafrost occurrence in mountains (Clague, 2008; Haeberli and Beniston, 1998; Harris et al., 2009; Harris et al., 2001a). The rate and magnitude of these changes, which influence the physical, hence mechanical, properties of the ground more or less directly, and consequently the stability and safety of structures, are difficult to predict but must be taken into account in the design of infrastructure (Hayley and Horne, 2008; Instanes, 2005; U.S. Arctic Research Commission Permafrost Task Force, 2003).

The general term 'mountain infrastructure' used in this chapter refers to structures with foundations (e.g. buildings, pylons for power lines or cable cars and defence structures), roads and railways, dams (e.g. hydroelectric, avalanche and rock fall retention), water pipes (e.g. sewage pipes), underground access tunnels, ski runs and technical snow production systems. Mountain infrastructure is either located in/on frozen bedrock (containing ice in pores and fissures) or debris accumulations such as scree slopes, moraines or rock glaciers (all containing varying amounts of ice). In contrast to arctic regions, where infrastructure in permafrost includes entire communities (e.g. Instanes, 2005), there are generally no large permanently inhabited settlements in the permafrost zone in mountain environments. However, densely populated settlements and transportation life lines are located at lower altitudes and can directly and/or indirectly be affected by processes occurring in permafrost terrain, requiring in situ engineering solutions such as retention dams (Keller et al., 2002) and the establishment of hazard maps for improved land-use management (Götz and Raetzo, 2002). Much of the infrastructure located directly on/in mountain permafrost pertains either to tourism, communication or power related industries and is of high economic and social significance, in particular in European mountains.

In areas other than the European Alps, hazard potentials related to mountain permafrost are mainly affecting lifelines, such as pipelines, power lines, railways or roads, hydro power infrastructure or mining activities at high altitudes (e.g., Kääb et al., 2005; Wei et al., 2006).

When designing infrastructure or assessing natural hazards in mountain permafrost environments, geoscientists, engineers and decision makers often do not have guidelines available for risk assessment and management, and to help to identify potential problems. Projects successfully completed in Alpine environments, including special anchoring techniques (Phillips, 2006; Rieder et al., 1980; Stoffel, 1995), cooling systems, building materials and innovative geometrical correction systems (Phillips et al., 2007) and site tailored solutions for Arctic infrastructure (Hayley and Horne, 2008) may help with ideas for new designs. Efficient, site- and project-specific monitoring, detailed observations and thorough site investigations developed in parallel have demonstrated the strengths of such joint engineering solutions (Keusen and Amiguet, 1987; Phillips et al., 2007; Steiner et al., 1996).

In this chapter, typical geotechnical hazards and challenges related to mountain permafrost, major geotechnical and geothermal characteristics are presented, as well as suitable field and laboratory investigations. Monitoring programmes are presented and advantages of the observational method are illustrated when building in such environments. Schematics and flow charts are introduced to provide a pathway through the design process. Finally, some examples of successful structures and adaptation methods are presented. By using the tools presented herein, the reader should be able to identify and manage problems, to assess special situations, and to introduce appropriate mitigation and adaptation strategies in a timely manner.

Mountain Infrastructure and Alpine Permafrost

Typical infrastructure in mountainous environments is listed above. Some structures may have to be tied back with anchors reaching into permafrost. Further hazards to infrastructure, settlements or reservoirs are evoked by slopes located in permafrost that may trigger mass movements during rainfall infiltration or thawing of snow or ice.

Due to the heterogeneity of the mountain environment, with the lower permafrost boundary changing significantly within short distances, structures are often located in zones that would be classified as sporadic permafrost. These are zones where patches of permafrost interchange with zones that are permafrost free. These changes in foundation properties create particularly vulnerable situations for the structures if some part of the foundation is located in permafrost, whereas the other is not.

Shallow or deep foundations are possible in permafrost environments. The design most suitable will depend on the actual situation and the predicted changes over the design life. In addition to geotechnical properties, geothermal properties and any thermal disturbances caused by the structure to be built must be considered during the design. Whenever ice is present in the ground, thermal and/or stress changes affect the ground as a function of time, hence the foundation condition does not remain the same over the lifetime of the structure.

Definition of Permafrost and Other Relevant Terms

According to the *Glossary of Permafrost and Related Ground Ice Terms* (van Everdingen, 1998 revised May 2005), permafrost is defined as:

"Ground (soil or rock and included ice and organic material) that remains at or below 0°C for at least two consecutive years."

Figure 1 shows additional definitions used in permafrost science and engineering. All these definitions are purely thermal and do not indicate whether ice is present or not. Changes in the geothermal gradient with depth further provide information on past temperatures and possible irregularities with depth, such as presence of taliks.

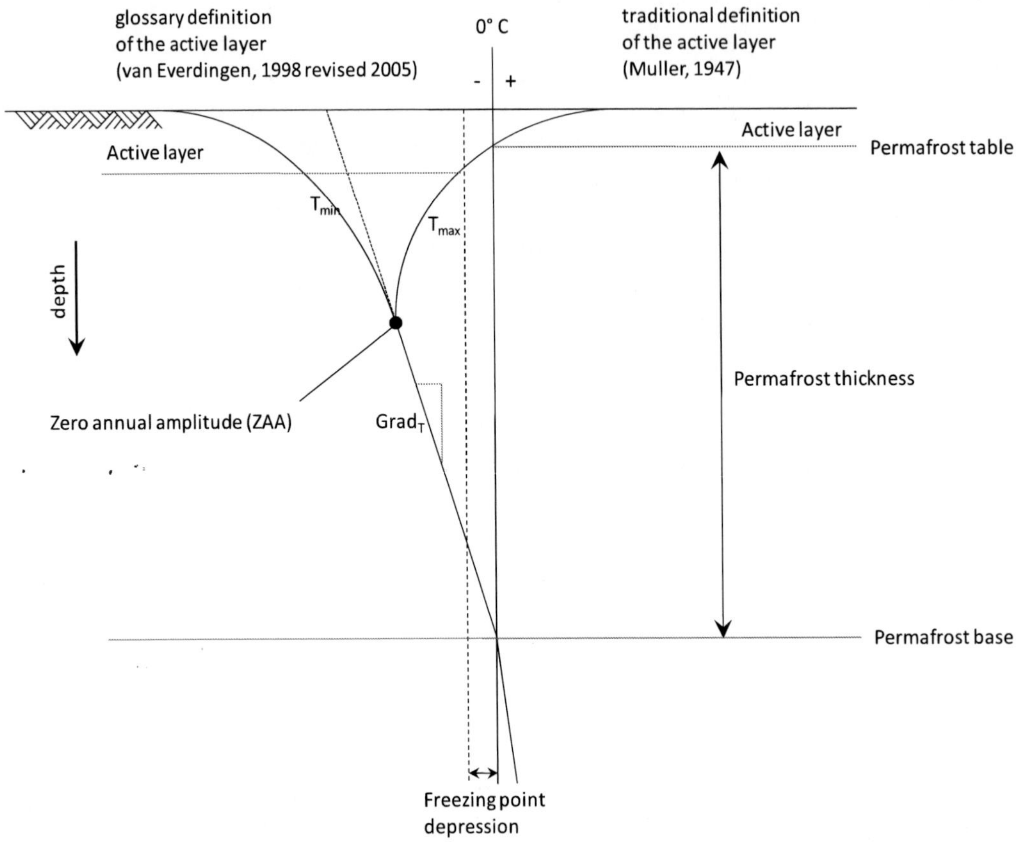

Figure 1. Definition of permafrost after van Everdingen (1998 revised 2005) and Muller (1947).

Frozen ground on the other hand is defined as:

"Soil or rock in which part or all of the pore water has turned into ice."

The second definition contains more information about the condition of a particular soil, but in contrast to the definition of permafrost, frozen ground is only a snapshot in time. Frozen ground can be found in temperate zones during cold months or induced artificially for ground freezing applications (e.g. Harris, 1995; Pimentel et al., 2007).

When designing and assessing permafrost conditions, it is therefore fundamental to assess the spatial ground conditions and relevant characteristics, keeping in mind that the ground is to be used as foundation material and its properties need to be known for geotechnical and geothermal analysis.

Active Layer

The active layer needs special attention because mechanical and thermal processes within this thin surface layer will often control the design of infrastructure. Ground movements or seasonal changes in strength require a good understanding of the thermo-hydro-mechanical response of this layer. According to the permafrost glossary (van Everdingen, 1998 revised 2005), the active layer is defined as (Fig. 1):

"The layer of ground that is subject to annual thawing and freezing in areas underlain by permafrost."

Arnold et al. (2005), for example, studied the stability of the active layer of a rock glacier in Switzerland. The interlocking effect in the coarse, elongated and angular particles was found to cause significant dilatancy at the low stresses pertaining near the ground surface and hence extremely high strength parameters were obtained. However, sliding over a possible massive ice layer at the permafrost table might be of concern because interface friction is reduced by a factor of two. Analysis of field data in respect of the hydrothermal processes developing within the active layer confirmed that this scenario might well occur (Rist and Phillips, 2005). During snow melt, the ground temperatures increase rapidly (0.6°C/day), which is partly caused by convective heat fluxes as melt water percolates to the permafrost table through the voids in the soil matrix. Seepage may induce down-slope displacements within the active layer during snow-melt.

Harris et al. (2008) and Kern-Luetschg et al. (2008) demonstrate the effect of the freezing of the active layer on solifluction. Progressive soil freezing from the surface down and from the permafrost table upwards forms a closed hydraulic system within the central zone of the active layer. Water migration towards the respective freezing fronts reduces moisture contents in the central zone, and pore water suction increases the effective stress and hence the unfrozen shear strength.

Issues / Hazards in Mountain Permafrost With Regard to Climate Change Uncertainties

Climate change is creating major challenges to the design of new and the assessment of existing infrastructure in mountain permafrost. The challenges are mainly related to the uncertainties that are associated with prediction of long-term climate change scenarios. On the one hand, designs must be economic but on the other hand some assumptions have to be made with respect to changing ground conditions. Climate variability not only involves changes in air temperature, but also changes in precipitation, e.g. rain and snow days, vegetation, e.g. evapotranspiration, wind (speed and directions) and solar radiation, i.e. radiant energy. These changes, in turn, affect the surface energy balance, hence the heat transfer into and out of the ground and therefore permafrost conditions. When incorporating long-term climate change effects, it is crucial that not only air temperature changes are considered, but also other climatic parameters that will influence the surface energy balance.

There are several hazards to mountain infrastructure that are related to permafrost and in particular to changes in climate conditions (Fig. 2). A range of typical hazards is listed below as an aide memoire. However, it is not exhaustive and additions may be required for individual projects.

Long-term changes in ice temperature and thawing

- o Major changes in soil properties
- o Thaw consolidation / settlements
- o Reduction in strength
- o Changes in soil deformation characteristics, creep rates
- o Possible changes in failure mechanisms, e.g. blocks becoming unstable from thawed joints

Change in precipitation

- o Changes in snow loads
- o Changes in snow drift
- o Changes in run-off, water pressures
- o Influence on ground temperatures
- o Changes in duration and timing of the winter snow cover

General ground warming

- o Change in active layer thickness and composition
- o Changes in the duration and timing of the active layer
- o Changes in intensity and duration of frost heave
- o Changes in soil strength properties
- o Effect on frost weathering

Slope stability related issues

- o Influence on unstable slopes, reduced slope stability
- o Effect of slope orientation
- o Effect on erosion processes
- o Possibly affected area, change in soil volumes for mass movements
- o Rate of movement
- o Effect on run-out zones

It is important to note that none of the above stated effects will occur alone, but the various combinations that influence each other may form the most critical condition for a specific structure.

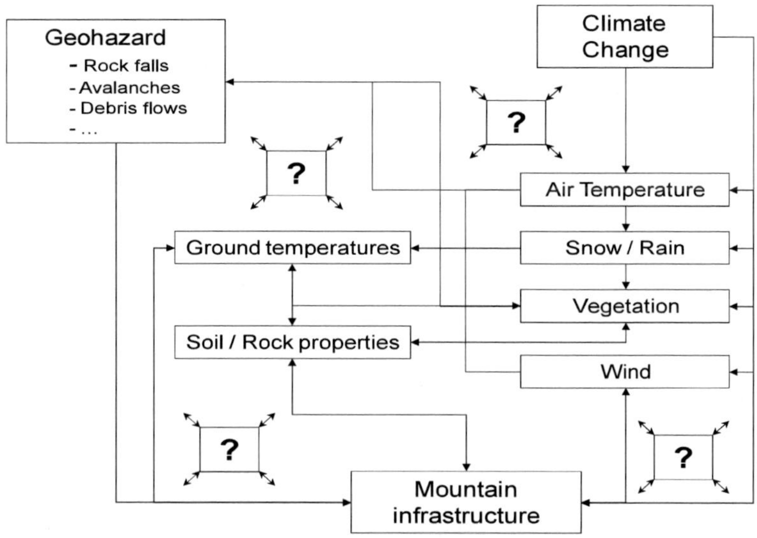

Figure 2. Interactions between climate change factors, geohazards and mountain infrastructure.

Some Aspects of Alpine Permafrost Hydrology

Mountainous permafrost zones can receive water from several sources including direct precipitation, runoff from adjacent slopes, avalanches, snow-melt and groundwater. Water output may occur through surface runoff, subsurface discharge, subsurface seepage, sublimation, vapour flux, evaporation and ground ice melt. Due to this hydrological complexity, water discharge from permafrost areas can not simply be attributed to melt of ice from the frozen ground alone. In fact, ground ice may be hundreds to thousands of years old, forming an important long-term groundwater storage that has no major influence on the yearly water balance.

To date, there have not been any long-term conclusive research programmes that have measured the full hydrological cycle. In addition to the water flow through partially saturated frozen ground, water might also flow within taliks, which are unfrozen zones that influence thermal and stress conditions of the ground. Detailed temperature records at depth help identifying such layers.

Seasonal characteristics of the hydrologic cycle may include:

- *Late winter*: water within the active layer is frozen and any observed discharge originates entirely from groundwater flow systems at the permafrost base or in taliks.
- *Late spring/early summer*: the thawing front penetrates the ground: melting snow and ice recharges the upper portion of the permafrost creating a seasonal aquifer perched on top of the frozen core.
- *Summer*: the majority of the free water within the upper portion of the permafrost has discharged and most of the discharge observed, other than run-off from rainfall, originates from the permafrost base or from taliks. Melt water from snow patches in depressions on the surface continues to run off through the summer.

- *Late summer/early fall*: the freezing front penetrates downward from the ground surface. Any remaining water is accumulated in the central portion of the active layer. Under cold permafrost conditions, two-sided freezing may occur, in which case the active layer also freezes from the permafrost table upwards (Harris et al., 2008).

GEOTECHNICAL PROPERTIES OF ALPINE PERMAFROST

Introduction

Arenson et al. (2007) present an overview of geotechnical properties of frozen soils and the many factors that have an influence on the mechanical response of permafrost soils. As high ice contents with excess ice and high air contents are often found in soils in mountainous environments, this section focuses on soils with these properties. Andersland and Ladanyi (2004) or Esch (2004) present thorough overviews for the remainder of the spectrum of frozen soils. An additional overview on most recent progress in permafrost geotechnics is presented in Springman and Arenson (2008).

Unfrozen Water Content

While water under atmospheric pressure freezes at 0°C, water in soil pores freezes at slightly colder temperatures as it is influenced by the soil skeleton. Mineralogy, particle size and pore water chemistry may influence the freezing point and the unfrozen water content so that at temperatures below zero degrees, liquid, unfrozen water and ice coexist (Anderson and Tice, 1972; Fish, 1985; Williams, 1967a; Williams, 1967b). In a mountainous environment with coarse- grained soils prevailing and generally no pore water salinity, the amount of unfrozen water below -2°C is negligible. To estimate the unfrozen water content, Smith and Tice (1988) and Tice et al. (1976) propose the following relationship.

$$w_u = \alpha \cdot \theta^{\beta},$$

where w_u is defined as weight of water divided by dry weight of soil expressed in percentage, θ is the temperature expressed as a positive number in degrees Celsius below freezing, and α and β are soil parameters. Typical values for gravelly materials found in mountainous environments are $\alpha = 2.1$ and $\beta = -0.408$ (e.g. Smith and Tice, 1988).

General Soil Strength Considerations

The strength of frozen soil is controlled by the interaction of soil particles similar to unfrozen soils and the cementation effect of the ice matrix. However, the strength properties of the matrix are strongly non-linear and temperature- as well as loading- and deformation-rate dependent. Generally, the strength of ice increases as the temperature decreases. At temperature close to the melting conditions, the specific surface of the soil particles affects the phase change. Because the unfrozen water affects the activation energy (e.g. Barnes et al., 1971; Fish, 1985), adopting the Arrhenius approach to account for variations in temperature will be invalid at temperatures close to the melting point and different approaches (e.g. Hivon and Sego, 1995) should be used.

When the ice content varies, a minor increase in strength is noted up to a volumetric ice content of approximately 40% (Arenson and Springman, 2005b; Goughnour and Andersland, 1968). However, research suggests that dispersed soil particles within dirty ice alters the failure mechanism, so that strength is slightly lower than in pure ice (Arenson and Springman, 2005b; Hooke et al., 1972; Yasufuku et al., 2003). Structural hindrance between the solid particles and dilation occurs as the volumetric ice content decreases, increasing the strength and reducing creep deformation significantly. At high relative soil densities, the resulting increase in strength over the same soil in its unfrozen state may be quantified as cohesion at zero stress (e.g. Arenson et al., 2004; Nater et al., 2008). However, at very large strains, ice-bonding fails, destroying the cohesive effect and the strength of the frozen material will be similar to the strength of the equivalent unfrozen soil.

Air voids also have a significant influence on the volumetric strain behaviour. Volumetric strains of more than 10% were recorded in triaxial compression tests at an axial creep strain of 20% for a sample with an initial air content of 25%. Experimental work on glacier ice containing air bubbles showed that the mean number of air bubbles per unit volume correlated inversely with the mean uniaxial compressive strengths (Gagnon and Gammon, 1995).

The loading regime further changes the failure mechanism and consequently the strength of the frozen soil. Fast loading of ice rich material or ice results in brittle failure, whereas low strain rates provoke a ductile response dominated by creep deformations. Arenson and Springman (2005a) present a schematic representation of this behaviour (Fig. 3).

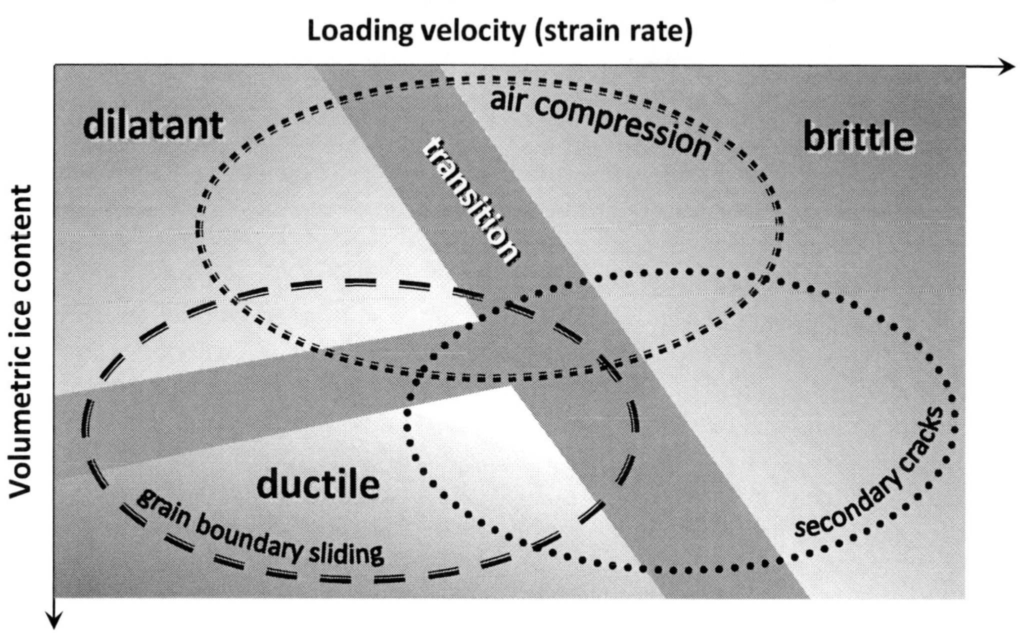

Figure 3. Sample response as a function of the loading rate and the ice content.

Strength parameters for frozen soils vary significantly and specific laboratory investigations are recommended for particular problems. The unfrozen strength of the frozen soil can be used as a lower boundary assuming that the cementing effect of the ice is no longer present and allowing for the effect of (unfrozen) groundwater.

Ground Characteristics

Rock and Soil Foundations

Generally, rock foundations in permafrost are no different to any other rock foundation. Therefore permafrost bedrock and rock walls only present an additional hazard when they are jointed and fissured and these discontinuities are filled with ice. Direct shear tests and centrifuge modelling on ice-filled joints imply that where the direction of dip of the joint planes is appropriate, the stability of a steep, jointed rock slope is maintained by the ice (Davies et al., 2001; Davies et al., 2000). The factor of safety decrease as the temperature increases and reached the melting point of the ice. This hypothesis implies that a jointed rock slope that is stable when there is no ice in the joints, and is also stable when ice in the joints is at low temperatures, may become unstable as the ice warms. The process can be attributed to the reduction in strength and stiffness of the cementing ice as it warms and the existence of water that may reduce the effective stress between the blocks (Günzel, 2008). Gruber and Haeberli (2007) suggest that the enhanced creep susceptibility of warm ice may have caused the increased rockfall activity at high altitude in Europe in the summer of 2003.

Because the existence of ice in rock joints controls the strength of the rock, it is crucial that it is reported in addition to common rock mass classification (e.g. Bieniawski, 1989).

Foundations on frozen soils are challenging because the mechanical properties of the ground varies as temperatures, ice content and loading regimes changes as indicated above. Creep deformations pose particular challenges and need special attention during the design because they can influence the serviceability of an engineered structure.

It is therefore important to properly report the ground ice conditions during soil and rock classification. The heterogeneity of various permafrost soils necessitate an appropriate sample size for the determination of the volumetric ice content that depends on the maximum grain diameter for soils or the non-fissured length in rocks (Fig. 4).

Creep of Frozen Soils and Rock Joints

Creep occurs as soon as excess ice is present in the soil. In a dense soil when only pores are frozen, structural hindrance inhibits any creep deformation. However, for low particle contents the matrix can deform and creep occurs. Even thin ice layers, e.g. from segregated ice or in rock joints, may be enough to trigger creep deformations. It is important that ice lenses are identified to judge the ground's susceptibility for creep movements even in soils with low ice contents and in rocks.

Creep laws and properties of for various frozen soils are presented in Andersland and Ladanyi (2004). Creep parameters for ice-rich soils can also be found in Arenson and Springman (2005a).

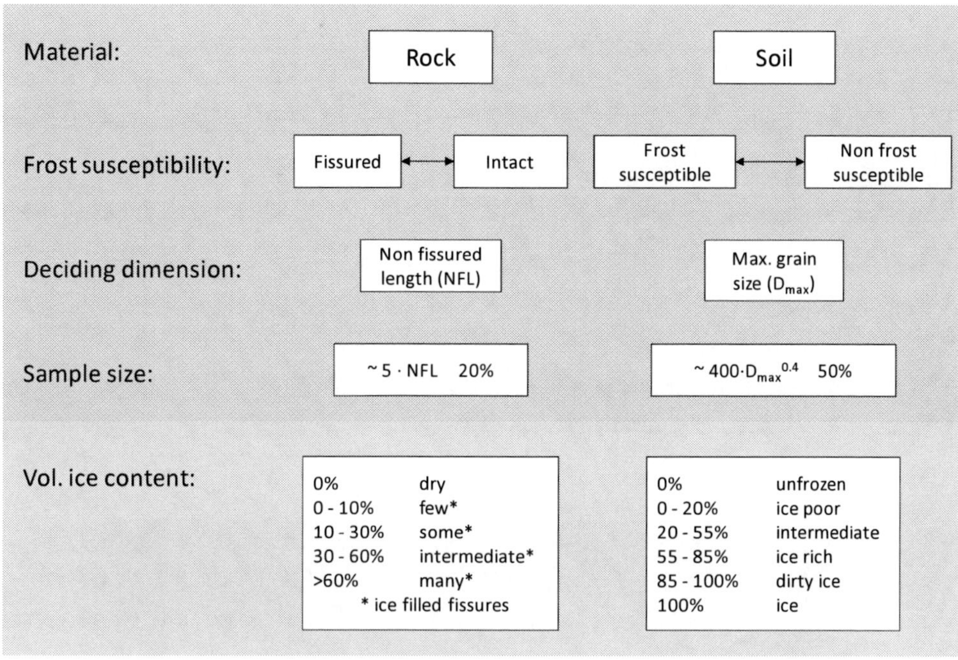

Figure 4. Determination of sample size and volumetric ice contents as a function of grain size diameter and non-fissured length for soils and rocks.

GEOTHERMAL CONSIDERATIONS IN ALPINE PERMAFROST

By understanding the geothermal processes involved during the freezing and thawing of permafrost soils, one can better judge the thermo-mechanical response of the ground to changes in temperature and stress.

Mountainous soil and rock properties and conditions are very diverse. Ground ice with a range of origins can be present within a mountain permafrost environment, including glacier ice, compacted snow, segregated ice, ice derived from adfreezing from rain or melt water during thaw, snow avalanches, etc. Consequently, alpine permafrost can be solid rock with ice-filled joints, fine grained soils with low ice contents, ice supersaturated gravels (excess ice), where not all particles are in contact, or dirty ice with some dispersed solid particles distributed within the ice (Fig. 5). Sampling from triple cored, air cooled drilling has shown that air contents of over 20% by volume can exist (Arenson and Springman, 2005b). In contrast to most glacier ice, the ice in permafrost, rock glaciers included, is much older because there is no distinct accumulation and ablation zone as would be found in mountain glaciers.

The time- and temperature dependent mechanical properties of the ice in the soil are the factors that commonly govern the geotechnical properties of frozen permafrost foundations. Knowledge about the ice (e.g. content, structure, distribution) in the ground is a major component in the design. Solid, dry, unfractured bedrock has similar geotechnical properties in a frozen state (e.g. at -10°C) as in an unfrozen condition (e.g. at +10°C). Saturated sand at +10°C and at -10°C on the other hand, with all the pore water frozen, behaves in a different manner.

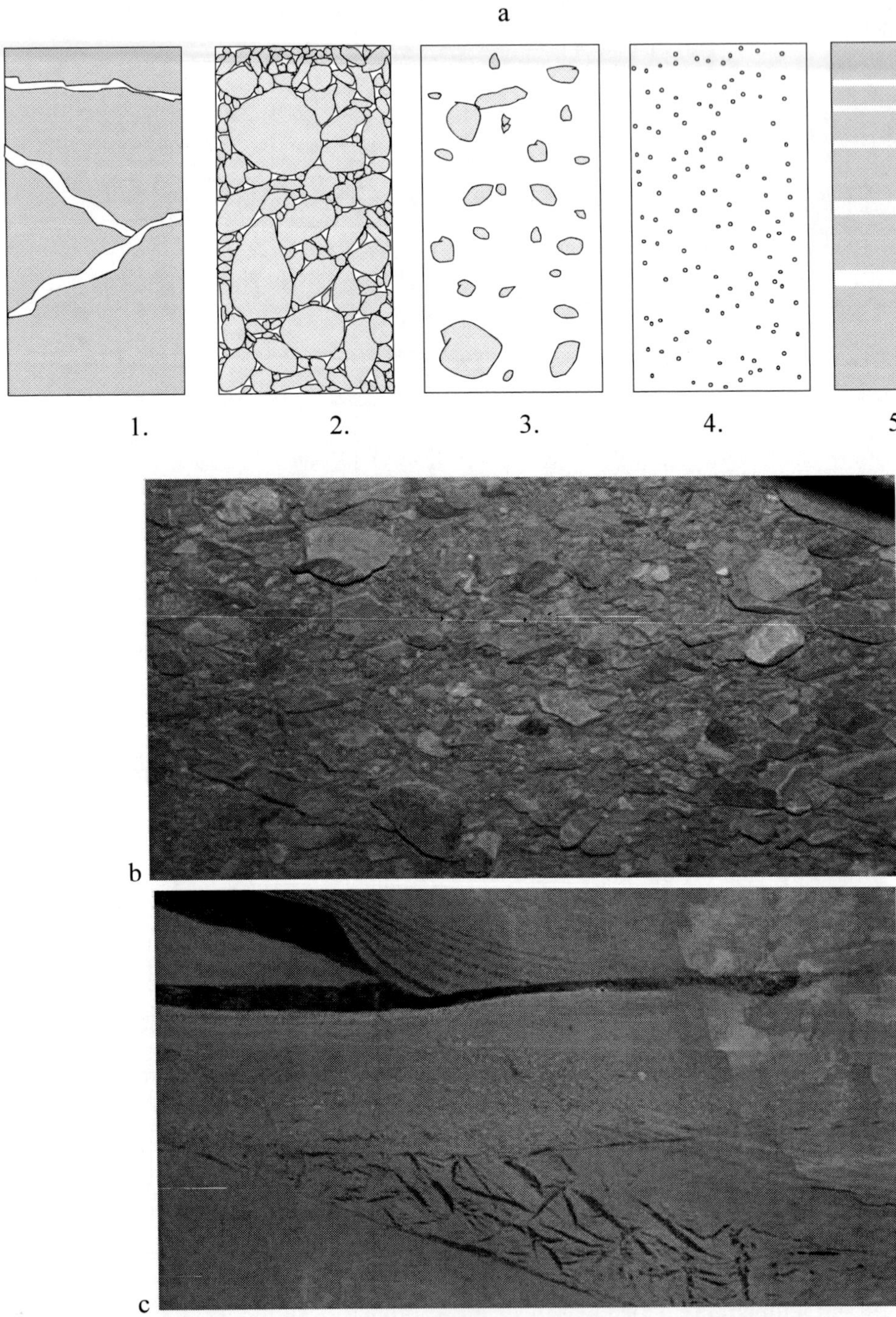

Figure 5. Different types of frozen ground (a) 1. Ice filled joints, 2. Compact frozen gravel, 3. Ice rich gravel, 4. Dirty ice, 5. Ice lenses in fine grained soils. (b) frozen gravel, CRREL permafrost tunnel, Anchorage Alaska (L. Arenson). (c) ice lenses in frozen silt, CRREL permafrost tunnel, Anchorage Alaska (L. Arenson).

The actual ice content is only one factor that influences the mechanical behaviour of a frozen soil. Its structure, i.e. the location, thickness and assembly of the ice lenses or ice-filled pores, is also important. Because permafrost and ice formation may have occurred over hundreds / thousands of years, ancient freeze-thaw cycles at greater depths can result in heterogeneous ice structures with ice rich layers that cannot be identified from the surface. The freezing processes that include the thermal, hydro-geological and geomorphological histories are the driving factors behind this ice lens formation, hence the final structure of a frozen soil.

Three-dimensional effects further influence the geothermal regimes in mountainous environments. In particular, thermal gradients with depth depend upon slope angle, orientation and content as well as on snow, glacier cover or groundwater flow (e.g. Gruber et al., 2004).

DESIGN PROCESS

Proper site investigations, long-term planning, redundancies and the ability for adaptations are crucial for structures located in mountain permafrost environments. The actual degree mainly depends on the project and the sensitivity of the structure (Bommer et al., 2008). In-depth site investigations that include climate data and ground temperatures are critical for the determination of the actual ground thermal conditions as well as predictions for future ground temperature trends.

The design process must start with a thorough screening process. In Canada, the Panel on Research and Development (PERD) developed such a process (Hayley and Horne, 2008; PERD, 1998), that can be adopted for mountain permafrost infrastructure. Figure 6 shows the schematics that result in an assessment for the consequences and the sensitivity of the structure to be built. These two assessments are combined in a Consequence – Sensitivity matrix that forms the basis not only for the required design, but also for the required site investigation. Some examples are given in Figure 7. A temporary road, built mainly on permafrost bedrock, requires less thorough site investigations with specific laboratory tests to determine soil properties than a dam for a hydropower project that is partially founded on frozen ground. The latter should include possibilities for future adaptations of the structures as improved data of future climatic trends and structural response become available, based on a detailed monitoring programme. Structural reassessments should be made mandatory as part of the licensing process in such a case.

Failure Mode and Effects Analyses (FMEAs: Nahir et al., 2005; Robertson and Shaw, 2005) are an additional important pillar during a design process. The result of the analysis is to be incorporated into the screening process as it identifies sensitivities as well as consequences. This should be an unreserved process with various experts involved. The earlier an FMEA is carried out, the better it can be used to guide the project design.

Fischer and Huggel (2008) present a methodical design for stability assessments of permafrost affected rock walls that provides additional information when dealing with structures in rock.

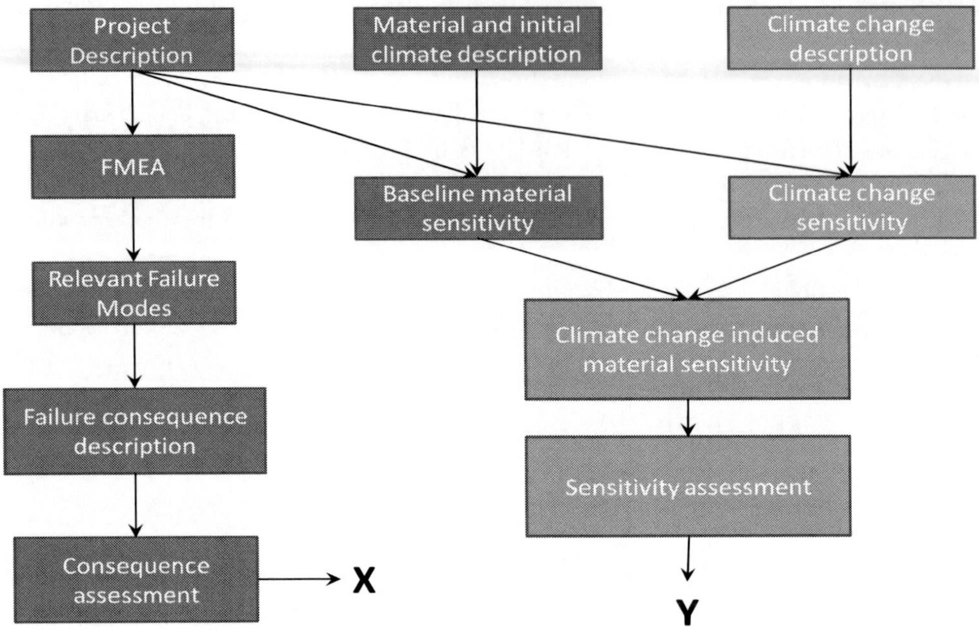

Sensitivity matrix:

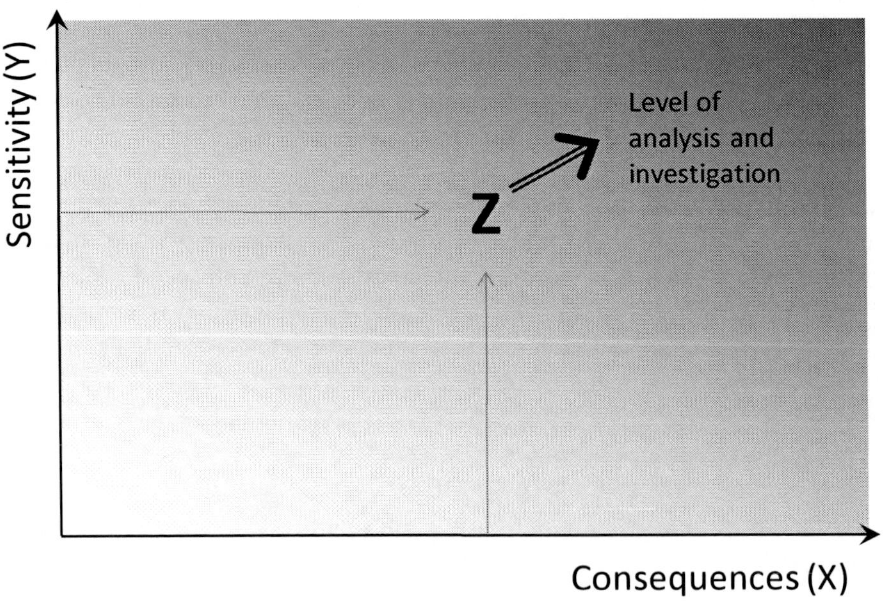

Figure 6. PERD process flow chart (after PERD, 1998).

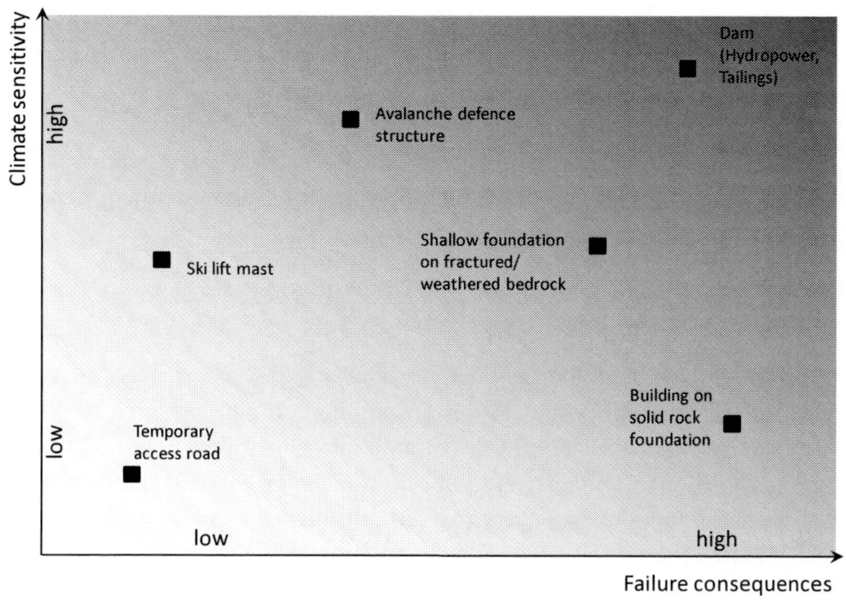

Figure 7. Examples of mountain permafrost infrastructure.

Failure Mode and Effects Analyses (FMEAs: Nahir et al., 2005; Robertson and Shaw, 2005) are an additional important pillar during a design process. The result of the analysis is to be incorporated into the screening process as it identifies sensitivities as well as consequences. This should be an unreserved process with various experts involved. The earlier an FMEA is carried out, the better it can be used to guide the project design.

Fischer and Huggel (2008) present a methodical design for stability assessments of permafrost affected rock walls that provides additional information when dealing with structures in rock.

Adaptational Design

Design for adaptation is an important element in the design of structures when either high sensitivity or high consequences are expected, or even both. As it is not economical to design for all possible scenarios, best estimates of changes in soil properties and climate boundary conditions should be investigated and accounted for in the design. A "high" case and extreme years should also be looked at, as will be described below. Although climate models are improving, it is practically impossible to predict the foundation conditions in 50+ years time. Measures should therefore be incorporated into the design so that they can be implemented based on monitoring results and pre-defined thresholds. Such measures could include anchor redundancies, access for anchors to be installed if rock masses could become unstable, or the installation of pipes that can be used for later installation of thermosyphons or an active refrigeration system. Flexible systems need to have ample room so that possible movements of the foundation can be adjusted accordingly.

In his Rankine lecture, Professor Ralph B. Peck (1969) made the following comments: "if the governing phenomena are complex, or are not yet appreciated, the engineer may measure the wrong quantities altogether and may come to dangerously incorrect conclusions". This is a situation often encountered in alpine environments and where the observational method, as introduced by Terzaghi and Peck (1969), is an ideal approach. However, in contrast to its original concept, monitoring, evaluating and modifying or adapting is not limited to the construction phase but more importantly it should be carried out continuously during the service life of the structure and even for a certain time after abandonment depending on the design purpose (e.g. tailings facilities).

Accounting for Climate Change

Box 11.3 of the most recent IPCC report on regional climate projections provides a brief overview of expected climate change in the mountains (Christensen et al., 2007). One of the most challenging questions to be answered during the design process, and one that has rarely been carried out in mountainous permafrost environments is how to account for climate change. Good climate records for a specific site are imperative for the best possible projection of future conditions. The site must be compared with available data from nearby long-term stations to obtain past average conditions. Offsets can be determined based on comparisons between local records and data from such a long-term station. Generally, conditions from the past 30 years provide an average that is utilised for modelling initial conditions. During the calculation of the past average conditions, extreme events, such as temperature anomalies caused by the El Niño Southern Oscillation (most recently 2006-07, or a major event in 1997-98), must be judged accordingly.

When selecting a climate scenario for predicting future climate conditions, the IPCC suggests five criteria that should be met if they are to be useful for impacting on researchers and policy makers (IPCC, 2007):

- *Criterion 1*: Consistency with global projections. They should be consistent with a broad range of global warming projections based on increased concentrations of greenhouse gases. This range is variously cited as 1.4°C to 5.8°C by 2100, or 1.5°C to 4.5°C for a doubling of atmospheric CO_2 concentration.
- *Criterion 2*: Physical plausibility. They should be physically plausible; that is, they should not violate the basic laws of physics. Hence, changes in one region should be physically consistent with those in another region and globally. In addition, the combination of changes in different variables (which are often correlated with each other) should be physically consistent.
- *Criterion 3*: Applicability in impact assessments. They should describe changes in a sufficient number of variables on a spatial and temporal scale that allows for impact assessment.
- *Criterion 4*: Representative. They should be representative of the potential range of future regional climate change. Only in this way can a realistic range of possible impacts be estimated.
- *Criterion 5*: Accessibility. They should be straightforward to obtain, interpret and apply for impact assessment.

General Circulation Models (GCM) represent physical processes in the atmosphere, ocean, cryosphere and land surface. They are the most advanced tools available to date for simulating the response of the global climate system to increasing greenhouse gas concentrations. GCMs, possibly in conjunction with nested regional models (fine-resolution regional climate models - RCM), have the potential to provide geographically and physically consistent estimates of regional climate change, which are required in the impact analysis. However, even the selection of all available GCM experiments would not guarantee a representative range, due to other uncertainties that GCMs do not fully address, especially the range in estimates of future atmospheric composition. The GCMs are to be used to calculate median values describing base case climate change scenario as well as a high case scenario, which could be median plus one standard deviation. Since climate change does not only affect air temperatures, changes in precipitation, vegetation cover and wind speed should also be considered to the best possible degree, since they have an effect on the surface energy balance that can not be ignored. Perhaps the most important factor influencing permafrost in mountain areas is the snow cover, which is present for long periods of the year and can have a cooling or insulating effect, depending on the timing and duration of the snow cover. Long-term snow data in the Swiss Alps show a distinct step-like reduction in the number of snow days at altitudes below 1800 m asl since the mid 1980's (Marty, 2008), yet no long-term trends can be distinguished at altitudes above 2100 m asl, where permafrost is found. It is therefore difficult to predict future snow scenarios at high altitudes. Potential changes and trends should, however, be taken into account for structure design.

Finally, seasonal variations in the warming trend also must be incorporated in long-term predictions. As indicated in the IPCC reports (IPCC, 2007) and modeled by GCMs, warming during the winter months is higher than average warming, whereas air temperatures during the summer months increase at a rate lower than average. Therefore heat extraction during the cold winter months is lower than an average would predict. Because several non-linearity processes are involved, e.g. surface energy balance, not considering seasonal trends in the long-term models would not be adequate.

A deterministic / probabilistic approach is recommended due to all these uncertainties, variabilities and sensitivities when accounting for climate change, similar to seismic designs for earthquakes (e.g. Klügel, 2008).

Serviceability Limit State (SLS)

Long-term settlements and ground movements are often the deciding factors in the design of infrastructure in mountain permafrost environments. The temperature, time and stress dependence of frozen ground must therefore be evaluated. When carrying out laboratory tests to determine these properties, laboratory boundary conditions must be chosen according to the expected field conditions. Because the long-term behaviour of frozen soils is highly non-linear, data extrapolation is not recommended. Caution is also to be used when interpolating test data. Sample size effects must be considered in the selection and possible correction of the data when using laboratory data for field designs. Data and relationships for permafrost soils are presented in Andersland and Ladanyi (2004), or in Arenson and Springman (2005a) and Arenson et al. (2004) for ice rich alpine permafrost samples.

Ultimate Limit State (ULS)

Bearing capacity and slope stability analysis may be carried out using a complex plasticity based model (Clarke et al., 2008) that requires many parameters, or a simple Mohr – Coulomb failure criterion (e.g. Nater et al., 2008). Similar to the SLS state, it is important that the soil parameters are determined in the laboratory for the ground and loading conditions found in the field. Because ground temperatures will change in the future, ultimate limit state analyses are not limited to one particular point in time, but require transient analysis, resulting in predictions of the evolution of the safety factor with time. In order to account for all uncertainties involved when designing for long-term ULS in permafrost, it is suggested that the desired safety factor is increased by 25%, as an initial approximation. However, this change may have to be adjusted for different projects, elements and quality of climate parameters available.

Geothermal Modelling

Geothermal modelling should form a standard part in any permafrost design. Three aspects of geothermal modelling need special attention and must be subject to sensitivity analyses:

- Climate boundary conditions,
- Geothermal properties, and
- Initial conditions.

The more measured climate data, including air temperature, snow depth, solar radiation, wind speed and direction, and precipitation are available, the more accurate soil thermal properties can be calibrated and initial conditions established. Three-dimensional effects (Gruber et al., 2004) in mountainous terrain can be particularly challenging to incorporate accurately and therefore sufficient data must be available to obtain the required confidence in the model.

A long-term, past climate average should be utilised to obtain steady state conditions before measured climate conditions are applied. The measured ground temperatures as well as active layer thicknesses are to be compared with the modeled data and the model parameters adjusted accordingly. Without proper determination of initial conditions, predictions of subsequent responses are very uncertain.

Different scenarios should be adopted for long-term geothermal modelling that account for uncertainties in climate change predictions. Not only will air temperatures change in the future, additional climate factors must be represented, too. These form the basis for the sensitivity analysis that is crucial for any structure that depends on the integrity of a foundation where future changes are difficult to forecast. Possible long-term changes in the ground geothermal properties must be included in the transient modelling. It is recommended to alter initial conditions to study their effect on long-term behaviour. This step is important when no or only little data are available to calibrate the model and obtain initial conditions. In summary, the fewer climate and ground temperature data that are available, the greater the variations and range must be evaluated during the geothermal modelling phase.

SITE INVESTIGATION

A site inspection with experts will reveal valuable information on possible permafrost existence and extent. Surficial geomorphological, geological and hydrological features help in site assessment. Simple tests, such as measuring temperatures in surface water or digging into the ground will further provide information on possible permafrost existence. Remote sensing using aerial photogrammetry, airborne LiDAR, InSAR and other technologies (Kääb, 2008), and permafrost distribution modelling (Riseborough et al., 2008) are important tools during the early design process and should also be used for planning the site investigations. However, site visits and desk studies using remote sensing data are only the beginning of a thorough site investigation programme that is required for structures to be built in or on mountain permafrost.

In-Situ Site Investigation (Geomorphology, Geology)

The analysis of the geomorphological features at a given site provides a preliminary diagnosis of whether permafrost is present or not. In addition to the analysis of topographical maps and aerial photographs, in-situ site investigations are carried out to identify potential permafrost-related features, prior to more complicated and costly geophysical and geotechnical investigations. The use and relevance of geomorphological analyses is unfortunately often underestimated, although the method is a low-cost and simple one. A solid understanding of alpine morphology is nevertheless required for the successful interpretation of the features observed and therefore expert knowledge is to be sought.

Various geomorphological features and processes can be discerned; as a result, ranges of ice contents may be estimated and an overview of the occurrence and potential for permafrost related processes such as erosion or mass movements may be established. Whereas permafrost terrain at high latitudes is characterised by a large number of clearly recognisable features, it is more difficult to determine the presence of high altitude permafrost on the basis of morphology alone – as is reflected in books on the periglacial environment, where only a few pages are dedicated to mountain permafrost, as it is often primarily associated with rock glaciers (e.g. French, 2007).

Other features that are commonly found in mountain permafrost regions, and can be of assistance in the diagnosis of permafrost (Fig. 8), include ice-rich moraines (Fig. 9), perennial snow patches (Fig. 10), creeping scree slopes and thermokarst depressions (Fig. 11). Rock glaciers (Fig. 12) are the most reliable diagnostic feature and can be used to indicate the lower altitudinal limit of discontinuous permafrost. Active rock glaciers contain ice and creep downslope - they are characterised by having steep flanks and snouts (>38°) as well as a bulging shape, with transverse ridges and hollows. Little vegetation (Fig. 13) or lichens grow at the surface. Springs at the snout are generally colder than 3°C.

Figure 8. Mountain permafrost environnement (Brooks Range, Alaska; D. Kinakin).

Figure 9. Ice-rich lateral moraine, Morteratsch glacier, eastern Swiss Alps. Ice is visible in the scars (M. Phillips).

Figure 10. Perennial snow patches (avalanche debris) at 2800 m asl above Pontresina in the eastern Swiss Alps (M. Phillips).

Figure 11. Recently formed and rapidly growing thermokarst depression on a ski run at 2600 m asl in the western Swiss Alps (Depth of hole: 3 m, Summer 2007, M. Phillips).

Figure 12. Rock glacier Sesvenna in the Eastern Swiss Alps (M. Phillips).

Figure 13. Stunted trees (80-150 years old) on a low-altitude permafrost scree slope in the Swiss Jura (M. Phillips).

Ice-rich moraines are generally found in recently deglaciated terrain. High ice contents can cause them to be very steep (>38°). Ice can be visible in fresh scars on the steep flanks and creep-induced features such as gelifluction lobes can occur (Fig. 8). Scree slopes are one of the most frequently occurring features in alpine environments - sometimes creep-induced features can be visible, which may however be due to solifluction. Ice is very rarely visible – and only seen in fresh scars (e.g. due to stream erosion at the base of a slope). Any vegetation on alpine permafrost is azonal or stunted, helping to distinguish the features from non-permafrost ones. Perennial snow patches are present for long periods of the summer on a regular basis and simply indicate that the underlying ground temperature is at 0°C or colder. They often consist of avalanche debris at the base of steep slopes or of windblown snow in hollows. Thermokarst depressions and slumps etc. can occur in any kind of frozen ground and are an indicator of permafrost degradation. Thaw subsidence due to the melting of ground ice leads to the formation of fresh depressions that often grow rapidly, in the course of one summer. Although rock walls look similar in any type of terrain, fresh rock fall scars containing ice can sometimes be seen in permanently frozen rock walls.

The geological site investigation gives a first indication of the composition, structure, stability and potential ice/water content of the ground. In addition to the standard geological observations, it is important to be able to establish the duration and type of the last glaciation at the site – which has an influence on ground temperature. The thickness and origin of ground surface material should be determined, and water/ice contents estimated. In order to establish a detailed stratigraphy and identify the type of ice (e.g. segregated ice lenses, massive ice layers, ice in cracks or covered ice), drilling is necessary. Geophysical measurements can deliver information on the layering of the ground and the location of ice over a large surface area but does not provide detailed geological and geotechnical information. Potential instabilities, the mass of creeping bodies and the mechanisms of erosion in the vicinity of the construction site should be identified in the geological analysis. Geological maps and field investigations also allow the presence of aquicludes to be determined – this helps to forecast where water and ice may accumulate due to construction activity or to the presence of an infrastructure. Ultimately, the geological report should state prohibited and recommended anchor directions.

Trench / Sampling

Digging a test trench gives valuable information about active layer thickness and properties of near-surface permafrost soils. For technical reasons, the maximum trench depth in alpine terrain mostly does not exceed 4-5 metres. It is, however, ideal to sample soils for additional laboratory investigations (block samples). Grain size distributions and ice contents can be estimated within a trench. Because exposed ice thaws rapidly, mapping has to be carried out immediately after the excavation. It is also important to record even fine ice lenses because the layering contains valuable information on the frost susceptibility of the active layer. However, it is only a shallow investigation and because the soil properties generally change significantly with depth, further investigations are required that give information on deeper ground sections. Not only is this necessary for deep foundations, but also for shallow ones since thermal disturbances at the surface affect the ground at greater depth. If thaw sensitive soils exist at greater depth, long-term thermal changes may cause thaw settlements or instabilities here.

Geophysics

Kneisel et al. (2008), Maurer and Hauck (2007), Musil et al. (2002) and Hauck (2001) present thorough overviews on geophysical methods for detecting permafrost in high mountains. The most important methods used for 2D or even 3D determination of the depth and lateral variability of permafrost are:

- Electrical resistivity tomography (ERT)
- Capacitively coupled ERT
- Frequency-domain electromagnetic induction (FEM) mapping
- Time-domain electromagnetic induction (TEM) sounding
- Seismic refraction tomography
- Ground penetrating radar (GPR)

It is recognised that in most cases one method is not sufficient and at least one additional method is needed in order to interpret the results unambiguously in terms of the existence of permafrost or not. A typical example in DC resistivity surveys is the difficulty in differentiating between isolated rock, ice and air occurrences, each resulting in anomalously high resistivity values. In combining DC resistivity with refraction seismic surveys, this ambiguity can be resolved as the seismic velocities of the three materials are markedly different.

In addition to surface geophysics, geophysical methods can also be used in boreholes for permafrost detection (Arenson, 2002; Vonder Mühll and Holub, 1992). By using cross-hole geophysical methods, 3D soil models can be created (Maurer and Hauck, 2007; Musil et al., 2003) that provide important information on the heterogeneity of the permafrost distribution. Borehole stability is often an issue in coarse, temperate frozen soils, and a casing, that influences the geophysical readings, may be required. Further methods that require the borehole to be filled with water might not be possible.

Note that geophysical investigations detect change in the physical properties of the ground, i.e. ice, water, mineralogy, or density. However, permafrost is thermally defined and if there is no change in the ground's physical properties, such as frozen pore water, no geophysical investigation method is capable of detecting permafrost. Temperature measurements with depth are therefore indispensable. Calibration of geophysically determined stratigraphies using borehole information is strongly suggested.

It is recommended to extend geophysical survey lines into non-permafrost if these zones are known, e.g. over the edge of a rock glacier. Such an extension can be used for reference when looking for permafrost features in the geophysical profiles.

Boreholes

Drilling boreholes is an important part of any permafrost site investigation and should be carried out with ample lead time to any construction. For low sensitivity, low consequence projects, uncored boreholes that are later used for ground temperature monitoring may be sufficient. Drilling chips already provide some information about the structure of the ground, i.e. ice contents, bedrock, etc. In addition, in-hole geophysics may be performed revealing data about variabilities in water content or density with depth.

Visual inspections of the borehole using downhole cameras can provide valuable information about possible existence of ice lenses or large voids (e.g. Arenson, 2002). If borehole instability prevents keeping the borehole uncased for longer periods, periodical inspections of the borehole bottom are recommended, even though it slows the drilling process.

If either the failure consequences or foundation sensitivity is high, coring of undisturbed permafrost samples is strongly recommended. Coring in mountainous permafrost is very challenging due to the heterogeneity of the soil where large boulders can be locked into an ice matrix. Arenson (2002) or Vonder Mühll (1996) provide some information on drilling and sampling in rock glaciers. The soils at the permafrost base can be coarse and large voids are present where liquid drilling mud would get lost, which would cause further environmental problems. Therefore cold air flushing is to be used. It is crucial that the air is cold enough to extract any heat that is generated during the drilling process. A triple tube system for coring, which results in the least sample disturbance, is the best choice in permafrost.

Long-term monitoring instruments should be installed in all boreholes drilled on site. The actual instrument depends on the project, and some can be combined, such as those to measure temperature and deformations. However, no borehole should be left unused, particularly as boreholes offer an opportunity to make line measurements with depth for a specific point on the surface and several therefore give a three-dimensional overview of ground conditions, which can vary strongly at a given site in mountain permafrost.

Additional Testing

Information on creep properties can be obtained from in-situ pressuremeter tests (Arenson et al., 2003; Ladanyi and Melouki, 1993). However, preboring is required, which may disturb the ground and influence the test results.

SPT (standard penetration testing) has been used in Arctic regions to measure the strength of the frozen ground. However, mechanical and thermal disturbances affect the ground and it is expected that the strength of the ground is higher than that ascribed empirically based on penetration rates. Cone penetration testing (CPT) in mountainous terrain is very problematic due to the coarse nature of most grounds. The probe is likely either be damaged or not to be able to penetrate the ground at all.

LABORATORY TESTING

Laboratory testing is an essential addition to field investigations. Due to lack of finances, poor planning and absence of specific permafrost knowledge this is hardly ever carried out for mountain permafrost projects, in particular in the Alps, except for simple soil classifications. It is, however, crucial when determining geothermal and geotechnical soil properties because these parameters can be measured in a controlled environment and extreme boundary conditions can be applied. It is moreover important to test the soil at the boundary conditions that are to be expected in the field. Laboratory testing includes:

- *Soil Classification*: Standard soil classification includes determination of ice content. Additional information, such as unfrozen water contents are recommended for high risk projects.
- *Thermal needle probe*: The thermal conductivity of a soil can be measured with a thermal needle (ASTM D5334-05, 2005). This parameter is needed for thermal modelling. Pushing the needle into a coarse frozen sample may be difficult to do.
- *Thermal plate*: Similar to the thermal needle, but no needle is pushed into the soil and the thermal conductivity can be measured from the sample surface.
- *Uniaxial compression test*: Uniaxial compression tests are the simplest way to determine creep or strength parameters for a frozen soil (ASTM, 2001; ASTM, 2006). However, the strengthening effect of the confinement can not be determined and the strain rates must be chosen carefully. Strain rates of 1%/min, as recommended by ASTM, may result in brittle behaviour that overestimates the long term strength.
- *Triaxial compression test*: Even though triaxial 'constant strain rate' and 'constant stress' (creep) tests are difficult and time consuming and are highly dependent on sample size and quality, currently they provide the best insight into deformation and failure mechanisms.
- *Direct shear test*: In order to study distinct shear interfaces, such as permafrost tables or rock joints, direct shear tests are highly recommended. These tests can be carried out in the laboratory (Yasufuku et al., 2003) or in the field (Springman et al., 2003).
- *Centrifuge modelling*: Centrifuge modelling (e.g., Harris et al., 2008; Harris et al., 2001b; Kern-Luetschg et al., 2008) can be a useful investigation tool for studying certain processes. Soil properties, however, can not be directly measured and scaling laws must be considered, in particular when modelling coarse soils.

The higher the risk (Consequence – Sensitivity matrix), the more knowledge about the soil's properties is required. A thorough laboratory testing programme that includes triaxial or direction shear testing is strongly advised. Depending on the project, even more complex studies that involve centrifugal modelling may be considered because they offer an opportunity to study failure mechanisms on a larger scale.

Monitoring of Environment and Structure

Good climate data are critical in the design of permafrost infrastructure. They form the basis for calibration work before any construction starts and they are important input data when assessing the reliability and performance in the future. Therefore a climate station that measures air temperatures, precipitation, snow depth, short wave and long wave radiation, and wind speed and direction, is to be installed as early as possible. Often there is an existing network of weather stations that can be utilised to determine the locally prevailing conditions (in particular air temperature and snow depth).

Reference measurements should be carried out at a certain distance from the structures. By comparing reference measurements (e.g., ground temperatures, deformations) with the data recorded within or close to the structure, various influences, e.g. due to climate change, can be analysed individually. By analysing such factors separately, better adaptation solutions can be designed in the future. The responsibilities of data analysis and management should be clearly determined and alarm values and reaction scenarios must be defined in advance.

Temperatures

In addition to snow depth, air and ground temperatures are the most important data to be monitored. Due to the heterogeneity of the ground conditions often encountered in mountainous environments, it is recommended to measure air temperatures and ground temperatures at location with different elevations and slope aspects. Wide-ranging active layer thicknesses and ground surface temperatures help in outlining permafrost conditions. Changes in temperature gradients below the depth of the zero annual amplitude further provide valuable information about past climatic conditions and the geothermal phase the permafrost is in, i.e. long-term cooling, warming or steady state (Fig. 14). Generally, changes in the temperature gradients below the depth of zero annual amplitude are good indicators for past long-term temperature conditions. However, effects due to the mountainous topography as indicated above must be included in the interpretation of the borehole temperature data.

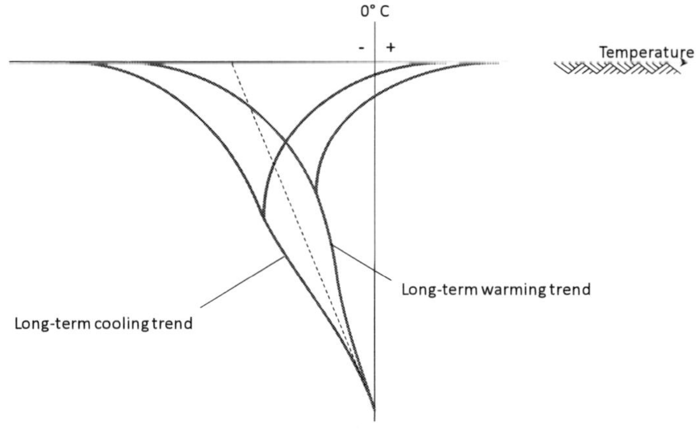

Figure 14. Effect of long-term surface temperature changes on the geothermal gradient. Warming periods result in temperatures below the depth of zero annual amplitude (ZAA) that are warmer than the deep temperature gradient would imply (red) and vice versa for long cooling periods (blue).

Ground temperatures at depth are to be measured within boreholes using thermistor strings. Close thermistor spacing at shallow depth is important to determine active layer thicknesses accurately. Temperatures should further be recorded at three to four locations below the permafrost base to obtain the local geothermal gradient, which is to be used in the geothermal model. Close thermistor spacing is also recommended at the permafrost base where thermal irregularities due to groundwater or air flow may occur (Vonder Mühll et al., 2003; Vonder Mühll et al., 1998). More information on installing thermistor strings can be found in Vonder Mühll et al. (2004).

Spatial distributions of the bottom temperature of the snow cover (BTS) allow further information about the permafrost distribution (Haeberli, 1973). Miniature temperature loggers (Hoelzle et al., 1999) are ideal for such temperature measurements and should be buried at 5cm depth. Interpretations of the ground temperatures are easier if the snow depth distribution is known for the site.

Infrared thermography is a useful tool when controlling insulation efficiency in buildings or the efficiency of active/passive cooling systems. Infrared cameras are widely available to date and any thermal anomalies can be visualised in real time.

Hydrology

The hydrology of mountain permafrost can be very complex (Giardino et al., 1992; Krainer and Mostler, 2002; Schrott, 1996; Schrott, 1998; Woo et al., 2008) and affect both the thermal regime due to convective heat transfer, and slope stability due to changes in the effective stresses. Water may flow within the active layer, below the base of the permafrost, through taliks and even through open channel systems in the frozen soil. To determine the influence and relevance of any permafrost body in a study area as a hydrologic resource, it is necessary to establish a water balance for the watershed area that the permafrost features encompass. Some tasks to establish hydrological information include:

- Measure climate data as indicated above that determine the amount of precipitation, water storage and storage capacities including water retard.
- Measure water flow at several locations with groundwater wells (flows into and out of permafrost areas are expected to be predominantly subsurface) including upslope run-offs.
- Measure snow depth and snow water equivalent (SWE) along a series of transects through the watershed prior to and at least once during snow ablation. If snow depths are known, SWE can also be calculated with reasonable accuracy for regions with long measurement records.

The objective of the measurements is to isolate a base flow component that could be attributed to slow release of water contained within the body of frozen ground and to groundwater originating from different sources located above.

Hydraulic properties of the various soil types encountered can be measured by carrying out laboratory tests from collected soil samples, or from slug or pump tests carried out in the field. Detailed site investigations are needed to determine the number and location of groundwater wells required for a specific problem.

Several different probes are available to measure moisture contents and pore pressures. Rist and Phillips (2005), for example, carried out field tests using TDR probes, vapour traps and lysimeters to measure suction and spatial variability of water infiltration within the active layer during freeze-thaw cycles. Other suction probes, such as tensiometers, are available to measure in situ suction (e.g. Ng et al., 2008; Springman et al., 2003; Thielen and Springman, 2005) in which water is replaced as the saturating medium in winter by an alcohol (circa 2% by volume) / water mixture.

Deformations

Monitoring ground and structural deformation is crucial for the serviceability of any structure in an alpine environment. The observational method approach and adaptation strategies rely on these measurements. Depending on the structure and situation, different deformation techniques can be used. Ground based surveying is the easiest way to measure surface deformations. Automatic total stations can be used for continuous monitoring in remote areas. Additional surface monitoring methods include photogrammetry, InSAR or LiDAR technologies (Kääb, 2008; Metternicht et al., 2005; Roer et al., 2005; Strozzi et al., 2004).

Borehole deformation measurements help to understand complex deformation profiles (e.g. Arenson et al., 2002; Haeberli et al., 1998). Horizontal deformations can be measured using slope inclinometers and vertical deformations with extensometers. Borehole inclinometers should cover the whole thickness of the permafrost layer and be fixed into non-moving ground, e.g. bedrock. In addition, the top of the borehole should be included into the terrestrial survey as a control point and to identify movements if the bottom is not fixed.

Time domain reflectometry (TDR) can also be used to measure ground deformations with depth (O'Connor and Dowding, 1999). TDR cables should only be used if distinct shear zones are present (Arenson, 2002). In contrast to inclinometers and extensometers, TDR deformation measurements have a much finer resolution in terms of the depth of a particular shear zone, however, the actual direction and extension of the deformation can only be estimated.

Various additional deformation monitoring instruments exist that may be suitable for particular projects, including tiltmeters, crackmeters, strain gauges, or pressure cells.

Construction Challenges in Mountain Environments

Using permafrost as a foundation and accounting for permafrost in the environment that may cause slope instability hazards are only part of the challenge when constructing in mountainous environments. There are many additional ones that need timely planning. The remote location of the construction is demanding on the logistics. Electricity and water may not be available and may have to be supplied to the site. Accommodation and food has to be planned for the site, and the effects of altitude on workers and machinery must be taken into consideration.

Severe weather conditions often shorten the construction season and simply slow down any construction activity. Cold temperatures are tough on machinery and increased abrasion and wear must be considered. These facts are often not accounted for enough but rapidly cause increasing construction costs.

Steep terrain and rugged conditions also limit access options. Special transport cable-cars are often installed just for the duration of the construction period. Access by air might be the only way to get to the construction site, which limits the weight of the equipment that can be used, is highly weather dependent, and very costly. When road access is possible, permafrost conditions are to be investigated since they might also limit access of heavy machinery in particular during summer months. Due to surface mass movements, e.g. solifluction, access roads might only be accessible for a limited time or have to be regarded regularly (Fig. 15).

Figure 15. Solifluction on temporary access road in the Brooks Range, Alaska (D. Kinakin).

The optimal time for construction also has to be determined. Preferably exposure of frozen soils is to be done during winter. This limits the thermal disturbance caused by the construction activity. In fact, by exposing the foundation to the cold winter temperatures and placing insulation at the end of the winter, initial foundation temperatures can be obtained that are colder than average. In addition, access on frozen terrain is easier than on thawed ground. However, winter construction is tougher on the machinery and workers due to climatic conditions (cold temperatures, snow), the period of sunlight is much shorter, and the construction site or the access route may be located in avalanche zones.

One problematic aspect of construction in permafrost is the introduction of concrete or grout into the ground and the resulting hydration heat given off during the curing process, which lasts at least a month. The hydration heat of 1 m^3 of concrete is between 65,000 and 162,600 kJ (e.g. PCA, 1997), depending on the cement content, compared to the 333 kJ required to melt 1 kg of ice. Hence, the heat produced during the curing of 1 m^3 of concrete can melt large quantities of permafrost ice. When smaller amounts of concrete or grout are involved, e.g. for micropiles, subzero ground temperatures can cause it to freeze before attaining the appropriate bearing capacity, implying that anti-freeze additives or so called 'retarders' are required to ensure that the curing process can be achieved successfully (Moser, 1999). The Swiss guidelines for the construction of avalanche defence structures also stipulate that warm water be used for the grout mixture injected into anchor boreholes in order to delay freezing (Margreth, 2007; Thalparpan, 2000). Whereas both anti-freeze additives and warm water have positive effects on the final bearing capacity of the grout and anchors, they will disturb the thermal regime of the permafrost. These aspects must therefore be considered carefully in terms of the effect on the ice content and long term stability of the ground.

Additional challenges for building in mountain permafrost include:

- *Blasting adaptations*: Due to the monolithic nature of frozen ground, more explosives are needed in frozen soils with ice.
- *Special machinery*: The frozen nature of the soil requires special equipment for excavation that is usually heavy.
- *Protection of excavation*: Changes in the soil strength due to thaw in excavation slopes must be considered, in particular if the excavated frozen slope is exposed to the warm summer temperatures. Fleeces for thermal insulation and optimising trench angles may be considered.

Case Histories, Problems and Technical Solutions

Heated Infrastructure

Heated infrastructure directly influences the thermal regime of the ground if the structure-ground contact surfaces are insufficiently insulated (Instanes et al., 2005). Active layer thickening and subsidence occurred, for example, beneath the northern tower of the

Kulmhotel Gornergrat (Zermatt) due to heating of the tower basement for over 20 years. Concrete was injected to stop the differential settlement, with apparent success (Hof et al., 2003). Bedrock instability due to loss of ice (possibly also induced by climate change) has been observed in the vicinity of several cable car stations and mountain huts (Dätwyler, 2004), causing damage to the buildings in recent years. Attempts have been made to stabilise the ground and buildings by pumping concrete into fissures and boreholes (e.g. approximately 200 tonnes of concrete were injected into the ridge under the Swiss Gemsstock cable car station in 1993, personal communication F. Nager, see also Fig. 16) or by inserting rock anchors (e.g. in 2001-2003 under the Erzherzog Johann hut on the Grossglockner, Austria). Even unheated structures can modify the thermal regime of the ground and induce subsidence, which has recently happened under garages on the Schilthorn (personal communication P. Feuz) and Gemsstock summits (Fig. 16). Tunnels can also have a disturbing effect, as air or water can warm frozen rock zones, as was observed during the drilling of access tunnels in Chli Titlis (Haeberli et al., 1979), Klein Matterhorn (Rieder et al., 1980) or in the Jungfrau east ridge (Wegmann, 1998). The heat transferred into the tunnel of the Klein Matterhorn tunnel by 490,000 visitors every year and generated by about 70,000 elevator movements in transporting people to the mountain top has caused the bedrock temperatures in the tunnel to rise from -8°C in 1999 to -3°C at present, as well as the refreezing of meltwater in the lift shaft, requiring remedial ventilation measures (King and Kalisch 1998, Baumann et al. 2005). Another current problem is thermokarst settlement in ground around water pipes for technical snow systems, leading to the formation of deep holes in ski pistes and damage to the pipes. Although steel trellis structures such as pylons and snow-supporting structures warm up to around 30°C diurnally under the influence of direct solar radiation, temperature measurements on micropiles below steel snow-supporting structures and in the surrounding ground have shown that the brevity of the warming periods prevents heat transfer into the ground (Phillips et al., 2000; Phillips and Schweizer, 2007). The main problem affecting these types of structure are temperature- and water dependent creep movements of the ice rich permafrost soils in which they are anchored (Phillips and Margreth, 2008; Rieder et al., 1980).

Figure 16. Fleece-covered snow ramp on the summit of Gemsstock, Swiss Alps (C. Danioth). To the right of the ramp, the rock wall was also covered and temperatures are being monitored. Concrete was injected into the ridge beneath the summit station in 1993 to increase the bearing capacity of the ground.

Modification of the Snow Cover

Infrastructure in snowy, windy areas can modify snow cover distribution (Hestnes, 2000; Thiis and Jaedicke, 2000) and thus affect the thermal regime of the ground. Obstacles above the ground surface tend to cause the formation of local snow drifts and wind scoops, whereas avalanche defence structures and technical snow systems cause widespread artificial change in snow depth over large surface areas through prevention of avalanche formation or addition of snow, inducing a delay of snow melt in summer (Fig. 17). In the case of snow-supporting structures, a heterogeneous snow cover is formed because snow tends to accumulate above the structures and wind scoops form below. Modelling the long-term effects of this phenomenon shows a perturbation of the annual temperature regime and overall cooling of the ground during the design life of the structures (Phillips et al., 2000). Modern techniques of ski run preparation involve the use of technical snow, which can lead to long-term cooling of ground temperatures due to the higher density of such snow (Fauve et al., 2002), which is used by approximately 10-15% of ski resorts in the Alps. Rixen et al. (2004) suggest that at alpine sites where mean ground temperature is close to 0°C, the additional temperature reduction may suffice to induce permafrost formation. In areas where permafrost already exists, mechanical snow grooming and use of technical snow could help to conserve the permafrost. On the other hand, the cooling effects of technical snow may be countered by the removal of blocky rock material from ski runs (Haeberli, 1992). Some ski resorts in the Alps have recently started covering high altitude parts of ski runs with insulating material to prevent the snow from melting at all in summer (Fig. 16). The long-term effects of this type of treatment on the underlying or adjacent permafrost are as yet unknown.

Figure 17. Delayed melt of technical snow on ski runs above St. Moritz, Switzerland (M. Phillips).

Snow fences might have similar effects on the snow cover and hence this may destabilise the permafrost (e.g. Hinkel and Hurd, 2006). On the other hand, they can be used in a design to protect critical areas from snowdrift and the accumulation of thick snow cover, so that an effective heat extraction can be obtained during winter.

Hydration Heat

During the construction of the midway station for a chairlift in Grächen (Swiss Alps) in 1997, hydration heat caused problems in a zone where no insulating material was placed between the concrete foundations and the underlying ground. This resulted in severe permafrost ice degradation and, as a consequence, settlement, cracking of the concrete and pronounced creep of the structure (5 cm during the first winter), which had to be replaced completely in 2003 (Phillips et al., 2007).

Ground Thermal Control

Changes in the ground thermal regime are the major cause of problems with permafrost infrastructure. Efforts should therefore be made to limit the changes in the ground temperatures and in particular to prevent phase changes in the ground. By moving the active layer into soils that are non-frost susceptible, frost heave and thaw settlements can also be avoided. Figure 18 shows various sketches of suitable foundations in permafrost environments. The various methods are described below.

Thermal Insulation

Special thermal insulation (e.g. Ulitsky et al., 2003) can be used in the design to protect the permafrost ground under a structure, e.g. roads or buildings. Cheng et al. (2004) recently presented a discussion on the existence of maximum and minimum embankment heights and the applicability of thermal insulation. Generally, insulating materials reduce the thermal flux due to their low thermal conductivity and thickness. The effect can either be obtained by using special materials, such as polystyrene (~0.03 W/m°C, e.g. Andersland and Ladanyi, 2004), foam glass (~0.04 W/m°C) or sufficient air space (air: ~0.02 W/m°C). Examples of insulation placement are given in VTT (1987).

Passive Cooling

Systems capable of extracting heat from the ground without additional energy can be incorporated into the design. Thermosyphons, thermoprobes or air-duct cooling systems offer such a possibility (McKenna and Biggar, 1998; Smith et al., 1991; Wen et al., 2005), however, they have mainly been installed in Northern regions and their efficiency in alpine environments is still to be tested. Passive cooling using gravity-driven air convection is an additional possibility (Fig. 19). Due to the difference in air density between warm and cold air, cold air sinks to the bottom of the coarse-grained layer, resulting in a cooling effect (Arenson et al., 2006; Goering, 1998; Goering et al., 2000; Ma et al., 2006). The main principles used in thermosyphons can also be incorporated into piles (thermopiles: Fig. 20) and mast designs.

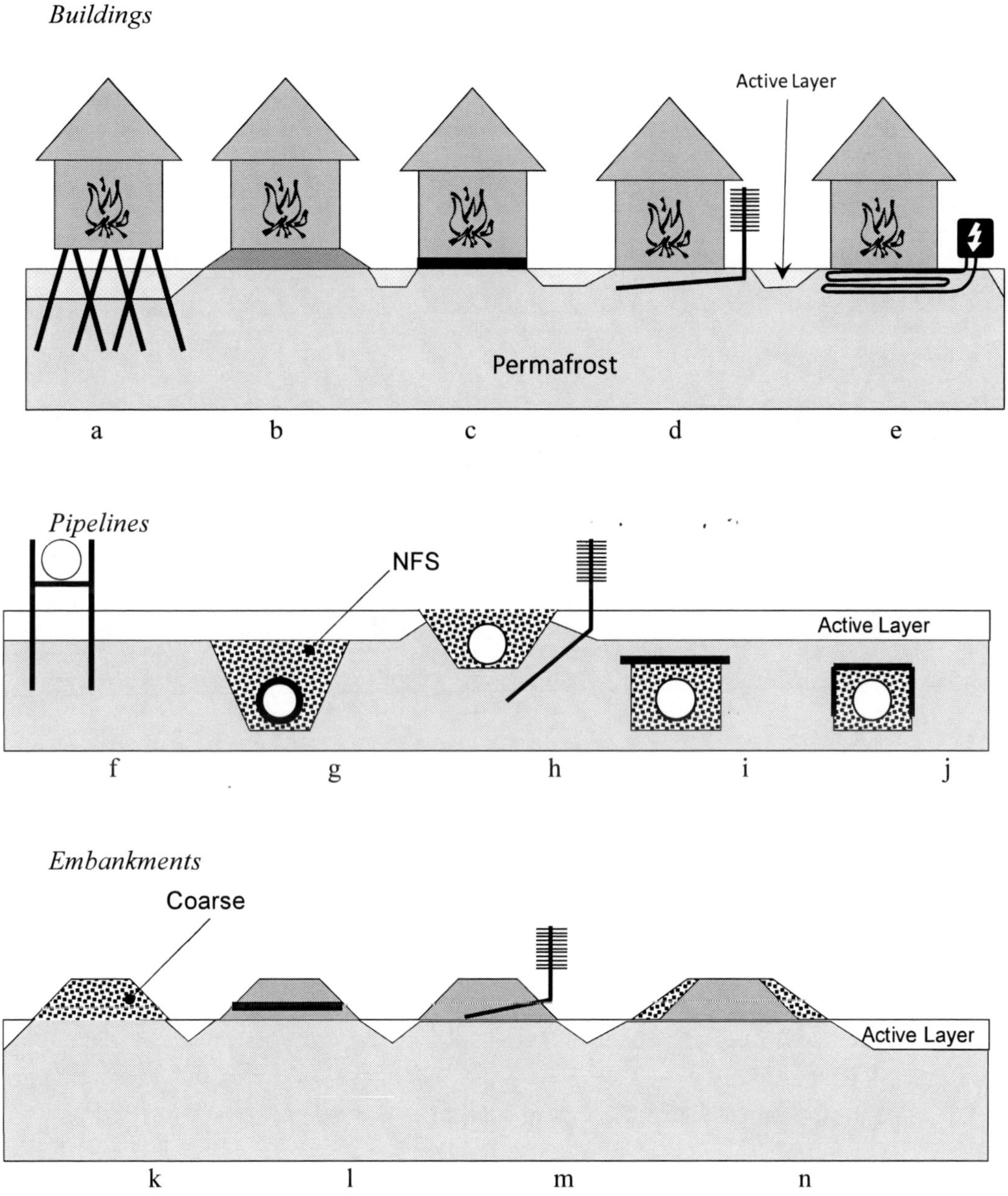

Figure 18. Various possible foundation methods in permafrost environments. Piles (a), elevated (b), insulated (c), thermosyphon (d), active cooling (e), elevated (f), buried and insulated pipe (g), buried and thermosyphon (h), non-insulated pipe buried with ground insulation panels (i, j), air convection embankment (k), insulation panels (l), thermosyphon (m), and convective shoulders (n).

Active Cooling Technology

Designs that involve active cooling should be considered as the last option. Similar to artificial ground freezing (Harris, 1995), heat is actively extracted from the ground in order to cool it and keep the temperatures below freezing point.

Significant amounts of cooling energy may be required, depending on the project. Convective heat fluxes from water and air must be considered and the design has to include redundancies as well as emergency scenarios and plans in the event of failure of the system. Active cooling may be a temporary option during construction or to accelerate cooling after thermal disturbances caused by construction activities.

Figure 19. Embankment shoulder protection using convective cooling. Fairbanks, Alaska (L. Arenson).

Figure 20. Thermopile foundation in Fairbanks, Alaska (L. Arenson).

Pipelines, Buried Conduits and Ducts

Thermal disturbances from buried pipelines can cause significant problems (e.g. Burgess and Smith, 2003; Huang et al., 2004; Palmer and Williams, 2003; Tart and Ferrell, 2002). Chilled pipelines may induce frost heave due to ice segregation and ice lens growth (Konrad, 1994), whereas warm pipelines induce additional thaw around the pipe, which results in thaw consolidation and settlement. In addition, deformation processes on steep slopes, such as creep or solifluction, may cause structural damage to pipelines (Tart, 2003). In order to minimise thermal disturbances, insulation systems are required for any buried structure to be built into permafrost containing ice as well as frost susceptible soils in an alpine environment. Only if no deformations are expected, can pipes without insulation be used. However, possible freeze up mechanisms within the pipe must be considered and modelled. Pipelines that are only used temporarily (e.g. water and sewage pipes in ski areas) should be emptied with dry, high pressurised air whenever possible. Flexible sleeves must be used at the joints.

Flexible Structures

Flexible systems without stiff connection elements to their foundation, such as snow-supporting nets (Phillips et al., 2003), or structures on point bearings, are designed to adapt to non-uniform ground movements. The crucial foundation geometry is not fixed and can therefore be corrected as creep and thaw settlements occur over time. One example of this type of structure is the chairlift midway station in Grächen, Switzerland Two concrete supports are carried by a T-shaped girder, which has three point bearings (two upslope and one downslope), all of which can slide horizontally to enable the entire midway station to find its optimal equilibrium position (Fig. 21). The two upslope bearings are fixed when not being repositioned, whereas the downslope bearing can move freely, allowing the downslope foundation to move independently. If settlements or displacements occur beyond a previously specified threshold, the T-girder is displaced or uplifted hydraulically and steel plates can be inserted. The point bearings can therefore be relieved and repositioned (Phillips et al., 2007). A similar type of system, also using three point bearings, was adopted in 2005 for the new 'Pardorama' restaurant above Ischgl in the Austrian Alps. Flexible systems need to be monitored at regular intervals to allow for timely geometrical corrections to be made.

Stabilization of Existing Structures / The Underlying Ground

Often, existing structures need to be stabilised because the original design did not account fully for all the problems related to permafrost conditions and/or forecasted thermal boundary conditions changed. The addition of anchors and concrete to the foundations of existing structures and into the ground has become widely used to stabilise structures and increase the bearing capacity of the underlying terrain (see examples mentioned earlier in this chapter). The costs involved are generally very high and the long term effectiveness is not yet known.

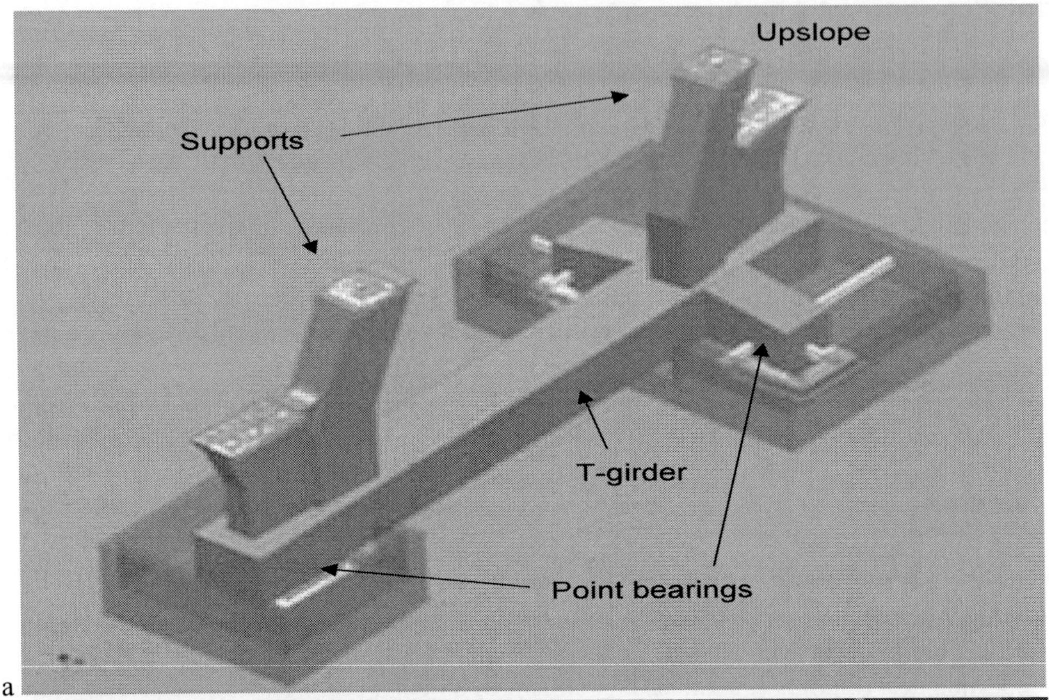

a

b

Figure 21. Chairlift station with three point-bearings (a), which can be displaced using hydraulic pumps (reprinted from Phillips et al. 2007. Structural design by Leitner AG). Hydraulic pump (b) under the T-girder allowing repositioning of the structure.

Pile Foundations and Grouting

Pile foundations in permafrost mobilise resistance mainly by side friction rather than end bearing and therefore only side friction is to be accounted for in a design. In consequence, load transfer between the pile and the surrounding ground is therefore crucial and thermal disturbances, as described above, must be minimised. Sego and Biggar (1990) present four ways that grout will cure adequately at sub-zero ground temperatures: 1) grout temperature may be artificially maintained above 0°C using external heat source; 2) use of cement with rapid rates of hydration; 3) use of salts or other additives; 4) grouts with very low water content. A series of different grouts has been tested for their applicability in permafrost in the laboratory as well as in the field (Biggar and Kong, 2001; Biggar and Sego, 1993a; Biggar and Sego, 1993b; Biggar and Sego, 1993c; Biggar et al., 1993; Biggar et al., 1996). Instead of using a sand slurry backfill, the authors used a cementitious grout (Ciment Fondu), which resulted in greater pile load carrying capacity as the failure surface was transferred into the grout annulus. The high alumina cement-based grout cures by rapidly emitting heat, which maintains it above 0°C until it hydrates and hardens. The hydration heat was minimised so that only minor thermal disturbance of the native ground was observed.

Close cell spray foam, which is easier to transport, may be used as an alternative to create polymer piles (Zaharko and Sego, 2008). As the resin mixture expands, voids are filled creating a bond between the pile and the ground. Because the heat production during the exothermal reaction is very low, very limited thaw occurs around the grout.

Design recommendations have recently been developed for the creep response and bearing capacity of helical piles in Arctic permafrost as foundations for lightweight structures (Zubeck and Liu 2003). While these piles could not be installed through a blocky active layer, there may still be applications within alpine permafrost for such an approach.

CONCLUSION

Designing infrastructure in mountainous permafrost environments is challenging and demanding. However, by acknowledging the special circumstances and working conditions encountered during construction in the mountains, as well as systematically managing uncertainties, in particular due to climate change, successful designs can be achieved. Monitoring plays a key role in this process because a solid data basis helps in establishing current ground conditions and properties to be determined, and further allows for more realistic future predictions to be made. Ongoing monitoring during the life span of a structure helps to design and implement adaptations in timely manner.

A systematic approach is presented that helps in determining the climate sensitivity and failure consequences. Based on the approach the necessary level of investigations and design details can be established. Due to the sensitive nature of some foundation in mountain permafrost, flexible structures that can be adjusted with time are required or elements have to be included that might later be used as the condition of the foundation change. By incorporating elements for structure adaptation at the beginning future rehabilitation costs can be minimised.

Geothermal modelling must be part of any infrastructure design process because it helps in evaluating the sensitivity of the foundation and the structure to changes in the ground thermal condition. In situ ground temperatures measurements are to be used for model calibration and therefore it is important that data are available over a period as long as possible. During the life-time of a structure, these simulations are to be revisited using data from ongoing monitoring programmes.

Designing and building infrastructure in mountain permafrost remains challenging. Complex, heterogeneous foundation conditions, unknown future climate conditions, difficult access for equipment and short construction periods add to the costs of the infrastructure and have to be considered in the design. Technologies and tools are available to engineers to successfully build in demanding environments.

References

Andersland, O.B. and Ladanyi, B., 2004. An introduction to frozen ground engineering. John Wiley & Sons, Inc., Hoboken, New Jersey, 363 pp.

Anderson, D.M. and Tice, A.R., 1972. Predicting unfrozen water content in frozen soils from surface area measurements. No. 393, Highway Research Record.

Arenson, L.U., 2002. Unstable alpine permafrost: a potentially important natural hazard - variations of geotechnical behaviour with time and temperature. Doctoral Thesis No. 14801 Thesis, Swiss Federal Institute of Technology, Zurich, 271 pp.

Arenson, L.U., Hoelzle, M. and Springman, S.M., 2002. Borehole deformation measurements and internal structure of some rock glaciers in Switzerland. Permafrost and Periglacial Processes, 13(2): 117-135.

Arenson, L.U., Johansen, M.M. and Springman, S.M., 2004. Effects of volumetric ice content and strain rate on shear strength under triaxial conditions for frozen soil samples. Permafrost and Periglacial Processes, 15(3): 261-271.

Arenson, L.U., Sego, D.C. and Newman, G., 2006. The use of a convective heat flow model in road designs for northern regions, Climate Change Technology Conference. Ottawa, Ontario, Canada: CD-Rom.

Arenson, L.U. and Springman, S.M., 2005a. Mathematical description for the behaviour of ice-rich frozen soils at temperatures close to zero centigrade. Canadian Geotechnical Journal, 42(2): 431-442.

Arenson, L.U. and Springman, S.M., 2005b. Triaxial constant stress and constant strain rate tests on ice-rich permafrost samples. Canadian Geotechnical Journal, 42(2): 412-430.

Arenson, L.U., Springman, S.M. and Hawkins, P.G., 2003. Pressuremeter tests within an active rock glacier in the Swiss Alps. In: M. Phillips, S.M. Springman and L.U. Arenson (Editors), Eighth International Conference on Permafrost. Zurich, Switzerland, A.A. Balkema: 33-38.

Arenson, L.U., Springman, S.M. and Sego, D.C., 2007. The rheology of frozen soils. Applied Rheology, 17(1): 12147-1 – 12147-14.

Arnold, A., Thielen, A. and Springman, S.M., 2005. On the stability of active layers in alpine permafrost, 11th International Conference and Field Trip on Landslides (ICFL). Trondheim, Norway, Taylor & Francis, Netherlands.

ASTM, 2001. D 5520, Standard test method for laboratory determination of creep properties of frozen soil samples by uniaxial compression.

ASTM, 2006. WK10696, New standard test method for laboratory determination of strength properties of frozen soil at a constant rate of strain.

ASTM D5334-05, 2005. Standard test method for determination of thermal conductivity of soil and soft rock by thermal needle probe procedure, ASTM, Philadelphia, Pennsylvania, USA.

Barnes, P., Tabor, D. and Walker, J.C.F., 1971. Friction and creep of polycrystalline ice. Proceedings of the Royal Society of London, Serie A, 324(1557): 127-155.

Bieniawski, Z.T., 1989. Engineering Rock Mass Classifications: A Complete Manual for Engineers and Geologists in Mining, Civil, and Petroleum Engineering Wiley-Interscience, New York, 272 pp.

Biggar, K. and Kong, V., 2001. An analysis of long-term pile load tests in permafrost from the Short Range Radar site foundations. Canadian Geotechnical Journal, 38(3): 441-460.

Biggar, K.W. and Sego, D.C., 1993a. Field Pile Load Tests in Saline Permafrost .1. Test Procedures and Results. Canadian Geotechnical Journal, 30(1): 34-45.

Biggar, K.W. and Sego, D.C., 1993b. Field Pile Load Tests in Saline Permafrost .2. Analysis of Results. Canadian Geotechnical Journal, 30(1): 46-59.

Biggar, K.W. and Sego, D.C., 1993c. The Strength and Deformation-Behavior of Model Adfreeze and Grouted Piles in Saline Frozen Soils. Canadian Geotechnical Journal, 30(2): 319-337.

Biggar, K.W., Sego, D.C. and Noel, M.M., 1993. Laboratory and Field Performance of High Alumina Cement-Based Grout for Piling in Permafrost. Canadian Journal of Civil Engineering, 20(1): 100-106.

Biggar, K.W., Sego, D.C. and Stahl, R.P., 1996. Long-term pile load testing system performance in saline and ice-rich permafrost. Journal of Cold Regions Engineering, 10(3): 149-162.

Bommer, C., Keusen, H.R. and Phillips, M., 2008. Engineering solutions for foundations and anchors in mountain permafrost. In: D.L. Kane and K.M. Hinkel (Editors), Ninth International Conference on Permafrost. Fairbanks, AK, U.S.A., Institute of Northern Engineering: 159-163.

Burgess, M.M. and Smith, S.L., 2003. 17 years of thaw penetration and surface settlement observations in permafrost terrain along the Norman Wells Pipeline, Northwest Territories, Canada. In: M. Phillips, S.M. Springman and L.U. Arenson (Editors), Eighth International Conference on Permafrost. Zurich, Switzerland, A.A. Balkema: 107-112.

Cheng, G.D., Zhang, J.M., Sheng, Y. and Chen, J., 2004. Principle of thermal insulation for permafrost protection. Cold Regions Science and Technology, 40(1-2): 71-79.

Christensen, J.H., Hewitson, B., Busuioc, A., Chen, A., Gao, X., Held, I., Jones, R., Kolli, R.K., Kwon, W.-T., Laprise, R., Rueda, V.M., Mearns, L., Menéndez, C.G., Räisänen, J., Rinke, A., Sarr, A. and Whetton, P., 2007. Regional Climate Projections. In: S. Solomon et al. (Editors), Climate Change 2007: The Physical Science Basis. Contribution of Working Group I to the Fourth Assessment Report of the Intergovernmental Panel on Climate Change. Cambridge University Press, Cambridge, United Kingdom and New York, NY, USA: 94 pp.

Clague, J.J., 2008. Effects of recent climate change on high mountains of Western North America. In: D.L. Kane and K.M. Hinkel (Editors), Ninth International Conference on Permafrost. Fairbanks, AK, U.S.A., Institute of Northern Engineering: 269-274.

Clarke, J., Fenton, C., Gens, A., Jardine, R., Martin, C., Nethercot, D., Nishimura, S., Olivella, S., Reifen, C., Rutter, P., Strasser, F. and Toumi, R., 2008. A multi-disciplinary approach to assess the impact of global climate change on infrastructure in cold regions. In: D.L. Kane and K.M. Hinkel (Editors), Ninth International Conference on Permafrost. Fairbanks, AK, U.S.A., Institute of Northern Engineering: 279-284.

Davies, M.C.R., Hamaz, O. and Harris, C., 2001. The effect of rise in mean annual temperature on the stability of rock slopes containing ice-filled discontinuities. Permafrost and Periglacial Processes, 12(1): 137-144.

Davies, M.C.R., Hamza, O., Lumsden, B.W. and Harris, C., 2000. Laboratory measurements of the shear strength of ice-filled rock joints. Annals of Glaciology, 31: 463-467.

Esch, D.C. (Editor), 2004. Thermal analysis, construction, and monitoring methods for frozen ground. Technical Council on Cold Regions Engineering monograph. American Society of Civil Engineers, Reston, VA: 492 pp.

Fauve, M., Rhyner, H. and Schneebeli, M., 2002. Pistenpräparation und Pistenpflege: Das Handbuch für den Praktiker, Eidgenössisches Institutes für Schnee- und Lawinenforschung (SLF), Davos, Switzerland.

Fischer, L. and Huggel, C., 2008. Methodical design for stability assessments of permafrost-affected high-mountain rock walls. In: D.L. Kane and K.M. Hinkel (Editors), Ninth International Conference on Permafrost. Fairbanks, AK, U.S.A., Institute of Northern Engineering: 439-444.

Fish, A.M., 1985. Creep strength, strain rate, temperature and unfrozen water relationship in frozen soil. In: S. Kinosita and M. Fukuda (Editors), Fourth International Symposium on Ground Freezing. Sapporo, A.A. Balkema, Rotterdam: 29-36.

French, H., 2007. The Periglacial Environment. Wiley, New York: 478 pp.

Gagnon, R.E. and Gammon, P.H., 1995. Triaxial experiments on iceberg and glacier ice. Journal of Glaciology, 41(139): 528-540.

Giardino, J.R., Vitek, J.D. and DeMorett, J.L., 1992. A model of water movement in rock glaciers and associated water characteristics. In: J.C. Dixon and A.D. Abrahams (Editors), Periglacial geomorphology: proceedings of the 22nd Annual Binghamton Symposium in Geomorphology. Wiley, New York: 159-184.

Goering, D.J., 1998. Experimental investigation of air convection embankments for permafrost-resistant roadway design. In: A.G. Lewkowicz and M. Allard (Editors), Seventh International Conference on Permafrost. Yellowknife, NWT, Canada, Collection Nordicana, vol. 57: 319-326.

Goering, D.J., Instanes, A. and Knudsen, S., 2000. Convective heat transfer in railway embankment ballast., International Symposium on Ground Freezing and Frost Action in Soils. Louvain-la-Neuve, Belgium: 31-36.

Götz, A. and Raetzo, H., 2002. Permafrost - mit dem Risiko umgehen. SIA, tec 21(17): 7-10.

Goughnour, R.R. and Andersland, O.B., 1968. Mechanical properties of a sand-ice system. Journal of the Soil Mechanics and Foundations Division, ASCE, 94(SM 4): 923-950.

Gruber, S. and Haeberli, W., 2007. Permafrost in steep bedrock slopes and its temperature-related destabilization following climate change. Journal of Geophysical Research-Earth Surface, 112(F2).

Gruber, S., King, L., Kohl, T., Herz, T.H., Wilfried and Hoelzle, M., 2004. Interpretation of geothermal profiles perturbed by topography: the Alpine permafrost boreholes at Stockhorn Plateau, Switzerland. Permafrost and Periglacial Processes, 15(4): 349-357.

Günzel, F.K., 2008. Shear strength of ice-filled rock joints. In: D.L. Kane and K.M. Hinkel (Editors), Ninth International Conference on Permafrost. Fairbanks, AK, U.S.A., Institute of Northern Engineering: 581-586.

Haeberli, W., 1973. Die Basis Temperatur der winterlichen Schneedecke als möglicher Indikator für die Verbreitung von Permafrost in den Alpen. Zeitschrift für Gletscherkunde und Glaziogeologie, 9(01. Feb): 221-227.

Haeberli, W., 1992. Construction, environmental problems and natural hazards in periglacial mountain belts. Permafrost and Periglacial Processes, 3(2): 111-124.

Haeberli, W. and Beniston, M., 1998. Climate change and its impacts on glaciers and permafrost in the Alps. Ambio, 27(4): 258-265.

Haeberli, W., Hoelzle, M., Kaeaeb, A., Keller, F., Vonder Mühll, D. and Wagner, S., 1998. Ten years after drilling through the permafrost of the active rock glacier Murtel, Eastern Swiss Alps; answered questions and new perspectives. In: A.G. Lewkowicz and M. Allard (Editors), Seventh International Conference on Permafrost. Yellowknife, NWT, Canada, Centre d'Etudes Nordiques, Université Laval,Quebec, PQ, Canada: 403-410.

Harris, C., Arenson, L.U., Christiansen, H.H., Etzelmüller, B., Frauenfelder, R., Gruber, S., Haeberli, W., Hauck, C., Hölzle, M., Humlum, O., Isaksen, K., Kääb, A., Lehning, M., Lütschg, M.A., Matsuoka, N., Murton, J.B., Nötzli, J., Phillips, M., Ross, N., Seppälä, M., Springman, S.M. and Mühll, D.V., 2009. Permafrost and climate in Europe: geomorphological impacts, hazard assessment and geotechnical response. Earth Science Reviews, 92: 117-171.

Harris, C., Davies, M.C.R. and Etzelmüller, B., 2001a. The assessment of potential geotechnical hazards associated with mountain permafrost in a warming global climate. Permafrost and Periglacial Processes, 12: 145-156.

Harris, C., Kern-Luetschg, M., Murton, J., Font, M., Davies, M. and Smith, F., 2008. Full-scale physical modeling of solifluction processes associated with one-sided and two-sided active layer freezing. In: D.L. Kane and K.M. Hinkel (Editors), Ninth International Conference on Permafrost. Fairbanks, AK, U.S.A., Institute of Northern Engineering: 663-668.

Harris, C., Rea, B. and Davies, M., 2001b. Scaled physical modelling of mass movement processes on thawing slopes. Permafrost and Periglacial Processes, 12: 125-135.

Harris, J.S., 1995. Ground freezing in practice. Thomas Telford; Distributed by American Society of Civil Engineers, London; New York, 264 pp.

Hauck, C., 2001. Geophysical methods for detecting permafrost in high mountains. Mitteilungen der Versuchsanstalt für Wasserbau, Hydrologie und Glaziologie der ETH Zürich. Eigenverlag der Versuchsanstalt für Wasserbau Hydrologie und Glaziologie, Zurich, 204 pp.

Hayley, D.W. and Horne, B., 2008. Rationalizing climate change for design of structures on permafrost: A Canadian perspective. In: D.L. Kane and K.M. Hinkel (Editors), Ninth International Conference on Permafrost Fairbanks, AK, USA, Institute of Northern Engineering: 681-686.

Hestnes, E., 2000. Impact of rapid mass movement and drifting snow on the infrastructure and development of Longyearbyen, Svalbard. In: K. Senneset (Editor), International Workshop on Permafrost Engineering. Longyearbyen, Svalbard, Norway: 259-282.

Hinkel, K.M. and Hurd, J.K.J., 2006. Permafrost Destabilization and Thermokarst following Snow Fence Installation, Barrow, Alaska, U.S.A. Arctic, Antarctic, and Alpine Research, 38(4): 530-539.

Hivon, E.G. and Sego, D.C., 1995. Strength of frozen soils. Canadian Geotechnical Journal, 32(2): 336-354.

Hoelzle, M., Wegmann, M. and Krummenacher, B., 1999. Miniature temperature dataloggers for mapping and monitoring of permafrost in high mountain areas: first experience from the Swiss Alps. Permafrost and Periglacial Processes, 10(2): 113-124.

Hooke, R.L., Dahlin, B.B. and Kauper, M.T., 1972. Creep of ice containing dispersed fine sand. Journal of Glaciology, 11(63): 327-336.

Huang, S.L., Bray, M.T., Akagawa, S. and Fukuda, M., 2004. Field investigation of soil heave by a large diameter chilled gas pipeline experiment, Fairbanks, Alaska. Journal of Cold Regions Engineering, 18(1): 2-34.

Instanes, A., 2005. Infrastructure: buildings, support systems, and industrial facilities. In: C. Symon, L. Arris and B. Heal (Editors), Arctic Climate Impact Assessment 2005. Cambridge University Press, Cambridge, New York, Melbourne, Madrid, Cape Town, Singapore, São Paulo: 907-944.

IPCC, 2007. Climate Change 2007: A report accepted by Working Group I of the Intergovernmental Panel on Climate Change but not approved in detail - Technical Summary. In: S. Solomon, D. Qin and M. Manning (Editors). WMO, Geneva, Switzerland: 74 pp.

Kääb, A., 2008. Remote sensing of permafrost-related problems and hazards. Permafrost and Periglacial Processes, 19(2): 107-136.

Kääb, A., Reynolds, J.M. and Haeberli, W., 2005. Glacier and permafrost hazards in high mountains. In: U.M. Huber (Editor), Global change and mountain regions: an overview of current knowledge. Springer: 225-234.

Keller, F., Haeberli, W., Rickenmann, D. and Rigendinger, H., 2002. Dämme gegen Naturgefahren. SIA, tec 21(17): 13-17.

Kern-Luetschg, M., Harris, C., Cleall, P., Li, Y. and Thomas, H., 2008. Scaled centrifuge modeling of solifluction in permafrost and seasonally frozen soils. In: D.L. Kane and K.M. Hinkel (Editors), Ninth International Conference on Permafrost. Fairbanks, AK, U.S.A., Institute of Northern Engineering: 919-924.

Keusen, H.-R. and Amiguet, J.L., 1987. Die Neubauten auf dem Jungfraujoch: Geologie, Felseigenschaften, Permafrost. Schweizer Ingenieur und Architekt, 30-31: 905-906.

Klügel, J.U., 2008. Seismic hazard analysis - Quo vadis? Earth-Science Reviews, 88(1-2): 1-32.

Kneisel, C., Hauck, C., Fortier, R. and Moorman, B., 2008. Advances in geophysical methods for permafrost investigations. Permafrost and Periglacial Processes, 19(2): 157-178.

Konrad, J.M., 1994. 16th Canadian geotechnical colloquium - Frost heave in soils - Concepts and engineering. Canadian Geotechnical Journal, 31(2): 223-245.

Krainer, K. and Mostler, W., 2002. Hydrology of active rock glaciers: Examples from the Austrian Alps. Arctic Antarctic and Alpine Research, 34(2): 142-149.

Ladanyi, B. and Melouki, M., 1993. Determination of creep properties of frozen soils by means of the borehole stress relaxation test. Canadian Geotechnical Journal, 30: 170-186.

Ma, W., Shi, C.H., Wu, Q.B., Zhang, L.X. and Wu, Z.J., 2006. Monitoring study on technology of the cooling roadbed in permafrost region of Quinghai-Tibet plateau. Cold Regions Science and Technology, 44(1): 1-11.

Margreth, S., 2007. Defense structures in avalanche starting zones. Technical Guideline as an aid to enforcement, SLF, Bern and Davos.

Marty, 2008. Regime shift of snow days in Switzerland Geophysical Research Letters, 35: L12501.

Maurer, H. and Hauck, C., 2007. Instruments and methods - Geophysical imaging of alpine rock glaciers. Journal of Glaciology, 53(180): 110-120.

McKenna, J.K. and Biggar, K.W., 1998. The rehabilitation of a passive-ventilated slab on grade foundation using horizontal thermosyphons. Canadian Geotechnical Journal, 35(4): 684-691.

Metternicht, G., Hurni, L. and Gogu, R., 2005. Remote sensing of landslides: An analysis of the potential contribution to geo-spatial systems for hazard assessment in mountainous environments. Remote Sensing of Environment, 98(2-3): 284-303.

Moser, K., 1999. Ankermörtel im Permafrost, EMPA, Swiss Federal Laboratories for Materials Testing and Research.

Musil, M., Maurer, H. and Green, A.G., 2003. Discrete tomography and joint inversion for loosely connected or unconnected physical properties: application to crosshole seismic and georadar data sets. Geophysical Journal International, 153(2): 389-402.

Musil, M., Maurer, H., Green, A.G., Horstmeyer, H., Nitsche, F.O., Vonder Mühll, D. and Springman, S., 2002. Shallow seismic surveying of an Alpine rock glacier. Geophysics, 67(6): 1701-1710.

Nahir, M., van Aanhout, M. and Reineche, S., 2005. Application of risk management to northern affairs program (NAP), Contaminated Sites in the North. Assessment and Remediation of Contaminated Sites in Arctic and Cold Climates: 189-198.

Nater, P., Arenson, L.U. and Springman, S.M., 2008. Choosing geotechnical parameters for slope stability assessments in alpine permafrost soils. In: D.L. Kane and K.M. Hinkel (Editors), Ninth International Conference on Permafrost. Fairbanks, AK, U.S.A., Institute of Northern Engineering: 1261-1266.

Ng, C.W.W., Springman, S.M. and Alonso, E.E., 2008. Monitoring the performance of unsaturated soil slopes. Geotechnical and Geological Engineering 26(6): 799-816.

O'Connor, K.M. and Dowding, C.H., 1999. Geomeasurements by pulsing TDR cables and probes. CRC Press, 402 pp.

Palmer, A.C. and Williams, P.J., 2003. Frost heave and pipeline upheaval buckling. Canadian Geotechnical Journal, 40(5): 1033-1038.

Panel on Energy Research and Delelopoment (PERD), 1998. Climate change impacts on permafrost engineeirng design, Environment Canada.

Phillips, M., 2006. Avalanche defence strategies and monitoring of two sites in mountain permafrost terrain, Pontresina, Eastern Swiss Alps. Natural Hazards, 39(3): 353-379.

Phillips, M., Bartelt, P. and Christen, M., 2000. Influences of avalanche-defence snow-supporting structures on ground temperature in Alpine permafrost terrain. Annals of Glaciology, 31: 422-426.

Phillips, M., Ladner, F., Muller, M., Sambeth, U., Sorg, J. and Teysseire, P., 2007. Monitoring and reconstruction of a chairlift midway station in creeping permafrost terrain, Grächen, Swiss Alps. Cold Regions Science and Technology, 47(1-2): 32-42.

Phillips, M. and Margreth, S., 2008. Effects of ground temperature and slope deformation on the service life of snow-supporting structures in mountain permafrost: Wisse Schijen, Randa, Swiss Alps. In: D.L. Kane and K.M. Hinkel (Editors), Ninth International Conference on Permafrost. Fairbanks, AK, U.S.A., Institute of Northern Engineering: 1417-1422.

Phillips, M., Margreth, S. and Ammann, W.J., 2003. Creep of snow-supporting structures in alpine permafrost. In: M. Phillips, S.M. Springman and L.U. Arenson (Editors), 8th International Conference on Permafrost. Zurich, Switzerland, A.A. Balkema: 891-896.

Phillips, M. and Schweizer, J., 2007. Effect of mountain permafrost on snowpack stability. Cold Regions Science and Technology, 47(1-2): 43-49.

Pimentel, E., Sres, A. and Anagnostou, G., 2007. Modelling of ground freezing in tunnelling, ITA-AITES World Tunnel Congress "Underground Space: the 4th Dimension of Metropolises", Prague.

Portland Cement Association (PCA), 1997. Portland cement, concrete and heat of hydration. Concrete Technology Today, 18(2): 1-4.

Rieder, U., Keusen, H.-R. and J.L., A., 1980. Geotechnische Probleme beim Bau der Luftseilbahn Trockener Steg - Klein Matterhorn. Schweizer Ingenieur und Architekt, 18: 428-431.

Riseborough, D., Shiklomanov, N., Etzelmüller, B., Gruber, S. and Marchenko, S., 2008. Recent advances in permafrost modelling. Permafrost and Periglacial Processes, 19(2): 137-156.

Rist, A. and Phillips, M., 2005. First results of investigations on hydrothermal processes within the active layer above alpine permafrost in steep terrain. Norsk Geografisk Tidsskrift - Norwegian Journal of Geography, 59: 177-183.

Rixen, C., Haeberli, W. and Stoeckli, V., 2004. Ground temperatures under ski pistes with artificial and natural snow. Arctic Antarctic and Alpine Research, 36(4): 419-427.

Robertson, A. and Shaw, S., 2005. Enviromine - Failure Modes & Effects Analysis (FMEA). http://robertsongeoconsultants.com/rgc_enviromine/Issues/cls_FMEA.asp, accessed: February 19, 2009.

Roer, I., Kääb, A. and Dikau, R., 2005. Rockglacier kinematics derived from small-scale aerial photography and digital airborne pushbroom imagery. Zeitschrift für Geomorphologie, 49(1): 73-87.

Schrott, L., 1996. Some geomorphological-hydrological aspects of rock glaciers in the Andes (San Juan, Argentina). Z Geomorph N.F., 104: 161-173.

Schrott, L., 1998. The hydrological significance of high mountain Permafrost and its relation to solar radiation. A case study in the high Andes of San Juan, Argentina. Bamberg Geographische Schriften, 15: 71-84.

Smith, L.B., Graham, J.P., Nixon, J.F. and Washuta, A.S., 1991. Thermal-Analysis of Forced-Air and Thermosiphon Cooling Systems for the Inuvik Airport Expansion. Canadian Geotechnical Journal, 28(3): 399-409.

Smith, M.W. and Tice, A.R., 1988. Measurement of the unfrozen water content of soils; a comparison of NMR and TDR methods. In: K. Senneset (Editor), Fifth International Conference on Permafrost. Trondheim, Norway: 473-477.

Springman, S.M. and Arenson, L.U., 2008. Recent advances in permafrost geotechnics. In: D.L. Kane and K.M. Hinkel (Editors), Ninth International Conference on Permafrost. Fairbanks, AK, U.S.A., Institute of Northern Engineering: 1685-1694.

Springman, S.M., Jommi, C. and Teysseire, P., 2003. Instabilities on moraine slopes induced by loss of suction: a case history. Géotechnique, 53(1): 3-10.

Steiner, W., Graber, U. and Keusen, H.-R., 1996. Construction in rock at 3550 meters elevation (Jungfraujoch, Switzerland). In: G. Barla (Editor), EUROCK'96 Prediction and Performance in Rock Mechanics and Rock Engineering. Torino, Italy, Balkema, Amsterdam: 543-550.

Stoffel, L., 1995. Bautechnische Grundlagen für das Erstellen von Lawinenverbauungen im alpinen Permafrost, Eidgenössisches Institut für Schnee- und Lawinenforschung (SLF), Davos, Switzerland

Strozzi, T., Kääb, A. and Frauenfelder, R., 2004. Detecting and quantifying mountain permafrost creep from in situ inventory, space-borne radar interferometry and airborne digital photogrammetry. International Journal of Remote Sensing, 25(15): 2919 - 2931.

Tart, R.G., 2003. Heave and solifluction on slopes. In: M. Phillips, S.M. Springman and L.U. Arenson (Editors), Eighth International Conference on Permafrost. Zurich, Switzerland, A.A. Balkema: 1135-1140.

Tart, R.G. and Ferrell, J.E., 2002. Performance of the Squirrel Creek slopes steep slopes on discontinuous permafrost, 11th International Symposium on Cold Regions Engineering: 39-49.

Terzaghi, K. and Peck, R.B., 1969. Soil Mechanics in Engineering Practice. Wiley.

Thalparpan, P., 2000. Lawinenverbauungen im Permafrost, Swiss Federal Institute for Snow and Avalanche Research, Davos.

Thielen, A. and Springman, S.M., 2005. First results of a monitoring experiment for the analysis of rainfall induced landslides. In: A. Tarantino, E. Romero and Y.J. Cui (Editors), International Symposium on Advanced Experimental Unsaturated Soil Mechanics—EXPERUS 2005. Trento: 549–554.

Thiis, T.K. and Jaedicke, C., 2000. Changes in the snowdrift pattern caused by a building extension - Investigations through scale modelling and numerical simulations. In: E. Hjorth-Hansen, I. Holand, S. Løset and H. Norem (Editors), Fourth International Conference on Snow Engineering. Trondheim, Norway, Balkema: 363-375.

Tice, A.R., Anderson, D.M. and Banin, A., 1976. The prediction of unfrozen water contents in frozen soils from liquid limit determinations. CRREL Report 76-8, U.S. Army Cold Regions Research and Engineering Laboratory.

U.S. Arctic Research Commission Permafrost Task Force, 2003. Climate Change, Permafrost, and Impacts on Civil Infrastructure. Special Report 01-03, U.S. Arctic Research Commission, Arlington, Virginia.

Ulitsky, V., Paramanov, V., Kudryavstev, S., Shaskin, K. and Lisyuk, M., 2003. Contemporary geotechnologies providing safe operation of railway embankments in permafrost conditions. In: W. Haeberli and D. Brandovà (Editors), Extended Abstracts of the Eighth International Conference on Permafrost reporting current research and new information: 167-168.

van Everdingen, R. (Editor), 1998 revised 2005. Multi-language glossary of permafrost and related ground-ice terms. National Snow and Ice Data Center/World Data Center for Glaciology, Boulder, CO

Vonder Mühll, D., 1996. Drilling in Alpine Permafrost. Norsk Geografisk Tidsskrift, 50: 17-24.

Vonder Mühll, D., Noetzli, J., Makowski, K. and Delaloye, R., 2004. Permafrost in Switzerland 2000/2001 and 2001/2002 Glaciological Report (Permafrost) No. 2/3 of the Glaciological Commission (GC) of the Swiss Academy of Sciences (SAS) and Department of Geography, University of Zurich.

Vonder Mühll, D.S., Arenson, L.U. and Springman, S.M., 2003. Temperature conditions in two Alpine rock glaciers. In: M. Phillips, S.M. Springman and L.U. Arenson (Editors), Eighth International Conference on Permafrost. Zurich, Switzerland, A.A. Balkema: 1195-1200.

Vonder Mühll, D.S. and Holub, P., 1992. Borehole logging in alpine permafrost, Upper Engadin, Swiss Alps. Permafrost and Periglacial Processes, 3: 125-132.

Vonder Mühll, D.S., Stucki, T. and Haeberli, W., 1998. Borehole temperatures in alpine permafrost: a ten year series. In: A.G. Lewkowicz and M. Allard (Editors), Seventh International Conference on Permafrost. Yellowknife, NWT, Canada, Centre d'Etudes Nordiques, Université Laval, Quebec, PQ, Canada: 1089-1095.

VTT (Valtion teknillinen tutkimuskeskus), 1987. Frost protection for house structures (in Finnish), Technical Research Centre of Finland, Geotech. Lab., Helsinki, Finnland.

Wei, M., Fujun, N., Satoshi, A. and Dewu, J., 2006. Slope instability phenomena in permafrost regions of Qinghai-Tibet Plateau, China. Landslides, 3(3): 260-264.

Wen, Z., Sheng, Y., Ma, W., Qi, J.L. and Wu, J.C., 2005. Analysis on effect of permafrost protection by two-phase closed thermosyphon and insulation jointly in permafrost regions. Cold Regions Science and Technology, 43(3): 150-163.

Williams, P.J., 1967a. Unfrozen water content of frozen soils and soil moisture suction. Publications of the Norwegian Geotechnical Institute, 72: 11-26.

Williams, P.J., 1967b. Unfrozen water in frozen soils. Publications of the Norwegian Geotechnical Institute, 72: 37-48.

Woo, M.-K., Kane, D.L., Carey, S.K. and Yang, D., 2008. Progress in permafrost hydrology in the new millennium. Permafrost and Periglacial Processes, 19(2): 237-254.

Yasufuku, N., Springman, S.M., Arenson, L.U. and Ramholt, T., 2003. Stress-dilatancy behaviour of frozen sand in direct shear. In: M. Phillips, S.M. Springman and L.U. Arenson (Editors), Eighth International Conference on Permafrost. Zurich, Switzerland, A.A. Balkema: 1253-1258.

Zaharko, B. and Sego, D., 2008. Foamed gravel backfill for piles in permafrost. In: 61st Canadian Geotechnical Conference and the 9th Joint CGS/IAH-CNC Groundwater Conference held at the Westin Edmonton, Alberta: 925-932.

In: New Permafrost and Glacier Research
Editors: Max I. Krugger and Harry P.Stern

ISBN: 978-1-60692-616-1

Chapter 2

PERMAFROST MODELING IN WEATHER FORECASTS AND CLIMATE PROJECTIONS

***Nicole Mölders*[*1,2] *and Gerhard Kramm*[*2]**

[1]Department of Atmospheric Sciences, Geophysical Institute,
[2]University of Alaska Fairbanks, Geophysical Institute,
University of Alaska Fairbanks, College of Natural Science and Mathematics,

ABSTRACT

This chapter briefly reviews current state-of-the-art in modeling permafrost in numerical weather prediction models (NWPMs), chemistry transport models (CTMs) and in general circulation models (GCMs) and earth system models (ESMs) for projecting the global climate. Pros and cons of various methods are assessed. Deficits of GCM/ESMs permafrost modeling practice are discussed based on gridded observed soil-temperature data; deficits of the treatment of permafrost in NWPMs and CTMs are elucidated by examples of site-by-site evaluations. In addition, the uncertainty in simulated soil moisture and heat fluxes due to uncertainty in soil physical and plant-physiological parameters is illustrated. The consequences of incorrect simulation of or even neglecting of permafrost processes for simulated weather and climate are discussed. Extreme changes in permafrost distribution and active layer depths, as they are associated with wildfires/fires and their impact on the simulated atmospheric conditions, are addressed as well. Finally the great challenges for improving permafrost simulations (grid resolution, lack of horizontally and vertically high resolved soil data, uncertainty in soil parameters, organic soils) in GCMs, ESMs, CTMs and NWPMs and how to address these challenges is outlined.

*903 Koyukuk Drive, Fairbanks, AK 99775, USA[1],+1 907 474 7910, molders@gi.alaska.edu
* P.O. Box 757320, Fairbanks, AK 99775-7320, USA[2], +1 907 474 5992, kramm@gi.alaska.edu

INTRODUCTION

At high latitudes and at high altitude of mountainous terrain, permafrost (defined as soil in which temperatures remains continuously at or below freezing point for, at least, two consecutive years) and the active layer (which thaws seasonally) are the primary subsurface components of the land-atmosphere system. Permafrost restricts the mobility of soil-water, and infiltration. Thus, capillary action, infiltration, and percolation are rather inefficient in permafrost. An important aspect of permafrost is the local temporal equilibrium between the ice, gaseous, and liquid phase of water within the soil. Any changes in heat diffusion and conduction caused by a change in snow thickness, insolation at the soil surface or infiltration affect all three water phases in the soil and soil temperature simultaneously. Any change in soil temperature results in freezing, thawing, sublimation or water vapor deposition and a release of latent heat or consumption of energy, again altering soil state variables (soil temperature, soil volumetric liquid water, ice and water vapor content) and fluxes (e.g., soil heat flux, soil water flux, soil water vapor flux). Thus, freeze-thaw cycles affect the thermal and hydrological properties of soil because of the release of latent heat and consumption of energy accompanied with phase transition processes.

Ice changes the dynamics of soil thermal fluxes through the dependence of soil volumetric heat capacity and thermal conductivity on soil volumetric water and ice content. The specific heat capacity of water is twice that of ice and the thermal conductivity of ice exceeds that of water about four times.

Regions with permafrost are characterized by low winter air-temperatures, low saturation pressure of water vapor and frequently stable stratification of the atmospheric surface layer. All these conditions lead to low evaporation. Consequently, soil moisture remains stored in the frozen active layer. Thus, in spring the capacity of the soil to take up additional water will be limited even if some of the soil already thawed. In consequence of all these, this means that permafrost may enhance spring peak flood events (Cherkauer and Lettenmaier 1999). In summer, transpiration and evaporation depend on the active layer depth, soil and vegetation type as well as on meteorological conditions.

The hydrological and thermal surface conditions associated with permafrost and the active layer affect the near-surface atmosphere, and hence weather and climate, by the exchange of heat, moisture, and matter at the soil-atmosphere interface (e.g., Stendel and Christensen 2002, Mölders and Walsh 2004). At the same time, the active layer depth is sensitive to weather; on the long-term, permafrost temperature and stability and active layer depth are sensitive to climatic change (e.g., Kane et al., 1991, Lawrence et al., 2006, Mölders and Romanovsky 2006). Permafrost thawing not only can cause huge economic and infrastructure damages and ecosystem changes, it also releases water and trace gases; the related changes in trace gas and water cycle and ecosystems again can feedback to climate. (e.g., Esch and Osterkamp 1990, Cherkauer and Lettenmaier 1999, Oechel et al., 2000, Serreze et al. 2000, Romanovsky and Osterkamp 2001, Zhuang et al., 2001). The coupling between soil moisture and thermal processes is fundamental to high-latitude soil irritations. Therefore, this coupling has to be considered appropriately in numerical weather prediction models (NWPMs) to capture the annual soil-temperature cycle and 2m-temperatures in winter (e.g., Viterbo et al., 1999). For all these reasons permafrost, permafrost dynamics, and soil-water freezing and soil-ice melting have also to be considered in climate and earth system modeling and for climate impact assessment.

Apparently permafrost variations have yet to receive a concerted effort within the context of global climate and earth system modeling. Recently, Luo et al. (2003) examined the performance of 21 modern land-surface models (LSMs) using their standalone versions and soil temperature observations along with fluxes and snow data from the 18-year Valdai dataset, a site without permafrost, but with regularly frozen ground in winter. Their study revealed that explicit inclusion of soil-water freezing and soil-ice melting improves the prediction of soil temperature and its seasonal and inter-annual variability. To appropriately represent the heat, moisture, and matter exchange at the soil-atmosphere interface, modern NWPMs, General Circulation Models (GCMs) and Earth System Models (ESMs) require suitable LSMs to simulate frozen ground and permafrost dynamics. For GCMs and ESMs such LSMs are indispensable for investigations of permafrost-climate feedbacks. ESMs also need to consider permafrost dynamics for examination on climate-permafrost-ecosystem change feedbacks.

Over the last decades geologists and geophysicists have developed site-specific permafrost models for investigation of permafrost dynamics (e.g., Goodrich 1982, Nelson and Outcalt 1987, Kane et al. 1991, Romanovsky and Osterkamp 1997, Smith and Riseborough 2001, Zhuang et al. 2001, Ling and Zhang 2003). Due to their fine vertical grid increments ($\leq$0.05 m) these kinds of permafrost models usually consume huge amounts of computational time. Since investigations of permafrost dynamics are typically oriented at decades or even centuries and the geological processes associated with changes in permafrost distribution are relatively slow, these kind of models typically run at large time steps (Mölders and Romanovsky 2006). Furthermore, most permafrost models are site-specific and calibrated (e.g., Romanovsky et al. 1997, Osterkamp and Romanovsky 1999, Romanovsky and Osterkamp 2001), i.e. new data have to be collected for calibration when they are supposed to be applied elsewhere (Mölders and Romanovsky 2006). Such calibration involves that the majority of available data at a site serves to determine optimal soil-transfer parameters, leaving the rest of data for model evaluation.

Applying a typical calibration technique certainly would lead to better predictions than those that are typically obtained with soil models designed for use in NWPMs, GCMs or ESMs. However, performing such a calibration technique for these models would require consistent soil temperature data for calibration for the typical domains of NWPMs and worldwide for GCMs or ESMs. As of today no such dataset exists making usage of calibrated permafrost models in NWPMs, GCMs or ESMs technically impossible. Furthermore, it has yet to be determined whether calibration coefficients may be climate sensitive. Coupling a permafrost model with a NWPM, GCM or ESM would also be a challenge because NWPM, GCM or ESM simulations require input of water and energy fluxes to the atmosphere at time steps of less than a minute to several minutes; consequently any vertically highly-resolved permafrost model would have to be run with this time-step making such coupled simulations computationally unattractive and in the case of weather forecasting even prohibitive. For all these reasons, the weather forecast, climate and earth system modeling communities do without calibration. They instead have developed various physically based concepts for predicting permafrost, active layer and soil frost processes. In doing so, knowledge and well-accepted concepts from permafrost and atmospheric sciences have been combined to build suitable soil models considering frozen soil physics for use in NWPMs, GCMs and ESMs.

All modern numerical NWPMs, GCMs and ESMs apply soil models embedded in their LSMs to simulate the thermodynamic and hydrological surface forcing (e.g., temperature, specific humidity, fluxes of water vapor and sensible heat) at the soil-atmosphere interface. The atmospheric scientific community has developed these soil models based on the best knowledge, and spent great efforts to evaluate and improve them (e.g., Yang et al. 1995, Shao and Henderson-Sellers 1996, Lohmann et al. 1998, Dai et al. 2003, Mölders et al. 2003a). Incomplete knowledge of soil type and heterogeneity as well as soil initial conditions generally limits the predictability of the soil state, fluxes of heat, trace gases, water vapor, and water, and phase-transition processes within the soil and at the atmosphere-soil interface. Further errors in simulating soil conditions and fluxes result from the necessity to parameterize sub-grid scale processes, prescribe soil physical parameters, discretized partial differential equations and incorrectly simulated forcing (e.g., precipitation rate and amount, insolation).

In the following, the theory of soil physics, history and current state of modeling soil physics in atmospheric sciences is reviewed and evaluated; error sources for simulated permafrost quantities are discussed.

Theory of Soil Physics

The earth system has various interactions between its various spheres (lithosphere, cryosphere, atmosphere, ocean). These interactions occur at various temporal and spatial scales. Depending on the time scale one is interested in certain processes are so slow that they seem not even to exist and hence are negligible at this scale (Fig. 1). For instance, at the typical forecast range of NWPMs (up to 5 or 10 days) the spatial distribution of permafrost does not change, but the active layer depth may change notably; over a typical climate period of 30 years that is considered by ESMs or GCMs, however, the spatial distribution of permafrost may significantly (even in a statistical sense) change in response to atmospheric warming or cooling over this climate period.

In the absence of impermeable layers soil-water motion in the vertical is more distinct than lateral soil-water movements due to gravity forces. Typically lateral soil-water movement, V, is of the order of up to several centimeters per day. In any atmospheric model, the horizontal extension of model grid cells, L, is several hundred meters to several 100 kilometers. Scale analysis shows that on the typical time scale, T, of atmospheric models the lateral soil-water movement is several orders of magnitude smaller than vertical soil-water transport (see Fig. 1).

The vertical heat- and water-transfer processes and soil-water/soil-ice freezing/thawing can be expressed based on the principles of the linear thermodynamics of irreversible processes (e.g., de Groot 1951, Prigogine 1961) including the Richards-equation (e.g., Philip and de Vries 1957, Philip 1957, de Vries 1958, Kramm 1995, Kramm et al. 1996, Mölders 1999). The governing balance equations for heat and moisture including phase transition processes and water extraction by roots χ read (e.g., Philip and de Vries 1957, de Vries 1958, Sasamori 1970, Flerchinger and Saxton 1989, Kramm et al. 1994, 1996, Mölders et al. 2003a)

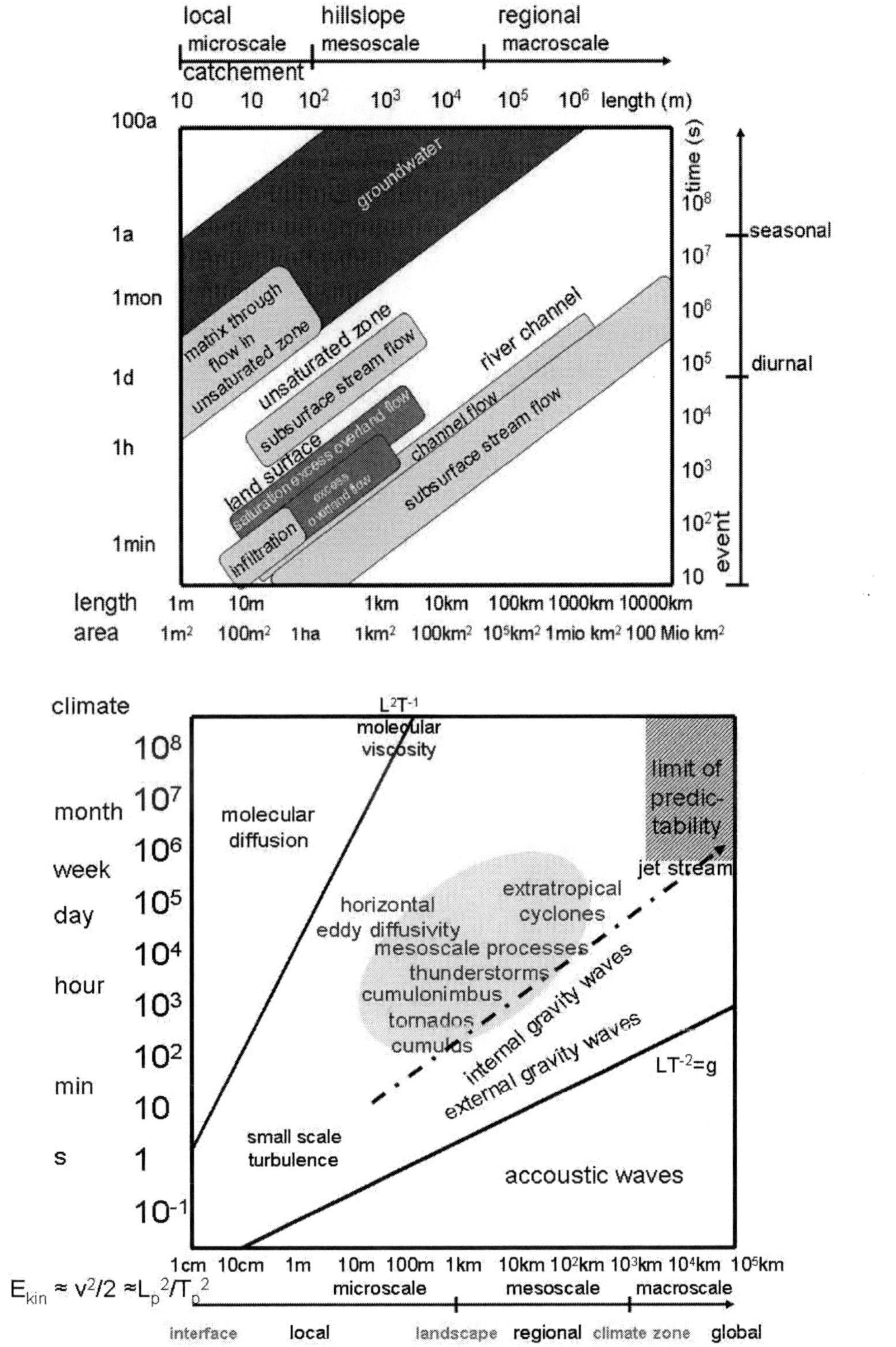

Figure. 1. Schematic views of hydrological (left) and atmospheric (right) scales modified after Bronstert et al. (2005) and Hantel (1997), respectively.

$$C\frac{\partial T_S}{\partial t}=\frac{\partial}{\partial z_S}\left(\lambda\frac{\partial T_S}{\partial z_S}\right)+\frac{\partial}{\partial z_S}\left(L_v\rho_w D_{T,v}\frac{\partial T_S}{\partial z_S}\right)+\frac{\partial}{\partial z_S}\left(L_v\rho_w D_{\eta,v}\frac{\partial \eta}{\partial z_S}\right)+L_f\rho_i\frac{\partial \eta_i}{\partial t} \tag{1}$$

$$\frac{\partial \eta}{\partial t}=\frac{\partial}{\partial z_S}\left(D_{\eta,v}\frac{\partial \eta}{\partial z_S}\right)+\frac{\partial}{\partial z_S}\left(D_{\eta,w}\frac{\partial \eta}{\partial z_S}\right)+\frac{\partial}{\partial z_S}\left(D_{T,v}\frac{\partial T_S}{\partial z_S}\right)+\frac{\partial K_w}{\partial z_S}-\frac{\chi}{\rho_w}-\frac{\rho_i}{\rho_w}\frac{\partial \eta_i}{\partial t} \tag{2}$$

Here z_S, λ, L_v, L_f, T_S, η, η_i, $D_{\eta,v}$, $D_{\eta,w}$ and $D_{T,v}$ are soil depth, thermal conductivity, latent heat of condensation and freezing, soil temperature, volumetric water and ice content, and the transfer coefficients for water vapor, water, and heat. Soil hydraulic conductivity $K_w = k_s W^{2b+3}$ depends on the saturated hydraulic conductivity k_s, pore-size distribution index b, and relative volumetric water content $W = \eta/\eta_s$ (e.g., Clapp and Hornberger 1978, Dingman 1994). The volumetric heat capacity of moist soil (Mölders et al. 2003a)

$$C = (1-\eta_s)\rho_S c_S + \eta\,\rho_w\,c_w + \eta_i\,\rho_i\,c_i + (\eta_s - \eta - \eta_i)\rho_a\,c_p \tag{3}$$

depends on the porosity of the non-frozen soil, η_s, the densities of dry soil, ρ_S, water, ρ_w, ice, ρ_i, and air, the specific heat of dry soil material, c_S, water, c_w, ice, c_i, and air at constant pressure. Soil volumetric heat capacity increases with increasing soil moisture for most of soils (e.g., Oke 1978). The thermal conductivity λ of unfrozen ground is a function of the soil-water potential $\psi = \psi_s W^{-b}$ also called matric potential, suction and tension head with ψ_s being the saturated water potential. Figure 2 exemplarily shows for various soil-types the dependence of thermal diffusivity on relative volumetric water content. At soil temperatures below $0^\circ C$, thermal conductivity depends on volumetric ice and water content. Thermal diffusivity more than doubles twice when relative volumetric water content increases from 0.5 to saturation (W=1; Fig. 2).

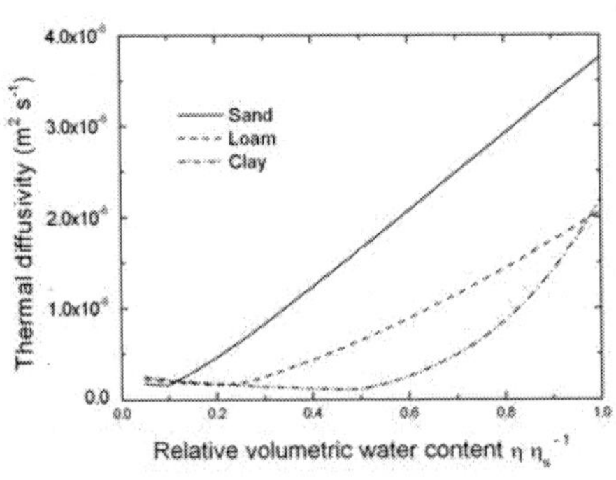

Figure. 2. Dependence of thermal diffusivity on relative volumetric water content for various soil-types. Modified from Mölders (2001).

The transfer coefficients for water vapor, water and heat are given by (Philip and de Vries 1957, Kramm 1995, Kramm et al., 1996)

$$D_{\eta,v} = -\alpha\nu D_w b \frac{\eta_s - \eta}{\eta} \frac{\rho_d}{\rho_w} \frac{g\psi}{R_d T_S} \tag{4}$$

$$D_{\eta,w} = -\frac{bk_s\psi_s}{\eta}\left(\frac{\eta}{\eta_s}\right)^{b+3} \tag{5}$$

$$D_{T,v} = \alpha\nu D_w (\eta_s - \eta)\frac{\rho_d}{\rho_w} \frac{L_v - g\psi}{R_d T_S^2} \tag{6}$$

Here g, α, ν, D_w, ρ_d, and R_d are gravity acceleration, a torsion factor that considers curvatures in the soil material due to roots (Sasamori 1970, Zdunkowski et al. 1975, Sievers et al. 1983, Kramm et al. 1996), a correction factor that is typically close to 1, the molecular diffusion coefficient of water vapor in moist air, and the density and gas constant for dry air, respectively.

In Eq. (1), the first term on the right side represents soil-temperature changes by divergence of soil-heat fluxes. The second term describes the divergence of soil-heat fluxes due to water-vapor transfer. The third term expresses how a soil-moisture gradient contributes to the soil-temperature change (Dufour effect), and the last term addresses soil-temperature changes due to freezing/thawing. In Eq. (2), the first two terms on the right side give the changes in volumetric water content caused by divergence of water vapor and water fluxes. The third term indicates how a temperature gradient contributes to the change in volumetric water content (Ludwig-Soret effect). The saturation vapor pressure is a function of soil-temperature. Consequently, a soil-temperature gradient leads to differences in saturation pressure and a water vapor flux that modifies soil moisture. This phenomenon is well known to exist in other porous media (e.g., snow). The fourth term gives changes due to hydraulic conductivity, the fifth considers water uptake by roots, and the last term represents changes due to freezing/thawing. The Ludwig-Soret and Dufour effects are cross-phenomena typically considered in the thermodynamics of irreversible processes.

If ice is present, soil-water potential Ψ, will remain in equilibrium with the vapor pressure over pure ice given by (Fuchs et al. 1978)

$$\Psi = \pi + \frac{L_f (T_S - 273.15)}{g\,T_S} \tag{7}$$

Here π is the osmotic potential. Osmotic effects due to solutes are typically omitted in NWPMs. However, they should be considered in chemistry transport models (CTMs), GCMs and ESMs in conjunction with solute chemistry because thawing of the active layer or permafrost releases traces gases (e.g., methane).

At any given soil temperature below 0°C all water in excess of (Flerchinger and Saxton 1989)

$$\eta_{max} = \eta_s \left\{ \frac{L_f \left(T_S - 273.15\right)}{g\, \psi_s\, T_S} \right\}^{-1/b} \tag{8}$$

freezes. Figure 3 exemplarily shows the dependence of maximum liquid water content on soil temperature for some selected soil-types. Considering the differences in volumes taken by water and ice, the volumetric ice content

$$\eta_i = \left(\eta_{total} - \eta_{max}\right) \frac{\rho_i}{\rho_w} \tag{9}$$

is proportional to the difference of the total water (liquid, solid, gaseous) within the soil layer minus the maximum liquid water content for temperatures below freezing point

Water extraction by roots and the following transpiration act as a soil-water sink. Soil-water uptake by roots, among other things, depends on vegetation-type, soil-physical and geologic characteristics, plant available soil-water, soil-temperature, aeration, competition or interaction with roots of other species, fertilizer, biologic and soil-chemical processes and transpiration. Various parameterizations have been developed with varying complexity (e.g., Gardner 1960, Cowan 1965, Federer 1979, Sellers et al. 1986, Martin 1990, Mölders et al., 2003a). The main differences between the various approaches are the assumptions on water-uptake restrictions, root-length, vertical distribution, whether or not root distribution varies with time, soil and/or vegetation type. Most recent LSMs used in atmospheric models assume equal distribution of roots in the root zone or only distinguish between the upper and lower root space (cf. Table 1). In the latter case, it is further assumed that the boundary between the two root spaces falls together with a soil-layer boundary; the same is true for maximum root length (e.g., Wilson et al. 1986, Martin 1990).

Simulating Frozen Ground

Since the cross-effects are very small under most conditions and since volumetric heat capacity and thermal conductivity of the substrate influence each other only marginally, decoupled equations to describe the energy- and water-transport within the soil are commonly used (e.g., Deardorff 1978, McCumber and Pielke 1981, Groß 1988, Dickinson et al. 1993, Schlünzen 1994, Jacobson and Heise 1982, Eppel et al. 1995, Chen and Dudhia 2002, Dai et al. 2003). However, this decoupling is realized in various ways as described in the following. Table 1 lists the various methods used by recent soil models of NWPMs, GCMs, and ESMs.

Table 1. Classification of soil-models used in NWPMs, GCMs and ESMs with respect to parameterizations used and model approaches. The symbol X indicates free choice of the number of layers, values in brackets are the typical choice in the models indicated.

Name	Reference	Purpose	Number of layers for			Treatment of			
			η, η_{ice}	T_s	root	moisture	temperature	roots	Soil frost
SECHIBA	Ducoudre et al. 1993, Polcher	GCM	2	2	2	Top-to-bottom filling			
NCEP	Chen et al. 1996	NWP	2	2	1	.	force-restore		
BATS	Dickinson et al. 1986, 1993, Yang et al. 1997	GCM	3	3	3	Richard's equation	heat diffusion	Federer-Cowan-model	Uniform freezing to –4C, limited thermal diffusivity
BEST	Cogley et al. 1990, Pitman		3	3	2	diffusion	heat diffusion		Explicitly, ice and water co-exist
BASE	Verseghy 1991, Desborough 1997, Pitman et al. 1991, Slater et al. 1998	GCM	3	3	3	Richard's equation	heat diffusion		Explicitly, ice and water co-exist
PROGMOD[1]	Ács et al. 2000	Substitutes surface flux data in diagnostic model	3	3	1	Richard's equation	force-restore	Hornet-approach, equal distributed	Bulk-heat capacity, freezing/ melting
SURF[2]	Claussen 1988	mesoscale modeling	2	5	0	Force-restore	heat diffusion	none	none
SURFW[3]	Claussen 1988, Mölders 1988, Fröhlich & Mölders 2002	mesoscale modeling	2	5	0	Force-restore	heat diffusion	none	none
HTSVS[4]	Kramm et al. 1994, 1996, Mölders et al. 2003, Mölders & Rühaak 2002, Mölders & Walsh 2004	Dry deposition, LSM in mesoscale modeling	X	X	X	Coupled diffusion equation including Richard's equation		Cowan-type, different in upper/lower root space	Explicitly, ice and water coexist

[1] Used by Universität Wien, Universität Budapest, Universität Bayreuth

[2] Used by GESIMA at GKSS, Universität Leipzig; similar LSMs are used by FITNAH at Universität Hannover or METRAS at Universität Hamburg, and Institut für Troposphärenforschung Leipzig

[3] Used by GESIMA at Universität Leipzig

[4] used by MM5 at EURAD Universität zu Köln, Universität Leipzig, (older Version at Universität Frankfurt)

Table 1. (Continued)

Name	Reference	Purpose	Number of layers for			Treatment of			
			η, η_{ice}	T_s	root	moisture	temperature	roots	Soil frost
OSULSM[5]	Chen & Dudhia 2001a,b	LSM in mesoscale modeling	X (4)	X (4)	3	Richard's equation	heat diffusion		none
MOSAIC	Koster & Suarez (1992)	GCM	3	2	2	Darcy's law	Force restore		
BUCKET	Manabe 1969, Robock et al. 1995, Schlosser et al. 2000	GCM	0	1	1	bucket	heat balance		
PLACE	Wetzel & Chang 1988, Wetzel & Boone 1995	Flexible	7	50	2	Richard's equation	heat diffusion		
SiB	Sellers et al. 1986, Xue et al. 1991, Sud & Moko 1999	GCM, mesoscale	3	2	2	diffusion	Force-restore	Federer-Cowan-model	
SEWAB[6]	Mengelkamp et al. 1999, 2001, Warrach 2001	Macroscale hydrological, mesoscale atmospheric models	6	6	1	Richard's equation	heat diffusion	different in upper/lower root space	
VEGMOD[7]	Schädler 1990, Grabe 2001								
ECHAM[8]	Dümenil & Todini 1992	LSM in GCM	1	5	2	bucket	heat diffusion	Equally distri-buted	
ECMWF[9]	Viterbo & Beljaars 1995	LSM in mesoscale modeling	4	4	3	Richard's equation	heat diffusion	Expon-ential decrease with depth	

[5] used by MM5 at EURAD Köln, LMU München, IFU Garmisch-Partenkirchen
[6] used by GKSS
[7] used by Universität Karlsruhe, Forschungszentrum Karlsruhe, IFU Garmisch-Partenkirchen
[8] used by MPI Hamburg
[9] used by ECMWF

Name	Reference	Purpose	Number of layers for			Treatment of			
			η, η_{ice}	T_s	root	moisture	temperature	roots	Soil frost
TERRA[10]	Jacobson & Heise 1982	LSM in mesoscale modeling	3	3	1	Force-restore	Force-restore	Resis-tance	no
WASIM[11]		Hydrology					none		
BIOME_ BCG[12]	Running & Hunt 1993.	Ecology	1	1	1	Top-to-bottom flow (no upwards flow)	Force-restore		
LSM_FU-Berlin	Blümel 2001	Determine H from satellite data	0	0	0	-.-	-.-	-.-	
ISBA[13]	Noilhan & Planton 1989, Mahouf & Noilhan 1991, Douville et al. 1995, Boone et al. 2000	LSM in GCM, mesoscale modeling, NWPM	2	2-3	1	Force-restore	Force-restore		Explicit, ice and water coexist
SWB	Chen et al. 1996, Schaake et al. 1995	hydrology	2	-.-	1	Richard's equation	-.-		
CLASS	Verseghy et al. 1991, 1993, Chatta & le Treut 1994, de Ronag & Polcher 1998	GCM	3	3	3	Darcy's law	Heat diffusion		Linear, temperature dependent freezing/melting
AMBETI[14]	Braden 1995	Agricultural consulting	3	3	3				
GISS	Abramopoulos et al. 1988, Lynch-Stieglitz 1994	GCM	6	6			Heat diffusion		

[10] Used by LM, DM at DWD, IfT, Universität Bonn, by REMO at MPI Hamburg, TU Cottbus
[11] used by ETH Zürich
[12] Used by Universität Bayreuth, BITÖK, PIK
[13] Used by FOOT3D, Universität zu Köln, Institut für Geophysik und Meteorologie
[14] used by DWD, Universität Bayreuth

Force-Restore Method

The force-restore method had been introduced by Deardorff (1978) and became standard for NWPMs in the mid-Eighties. A force-restore model (Fig. 4) considers at least a thin top layer of depth d_1 and a deep soil layer of depth d_2 for which the soil temperature and moisture states are calculated. The force-restore model considers two distinctly different time scales in soil. The conditions in the uppermost layer are governed by the rapid responses to atmospheric forcing (e.g., precipitation, evaporation, diurnal course of atmospheric heating). These changes are represented by the so-called force term. The deeper soil layer only responses slowly to the atmospheric forcing. It typically represents annual changes. The interaction between the upper and deeper soil is considered by the restore term that describes the supply of heat and soil moisture from the deep soil layer. Some versions of the force-restore model consider a third layer that considers decadal variation. In all layers, prognostic equations are solved to determine soil temperature or moisture conditions. In doing so, soil temperature and moisture conditions are assumed to be independent from each other except if freezing/thawing is considered. Then these phase transitions lead to a change is soil temperature.

NWPMs that use the force-restore method are limited in resolving the various soil horizons (Montaldo and Albertson, 2001). High latitude soils, however, frequently show a very heterogeneous vertical stratification because they were formed by during the ice age. Moreover, the force-restore methods does not permit for simulating the vertical distributions of soil processes like the diurnal variation of the boundary between an unfrozen upper and a frozen deeper soil layer because it works with only two or three reservoirs. However, surface-water and energy fluxes are extremely difficult to predict without knowing the exact depth of the freezing line.

NWPMs that use the force-restore method are, for instance, the Deutschland Model (DM) of the German Weather Service (e.g., Jacobson and Heise, 1982), the APREGE of Météo-France and the Spanish Weather Service that both use ISBA (Noilhan and Planton 1989, Mahfouf et al., 1995); GCMs using a force-restore method are CSIRO9, and the ARPEGE climate model (DéQué et al., 1994, Mahfouf et al., 1995). See also Table 1.

Multi-Layer Models

Multi-layer soil models (Fig. 4) are most suitable for permafrost simulation in NWPMs, CTMs, GCMs and ESMs because they permit for simulating the vertical distributions of soil processes like the diurnal variation of the boundary between an unfrozen upper and a frozen deeper soil layer. Consequently, huge efforts have been spent to enlarge multi-layer soil models by soil-frost processes (e.g., Koren et al., 1999, Boone et al., 2000, Warrach et al. 2001, Mölders et al. 2003a, Narapusetty and Mölders 2005, 2006). Koren et al. (1999), for instance, tested and evaluated a soil-frost model offline that now is included with modifications in the NCEP (National Center for Environmental Prediction) Eta model. Mölders et al. (2003a) included the physics of soil-water freezing and thawing of soil-ice into the soil-model of the Hydro-Thermodynamic Soil Vegetation Scheme (HTSVS; Kramm et al.

1994, 1996) that is used in several mesoscale meteorological models (e.g., GESIMA Mölders and Rühaak 2002; MM5 Mölders and Walsh 2004).

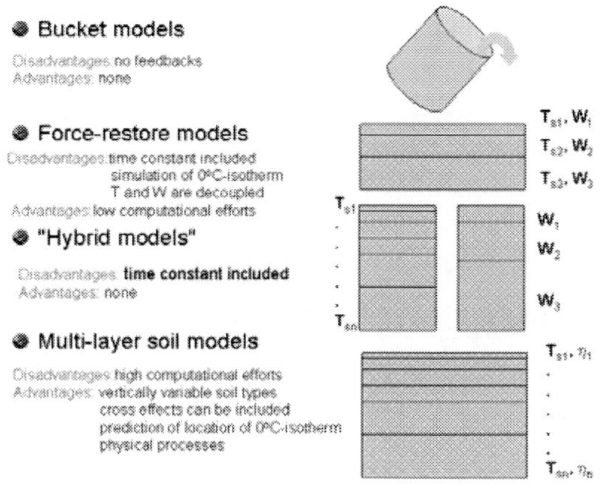

Figure 4. Schematic comparison of different concepts used for soil modeling in atmospheric models.

The impacts of soil-water freezing and soil-ice thawing in the active layer and the related processes have received little systematic study in the context of their influence on short-term weather.

Obviously, the coupled equation set (1) and (2) includes cross-effects like the Dufour effect (i.e., a moisture gradient contributes to the heat flux and alters soil temperature) and Ludwig-Soret effect (i.e., a temperature gradient contributes to the water flux and changes soil volumetric water content). Such a set of equation has either to be solved simultaneously by an iteration technique or must be simplified to avoid the iteration required by the coupling due to the cross-effects. Typically the interactions between the soil thermal and moisture regimes by the Ludwig-Soret and Dufour effect are neglected because they are negligible small under many circumstances. These interactions become noteworthy when chemicals are considered, for which they should be considered in CTMs and ESMs, when soil conditions suddenly switch from the dry to the wet mode, when soil temperatures vary around the freezing point, during snow-melt, and over the long-term these processes may gain influence on other processes or variables (Mölders and Walsh 2004). The Dufor-effect, for instance, was found to affect soil temperature up to 2 K, the Ludwig-Soret effect affects water recharge by 5 % of the total recharge over the long-term (Mölders et al., 2003b). Changes in soil temperatures and moisture caused by these cross-effects may alter the exchange of heat and moisture at the atmosphere-soil interface under these conditions.

The partial differential equations have to be discretized by a numerical scheme. Typically in LSMs of CTMs, NWPMs, GCMs and ESMs the Crank-Nicholson-scheme sometimes in conjunction with Gauß-Seidel-techniques are used (e.g., Kramm 1995). When using a Crank-Nicholson-scheme it is advantageous to introduce a logarithmic coordinate transformation into Eqs. (1) and (2) by $\xi = \beta \ln(z/z_D)$ before integrating to apply equal spacing and central differences for well appropriate finite difference solutions. Here, z_D is the lower boundary, and β is a constant which is to be chosen for convenience. Sensitivity studies showed that discretizing the partial differential equations by a type of Galerkin finite element scheme is advantageous for simulation of frozen soil physics (Narapusetty and Mölders 2006).

In most LSMs of NWPMS and CTMs thermal conductivity is assumed to be either constant or parameterized by using McCumber and Pielke's (1981) empirical formula (see also Kramm 1995, Kramm et al., 1996)

$$\lambda = \begin{cases} 419\exp(-(P_f + 2.7)) & P_f < 5.1 \\ 0.172 & P_f \geq 5.1 \end{cases} \tag{10}$$

With and $P_f = 2 + {}^{10}\log|\psi|$. Many state-of-the-art LSMs of NWPMS, CTMs, GCMs and ECMs use this parameterizations or variations thereof. For soil-temperatures below the freezing point the LSMs of many NWPMs, CTMs, GCMs and ESMs assume a mass-weighted thermal conductivity depending on the amounts of liquid and solid volumetric water content present

$$\lambda = \frac{\lambda_w \eta + \lambda_i \eta_i}{\eta + \eta_i} \tag{11}$$

Here the indices w and i stand for the liquid and solid phase of soil-water. In doing so, either a fixed or calculated value of thermal conductivity for the liquid and a value of 2.31 J/(msK) or so is used for the solid phase.

Predicted soil-temperature is highly sensitive to the thermal conductivity of the soil. Mölders and Romanovsky (2006) showed that the parameterization of thermal conductivity according to Eq. (4) provides much higher thermal conductivity values than typically found for permafrost soils; Eq. (4) also provides a decrease of thermal conductivity as the ground freezes, while observations typically indicate the opposite effect. In permafrost, thermal conductivity can be determined as Farouki (1981)

$$\lambda = \lambda_s^{(1-\eta_s)} \lambda_w^{(\eta_s - \eta_i)} \lambda_i^{\eta_i} \tag{12}$$

This formula is often applied in permafrost modeling (e.g., Lachenbruch et al., 1982, Riseborough 2002). Here, λ_s, λ_w(=0.57 W/(mK)), and λ_i(=2.31 W/(mK)) are the thermal conductivity of dry soil, water, and ice, respectively. Typical values for λ_s range between 0.06 and 0.25 W/(mK) (e.g., Pielke, 1984). In permafrost soils, typical values for λ range between 0.7 and 2.4 W/(mK) (e.g., Romanovsky and Osterkamp 2000).

Since permafrost soil pores are typically totally ice-filled, Farouki's formulation does not consider the possibility of partially air-filled pores because permafrost soils are usually saturated. Freezing of soil-water, however, also frequently occurs in mid-latitude winter or deserts where soil-pores are often partially filled with air. Since in NWPMs, GCMs and ESMs have also to be able to predict soil-temperatures accurately under these conditions, Mölders and Romanovsky (2006) enlarged the parameterization to include the impact of air

$$\lambda = \lambda_s^{(1-\eta_s)} \lambda_w^{\eta} \lambda_i^{\eta_i} \lambda_a^{(\eta_s - \eta - \eta_i)} \tag{13}$$

Here λ_a(=0.025 W/(mK)) is the thermal conductivity of air. This formulation is consistent with Eqs. (1) to (3) that explicitly consider water vapor fluxes (third and first on the right side of Eqs. (1) and (2), respectively) and air (last term of Eq. (3)). It leads to Farouki's formulation in the case of permafrost soils that are usually saturated meaning (Hinkel et al., 2001) $\eta_{air} = 0$, $\eta_s - \eta_i = \eta$, and $\lambda_a^{(\eta_s - \eta - \eta_i)} = \lambda_a^0 = 1$.

The empirical formulation with mass-weighted thermal conductivity values generally provides greater thermal conductivity values than Mölders and Romanovsky's (2006) parameterizations (e.g., Fig. 5, their Fig. 3). These authors report that thermal conductivity calculated with Eq. (10) ranges between 0.292 and 5.745 W/(mK), while values of about 2.2 W/(mK) and 1.5 W/(mK) were observed in the deeper and upper soil. Using the modified version of Farouki's formula yields thermal conductivity values between 0.149 (uppermost layer after dry episode) and 1.52 W/(mK) with about 1.1 W/(mK) on average. Note that an uncertainty analysis using Gaussian error propagating techniques identified Eq. (10) as a critical source of errors in predicted soil temperature because the natural variance in empirical parameters (pore-size distribution index, saturated water potential, porosity) propagates to great uncertainty in calculated thermal conductivity (Mölders et al., 2005). Uncertainty in parameters propagates less strongly when using the modified Farouki formula, for which parameter-caused statistical uncertainty in calculated thermal conductivity, soil temperatures, and soil-heat fluxes is lower than when using the mass-weighted formulation.

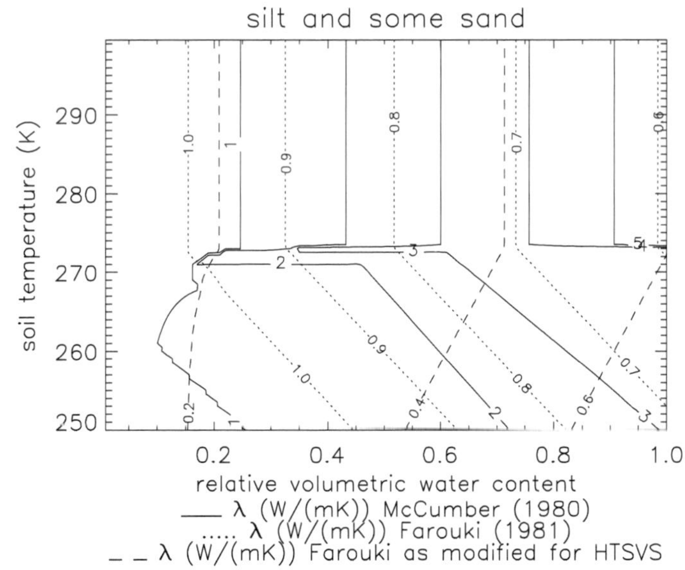

Figure. 5. Thermal conductivity as obtained by Eqs. (10) and (12). Figures for other soil types show similar basic pattern

Since the phase transitions alter soil temperature by release or consumption of heat diagnosis of soil ice has to be solved iteratively. Typically a first-order Newton-Ralphson-technique is applied (e.g., Mölders et al., 2003a).

NWPMs using multiple-layer soil models are, for instance, MM5 (Grell et al., 1994), WRF (Skamarock et al., 2005); GCMs and ESMs with multiple soil-layers are, for instance, the Community Climate System Model (CCSM) family, the ECMWF GCM and the Canadian GCM using CLASS (see also Table 1).

Hybrid Models

In hybrid soil models (Fig. 3), soil-wetness is determined by a force-restore-method (e.g., Deardorff 1978, Groß 1988, Schlünzen 1994, Jacobson and Heise 1982), while soil heat-fluxes and soil temperatures are calculated from a one-dimensional heat-diffusion equation (e.g., Claussen 1988, Groß 1988, Schlünzen 1994, Eppel et al., 1995). In these models, soil temperature layers typically differ from the two or three reservoirs used for soil moisture determination because the heat-diffusion equation is often solved for more than two or three layers to better capture the diurnal variation of soil temperature (e.g., Fig. 3). It is obvious that when soil temperature and soil moisture are calculated at different depths, permafrost hardly can be dealt with in this kind of soil model, for which they are not further discussed.

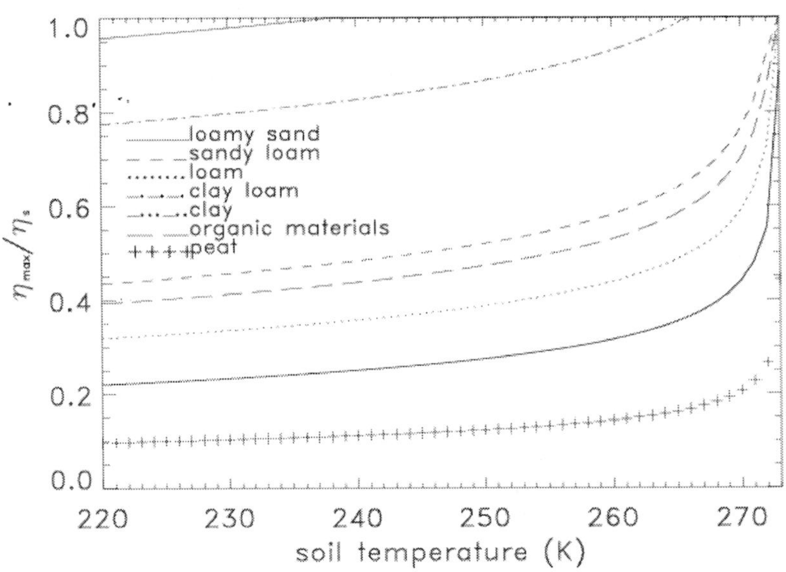

Figure. 3. Dependence of maximum liquid water content on soil temperature for some selected soil-types. From Mölders and Walsh (2004).

Vertical Resolution

In theory, fine soil-grid increments ensure accurate simulation of soil heat and moisture fluxes, temperature and moisture profiles. Unfortunately, global datasets of vertical distributions of soil type are not available. The of soil type and soil initial state data, and huge computational burden associated with a fine grid dictate the vertical grid resolution of soil models of CTMs, NWPMs, GCMs or ESMs. For reasonable turn-around times a compromise between efficiency and practical accuracy of soil-temperature is made. Modern CTMs and NWPMs typically use four to six (e.g., Smirnova et al., 1997, 2000, Chen and Dudhia 2001,

Skamarock et al., 2005, Grell et al., 2005, Mölders and Kramm 2007), GCMs and ESMs about ten logarithmically spaced soil layers that cover a depth down to 2 to 3m (e.g., Bonan et al., 2002, Stendel and Christensen 2002, Dai et al., 2003). Obviously, the number and position of grid nodes plays a role in how accurately the active layer depth can be captures (Fig. 6).

Boundary Conditions

The Earth's surface is the only physical boundary condition in atmospheric models. While the soil surface is part of the lower boundary with respect to the atmosphere, it is the upper boundary with respect to the soil. The lower boundary of any soil, i.e. the bottom of a soil model, is an artificial one. Ideally, it is put at a level of nearly constant soil temperature and moisture in 20 or 30 m depths or so. Doing so is especially important in permafrost soils, where decadal soil temperature variations exist even below 15 m depth (e.g., Romanovsky et al., 1997, Mölders and Romanovsky 2006). Most modern soil models used in atmospheric models have the lower boundary around 2 or 3 m depth (e.g., Kramm et al., 1995, Smirnova et al., 1997, 2000, Chen and Dudhia 2001, Dai et al., 2003, Mölders and Walsh 2004). NWPMs and CTMs typically assume climatologic soil temperatures that vary monthly and spatially at the bottom of the soil model for at least a month. Doing so introduces artificial sources and sinks for heat and moisture (e.g., Stendel and Christensen 2002). A constant soil temperature, for instance, will act as a heat source (sink) if the actual temperature is lower (higher) at that depth. While this shortcoming may be of minor impact when regarded over the short integration times of NWPMs and CTMs and if the soil temperature is appropriately set (e.g., Narapusetty and Mölders 2005), errors may accumulate over the long integration times (of at least 30 years=climate period) of GCMs and ESMs (Mölders and Romanovsky 2006). Therefore, soil models of GCMs and ESMs usually assume constant soil moisture and heat fluxes at their lower boundary (e.g., Dai et al., 2003, Oleson et al., 2004). Most soil models of GCMs or ESMs assume zero-flux conditions at the lower boundary (e.g., Oleson et al., 2004, Nicolsky et al., 2007). However, various observations (e.g., Zhang et al., 1996, Romanovsky et al., 1997, Mölders et al., 2003a, b) show non-zero heat and moisture fluxes at 2 or 3 m, the depth typically used as the lower boundary in soil models of GCMs or ESMs. Therefore, Mölders and Romanovsky (2006) performed simulations assuming zero-flux at 30 m depth where this assumption is generally fulfilled (see also Nicolsky et al., 2007). They found that a logarithmic grid-spacing with at least 20 layers is required to appropriately capture the diurnal cycle in the active layer (see also Fig. 6), the depth of the active layer, the annual soil temperature cycle, and the timing of thawing and freeze-up.

Heterogeneity Of Soil

Data from lysimeters filled with natural soil cores taken at the same site show evidence that the natural heterogeneity of soils may lead to notable differences in ground water recharge even on relatively short-term (Mölders et al., 2003a, b). It is obvious that such differences also impact soil-moisture and temperature condition. Such heterogeneity, however, is of subgrid-scale with respect to any soil model, and hence, not considered.

Other heterogeneity stems from the spatial variability of soils. Typically grid-cells of NWPMs and CTMs cover areas of several square-kilometers, while those of GCMs or ESMs encompass several 100 square-kilometers. Obviously soil type may vary or be even different over these areas. In most NWPMs and CTMs, the soil-type dominating within a grid-cell is assumed to be the representative one for the soil conditions within that grid-cell. This means soil temperature and moisture as well as heat and moisture fluxes are calculated using the soil parameters of the dominating soil. It also means that in areas of discontinuous permafrost either permafrost soil or no permafrost soil is assumed in a grid-cell. In both cases the neglecting of heterogeneity may lead to great errors in predicted soil temperature and hence active layer depth (see Fig. 7).

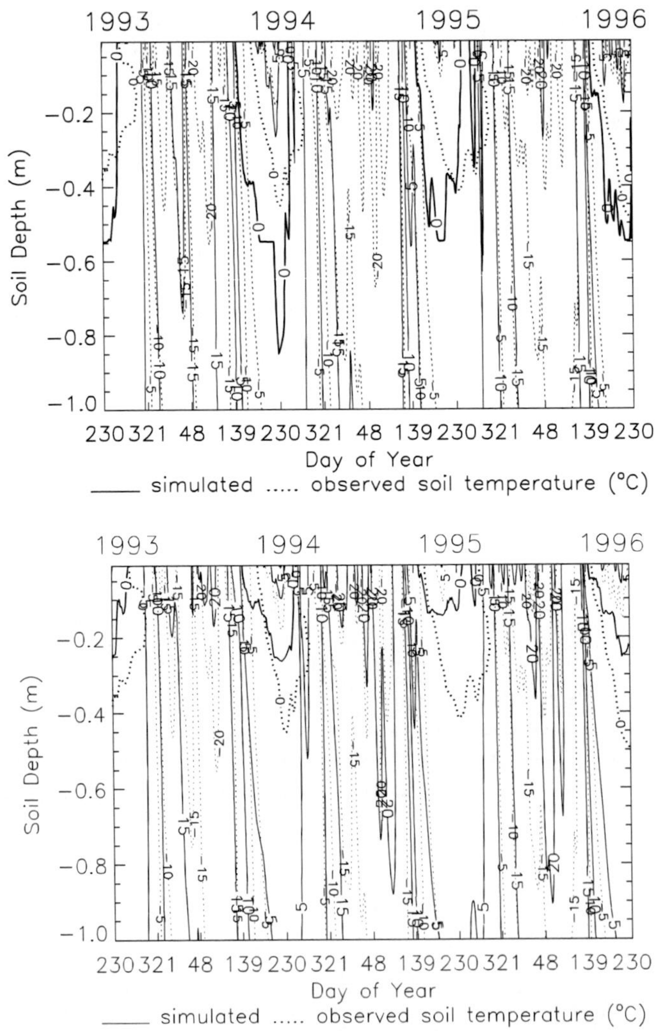

Figure. 6. Comparison of soil temperature as simulated at Barrow, Alaska with HTSVS and observed.In (a) 20 layers and in (b) 13 layers are logarithmically spaced to a depth of 30 m. Modified from Mölders and Romanovsky (2006).

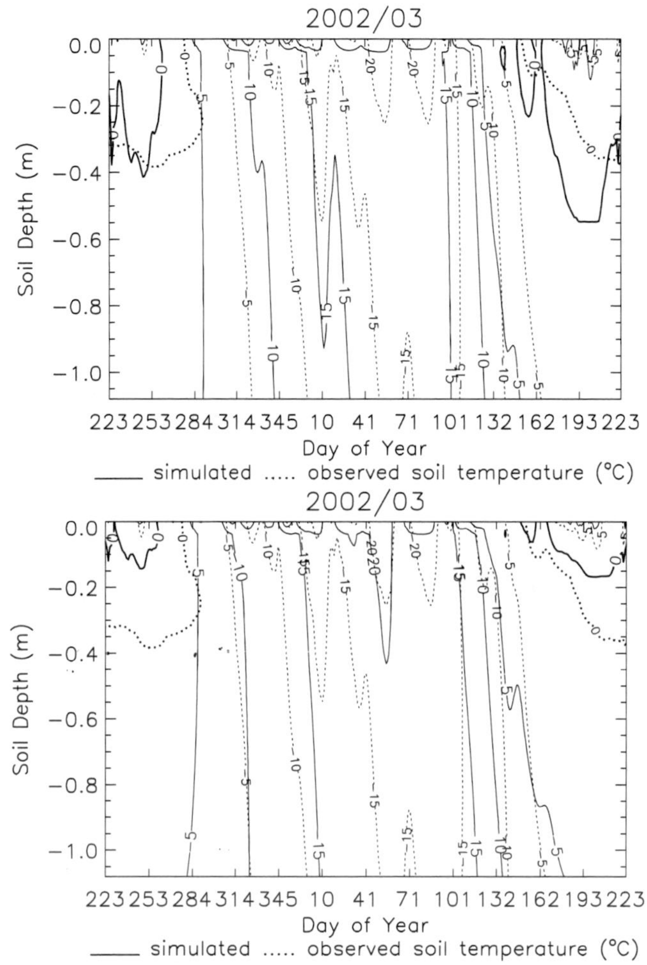

Figure. 7. Comparison of soil temperature as simulated with HTSVS and observed at Yakutsk, Siberia. In (a) soil type varies with depth, while in (b) a constant soil type, namely that of the uppermost layer is assumed. Observation data from Levine (2007; pers. communication

Some research meteorological models consider subgrid-scale spatial heterogeneity of soils by some kind of mosaic approach or subgrid-scheme (e.g., Mölders and Raabe 1996, Mölders et. al 1996). Considering subgrid-scale heterogeneity of soils can lead several Kelvin differences in soil temperature as compared to the strategy of dominant soil-type.

In modern GCMs and ESMs, the fact that soil type may vary horizontally in space is typically considered by some kind of mosaic or TOPMODEL approach (e.g., Dai et al., 2003, Essery et al., 2003, Oleson et al., 2005, Nui et al., 2005). Herein soil temperature and moisture conditions are determined for the various horizontal patches of different soil-type. The grid-cell soil temperature and moisture are then derived as an area-weighted average of the soil-temperatures of the various patches within the grid-cell.

Most modern soil models of NWPMs, GCMs and ESMs assume one soil-type for the entire soil column (e.g., Slater et al., 1998, Schlosser et al., 2000). Typically the uppermost soil-type is chosen to be representative for the entire soil column. The main reason is the lack of 3D-soil characteristic data. Nevertheless, some soil models of NWPMs (e.g., HTSVS in MM5) permit the user to consider vertical heterogeneity of soil for process research studies, rather than for general use in forecasts. Many soil models of modern GCMs or ESMs also are designed for consideration of vertically differing soil types, but basically make no use of the possibility due to the lack of 3D global distributions of soil-data. Examinations show that simulations without consideration of vertically varying soil characteristics miss many details in soil temperature and moisture patterns that result from the vertical profile of soil parameters (see Fig. 7). Mölders and Romanovsky (2006) found that even in the uppermost layer where the soil-type is the same, RMSEs between simulated and observed soil temperatures increased on average up to 0.3 K as compared to simulations with consideration of a vertically varying soil characteristic profile; moreover, simulations ignoring vertical soil characteristic profiles yield significantly different soil temperature variance than those considering it. Neglecting vertical soil characteristic profiles may yield to errors in predicting active layer depth and the timing of thawing and freeze-up of the active layer (e.g., Fig. 7).

Initialization Problem

One major problem is the initialization of soil moisture and temperature in NWPMs and CTMs. Unfortunately, global datasets of vertical distributions of soil temperature and moisture conditions do not exist.

In NWPMs and CTMs, usually the soil moisture and temperatures states obtained from the previous forecast are used as initial values for the following forecast. This procedure violates the assumption used to simplify the equations, namely that horizontal heat and moisture fluxes within the soil are negligibly small. In nature as in the model, mountains usually receive more precipitation than valleys (e.g., Müller et al., 1995). In nature, runoff on the short-term and lateral soil water fluxes on the long-term lead to moister valleys than mountains (except for glaciers where water is stored in the solid phase). Consequently, when initializing NWPMs and CTMs as described before the neglecting of lateral soil water fluxes leads to too high soil moisture in mountainous regions and too low moisture in the lower elevated terrain. In weather forecasts, these errors yield to incorrect prediction of local recycling of previous precipitation and hence wrong forecasts of convection, showers, and thunderstorms (e.g., Mölders and Rühaak 2002). In permafrost regions, some of the permafrost exists in the valleys and is fed by runoff from the mountains, i.e. in such cases cannot be appropriately captures due to the initialization method. These errors can be avoided by either inclusion of horizontal moisture transport (3D soil model), or coupling/using the soil model with a hydrological model (e.g., Mölders and Raabe 1997, Mölders et al., 1999, Mölders 2001, Walko et al., 2000, Mölders and Rühaak 2002).

Being aware that lateral soil-water movements may be important on longer time scales, in the nineties several authors (e.g., Kuhl and Miller 1992, Marengo et al., 1994, Miller et al., 1994, Sausen et al., 1994, Hagemann and Dümenil 1998) introduced parameterizations of different complexity to consider runoff in GCMs. Some kind of TOPMODEL-approach (e.g., Beven and Kirkby 1979) considers soil moisture heterogeneity (e.g., Dai et al., 2003, Essery et al., 2003, Nui et al., 2005) and permits also for consideration of discontinuous permafrost. GCMs and ESMs typically initialize soil temperature and moisture states homogenously worldwide and run the soil model with the forcing of one year for several centuries until an equilibrium soil state distribution is established.

UNCERTAINTY IN PERMAFROST MODELING

The atmospheric science community spend huge efforts on investigating uncertainty in modeling the soil conditions in atmospheric models because errors in simulated soil states and fluxes may propagate into errors in atmospheric state variables and fluxes. The following sources of error have been identified:

- Discretization, vertical and temporal resolution
- Initial and boundary condition
- Subgrid-scale heteorgeneity
- Forcing data
- Assumptions and/or parameterization concepts
- Uncertainty in soil physical parameters
- Data on soil type distribution

For a further discussion of the error sources mentioned in the first three bullets see also the respective subsections of section Simulating Frozen Ground.

Input of heat by precipitation, changes in insolation the soil surface due to cloudiness, changes in soil heat flux at the soil surface due to changes in wind speed can affect soil temperature, soil moisture, as well as soil moisture and heat fluxes (e.g., PaiMazumder et al., 2008). Since these changes in meteorological forcing occur on very short time scales, the temporal resolution like the vertical discretization has an impact on the accuracy with which diurnal change of soil temperatures and active layer depths can be predicted by soil models of atmospheric models. Figure 8 exemplarily shows results from simulations with different time steps and illustrates how temporal resolution can affect simulated soil temperature profiles on the long-term.

First of all, uncertainties in simulating soil temperature regimes may results from incorrectly simulated processes in the NWPM, CTM, GCM or ESM itself (e.g., Avissar and Pielke 1989, Calder et al., 1995, Mölders et al., 1996, 1997, Niu and Yang 2004).

Various sensitivity studies aimed at detecting error sources related to assumptions and/or parameterization concepts (e.g., Robock et al., 1995, Cuenca et al., 1996, Shao and Irannejad 1999). As aforementioned, the force-restore method, for instance, has only limited ability to resolve soil horizons (see Fig. 4) and to simulate the vertical distributions of soil conditions and processes (e.g., diurnal variation of the freezing line).

In any soil model in NWPMs, CTMs, GCMs and ESMs prescribed soil parameters (e.g., Table 2) represent different soil types. Ideally, the soil characteristics should be mapped as vertical and horizontal three-dimensional continuous distribution to capture the gradients and mixtures in soil type within a grid-cell or a patch of same soil type within a grid-cell. However, soils are spatially heterogeneous for which attributing a single soil type to an area or patch of several square-kilometers as it is required in atmospheric models can be ambiguous, and is a potential error source. Using a wrong soil type, for instance, can cause errors in predicted near-surface air temperatures and humidity of more than 0.5K and 0.5g/kg even in a 24-hour simulation (Mölders 2001). Assigning soil physical parameters to an area or patch is also ambiguous in GCMs or ESMs because soil surface properties can vary in time due to various events (e.g., burning of organic soils during wildfires, land avalanches, flooding, volcanic eruptions) or may be influenced by previous weather conditions (weathering) over centuries. Thus, prescribed fixed values of soil parameters for in nature time-dependent quantities may introduce uncertainty in climate and earth system modeling. Furthermore, the variability in some soil parameters is sometimes greater within the same soil type than across soil types (cf. Table 2). There is observational evidence from lysimeter studies that the heterogeneity within the same soil may cause differences in evapotranspiration and recharge of 112mm (14%) and 137mm (4%) in 5.6 years (Mölders et al.,, 2003b).

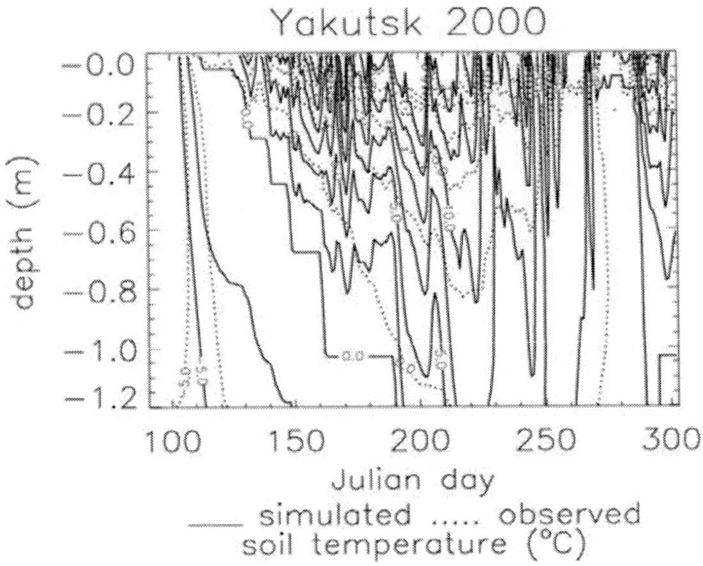

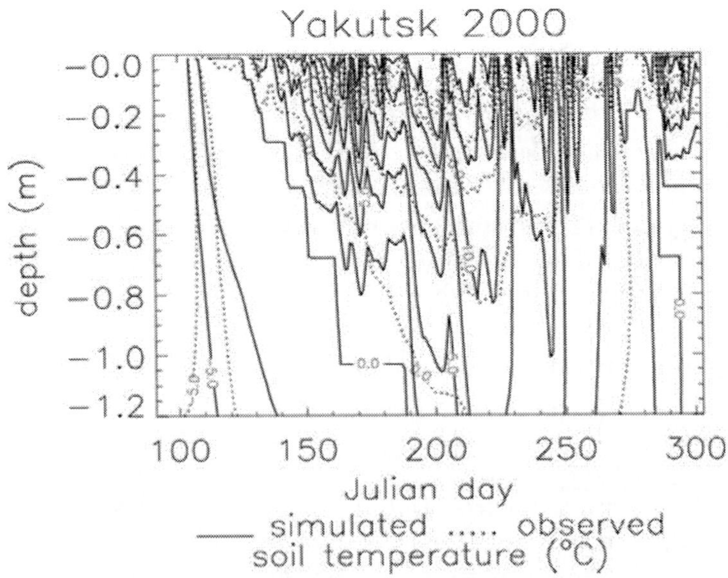

Figure. 8. Comparison of soil temperature as simulated with HTSVS and observed at Yakutsk, Siberia. The time step used in (a) is three times smaller than that used in (b). Observation data from Levine (2007; pers. communication).

Table 2. Typical mean values and standard deviations (in brackets) of soil characteristics. The symbols $k_s, \eta_s, b, \psi_s, \rho_s$ stand for the hydraulic conductivity at saturation, porosity, soil-pore distribution index, and density of the dry soil material. References are (a) Meyer et al. (1997), (b) Mohanty and Mousli (2000), (c) Schwartz et al. (2000), (e) Mendoza and Steenhuis (2003), (f) Kvaerno and Deelstra (2002), (g) Smith et al. (2003), (h) Parson (2001), (i) Wallace laboratories (2003), (j) Perfect et al. (2002), (k) Carey and Woo (1999), (l) Schlotzhauer and Price (1999), (m) Pielke (2001), (n) Grunwald et al. (2001), (o) Landsberg et al. (2003), (p) Calhoun et al. (2001), (q) Laurén and Heiskannen (1997), (r) Clapp and Hornberger (1978). Note that Cosby et al. (1984) provide slightly different values than Clapp and Hornberger (1978).

Soil-type	k_s 10^{-6} m/s	η_s m^3/m^3	b -.-	ψ_s m	ρ_S
Sand	176[r](43.9)[a]	0.395(0.056)[r]	4.05(1.78)[r]	-0.121(0.143)[r]	1580(90)[p]
Loamy sand	156.3[r](31.7)[a]	0.410(0.068)[r]	4.38(1.47)[r]	-0.090(0.124)[r]	1610(100)[n]
Sandy loam	34.1[r](13.7)[a]	0.435(0.086)[r]	4.90(1.75)[r]	-0.218(0.310)[r]	1520(140)[p]
Silt loam	7.2[r](6.2)[b]	0.485(0.059)[r]	5.30(1.96)[r]	-0.786(0.512)[r]	1400(90)[n]
Silt	2.81[r](1.325)[c]	0.476	5.33	-0.759	1420(70)[k]
Loam	7.0[r](3.028)[b]	0.451(0.078)[r]	5.39(1.87)[r]	-0.478(0.512)[r]	1350(110)[n]
Sandy clay loam	6.3[r](3.056)[e]	0.420(0.059)[r]	7.12(2.43)[r]	-0.299(0.378)[r]	1520(40)[n]
Silty clay loam	1.7[r](0.806)[f]	0.477(0.057)[r]	7.75(2.77)[r]	-0.356(0.378)[r]	1410(60)[n]
Clay loam	2.5[r](0.25)[g]	0.476(0.053)[r]	8.52(3.44)[r]	-0.630(0.510)[r]	1420(80)[n]
Sandy clay	2.2[r](8.333)[h]	0.426(0.057)[r]	10.40(1.64)[r]	-0.153(0.173)[r]	1570(120)
Silty clay	1.0[r](0.4)[j]	0.492(0.064)[r]	10.40(4.45)[r]	-0.490(0.621)[r]	1480(110)
Clay	1.3[r](0.569)[i]	0.482(0.050)[r]	11.40(3.70)[r]	-0.405(0.397)[r]	1470(140)[p]
Humus					
Peat	1.736 (0.938)[l]	0.923(0.342)	4.00(1.75)	-0.165(0.31)	106(243)
Moss	150 (400)[k]	0.900 (0.040)	1.00(1.75)	-0.120(0.310)	100(100)
Lichen	3356.5 (200)[q]	0.95 (0.060)	0.50(1.75)	-0.085(0.310)	120(30)

Various investigations using stand-alone versions of LSMs (e.g., Gao et al., 1996), NWPMs (e.g., Douville and Chauvin 2000), and GCMs (e.g., Wang and Kumar 1998) showed that initializing soil-moisture and temperature distributions is a huge source for errors in predicting the soil conditions correctly. Adjoint models and data-assimilation techniques can be applied for minimizing errors in initial soil conditions (e.g., van den Hurk et al., 1997, Callies et al., 1998, Reichle et al., 2001). Using this technique, however, is not possible for NWPMs, CTMs, GCMs or ESMs initialization due to lack of spatially continuous data.

The Project for Intercomparison of Land Surface Parameterization Schemes (PILPS) showed that LSMs strongly differ in accuracy because of, among other things, the choice of empirical parameters needed in parameterizations (e.g., Shao and Henderson-Sellers 1996, Slater et al., 1998). Typically, soil properties within a grid-cell or patch are expressed by assigning a mean value derived from laboratory or/and field studies thereby ignoring any variability. Consequently, predicted soil state variables and fluxes can differ over wide ranges in dependence of the parameter choice. Various parameter variation studies to assess whether slightly different parameters result in significant perturbations of soil temperature and moisture states. Such parameter-variation studies are subject to parameter interaction meaning that the parameter choice also affects simulated quantities that do not directly depend on the parameter. This fact makes optimal parameter choice difficult. Henderson-Sellers (1993), for instance, by using factorial experiments found that porosity is one of the most ecologically important parameters. Enhancing thermal diffusivities or volumetric heat capacities, for instance, may cool the soil and atmospheric boundary layer (locally more than 5 K and 1 K, respectively); enhancing volumetric heat capacities or thermal diffusivities may also affect atmospheric variables especially specific humidity, cloud and precipitation particles and may result in decreased maximum precipitation (Mölders 2001). Errors may also stem from incorrectly assigned soil types. An about 5 % change in soil-type distribution may alter daily averages of the soil-moisture fraction by 29 % with respect to the reference case, and surface temperature by 2.3 K (Mölders et al., 1997).

Besides systematic errors due to parameter choice, initialization, discretization, assumptions and physical parameterizations stochastic error is a source of uncertainty in predicted soil state variables and fluxes. As pointed out before, for describing soil heat and moisture transfer processes parameters have to be assigned that represent the soil characteristics. Herein stochastic errors result from the fact that the mean values of empirical soil parameters are in "error" by the amount of the standard deviation related to the natural (random) variability (Mölders et al., 2005). For many soil parameters, this variability expressed, for instance, by the standard deviation is of the same order of magnitude as the parameter itself (cf. Table 2). Consequently, any soil state variable or flux predicted with these parameters is "error"-burdened too. Such uncertainty may even reduce the trust in predicting permafrost dynamics in GCMs and ESMs. For NWPMs and CTMs, it may limit the ability to simulate the evolution of active layer depth which is important information for agricultural purposes and assessment of river runoff. For GCMs and ESMs, this uncertainty may complicate climate impact assessment.

Errors in soil state variables and fluxes related to parameter uncertainty are of random kind for which they can be evaluated with statistical methods, for instance, Gaussian error-propagation (GEP) principles. This method permits researchers to investigate the relative importance of soil physical parameters (e.g., porosity) in producing prediction uncertainty at various potential conditions. Using GEP Mölders et al., (2005), for instance, found that predicted distributions of soil temperature are less sensitive to uncertainty in thermal parameters than to uncertainty in hydraulic parameters. According to GEP results uncertainty in predicted soil-heat fluxes is within of the range as the typical errors in soil-heat flux measurements. They also found that the absolute value of soil-heat flux and its relative error decreases with increasing relative volumetric water content and concluded that soil-heat fluxes can be predicted with greater certainty after rain events or in the Tropics than under dry conditions or in dry regions.

Note that GEP can also be applied to examine how terms in the soil heat and moisture equations contribute to uncertainty in predicted soil temperature and moisture states. During phase transitions, the freeze-thaw term, for instance, can cause great uncertainty in volumetric water content and soil temperature (e.g., Mölders et al., 2005). Similar was found using other methods by Mölders and Romanovsky (2006).

Simulating Wildfire Impacts

Wildfires are a regular thread in many regions on Earth, so also to areas underlain by permafrost. Often the uppermost layer of permafrost contains huge amounts of organic material or completely consists of organic material like peat, moss or lichen (e.g., Beringer et al., 2001). Wildfires can burn this organic material. The degree to which this material is burned depends, among other things, on fire intensity, fire duration, and total soil water content of the material. Fires heat the soil and huge amounts of soil water evaporate during the fire. In permafrost, fire-induced changes in soil temperature go along with changes in total soil water content and the partitioning of the water phases (e.g., Hinzman et al., 2003). Consequently, infiltration, soil volumetric heat capacity and hydraulic conductivity before and after a fire differ appreciably. As compared to pre-fire soil conditions, post-fire soils are warmer. Such modified hydro-thermodynamic states of soil remain detectable long after the fire events. Due to their impact for soil temperature regime, active layer depth and soil surface temperature on the short and long-term it would be important to consider the impact of wildfire on soil temperature in the soil models of atmospheric models.

Currently, the impact of wildfires on permafrost is not considered in routine weather forecasts, CTMs, GCMs or ESMs. In CTMs, wildfire impacts on permafrost are currently neglected even when the CTMs are applied for wildfire smoke forecasts in areas underlain by permafrost. The neglecting wildfire impacts on permafrost in ESMs is despite some ESMs consider random aerosol release from wildfires and wildfire related land-cover changes in the biogeochemical cycles. One application that considered the impact of wildfires by land-cover changes and soil-temperature and moisture changes was performed with an NWPM by Mölders and Kramm (2007). Their results showed that the relatively warmer burned areas may increase atmospheric buoyancy and hence locally convection.

Challenges

The lack of horizontally and vertically high resolved soil data for organic and mineral soil, uncertainty in soil parameters, and organic soils are among the biggest challenges in modeling permafrost in atmospheric applications.

Due to the lack of 3D data on the soil type distribution most modern soil models used in NWPMs, CTMs, GCMs or ESMs assume one soil-type – typically that of the uppermost soil - for the entire soil column (e.g., Slater et al., 1998, Schlosser et al., 2000). Investigations show that this simplification/assumption results in missing many details in predicted soil-temperature patterns that result from the vertical soil type profile (Fig. 7). Due to neglecting vertical soil type profile characteristics the variance of simulated and observed temporal evolution of soil temperature can differ significantly with consequences for predicted active layer depth that may be dislocated about ±0.4 m or so (e.g., Mölders and Romanovsky 2006). Off-line simulations (i.e., the feedback processes between the atmosphere and surface are not considered) with different soil hydraulic models that were run with and without ensuring consistent soil hydraulic parameters, demonstrated that uncertainty in soil hydraulic parameters overwhelms that in the theory of soil hydraulic models (Shao and Irannjad, 1999).

EVALUATION

Evaluating soil temperature and moisture conditions simulated by NWPMs, ESMs, or GCMs has been a high priority of the third PILPS phase (e.g., Henderson-Sellers et al., 1995). PILPS demonstrated that results obtained from LSMs coupled to GCMs differ on the same order of magnitude as off-line PILPS experiments; differences in LSM complexity may cause statistically significant differences in temperature, pressure, and turbulent fluxes over land (e.g., Sato et al., 1989, Thompson and Pollard 1995, Yang et al., 1995, Qu and Henderson-Sellers 1998). The results of PILPS also suggested that a soil model must be able to capture soil-temperature conditions well when run offline with observed atmospheric forcing and known site-specific parameters (necessary condition), and it must be re-evaluated when being implemented in a NWPM, GCM or ESM (sufficient condition).

Soil models of NWPMs are typically evaluated by assuming that the soil temperature and moisture measurements at a site are representative for the grid-cell within which the site is located (e.g., Chen and Dudhia 2002, Narapusetty and Mölders 2005). It is well known that some discrepancies may arise due to the fact that the model grid-cell represents a volume-average condition for several square-kilometers of several centimeters thickness. For GCMs or ESMs, however, simulated soil temperature and moisture states represent even larger volumes. Due to the large area of several 100 square-kilometers covered by GCM or ESM grid-cells often several sites exist within the same grid-cell. Thus, a comparison like performed for NWPMs becomes highly ambiguous. Therefore, GCM and ESM simulations of soil regimes are typically evaluated using gridded climatologies that are derived from point observations projected on to a grid by some kind of interpolation methods (e.g., Li 2007, PaiMazumder et al., 2008).

Recently, the digital versions of the Ground Ice Conditions map and the International Permafrost Association (IPA) Circum-Arctic Map of Permafrost (known as IPA map) were combined with ancillary data sets of the Global Land One-kilometer Base Elevation data base and the global land-cover characteristics data base to provide a gridded distribution of northern hemispheric permafrost and ground ice (Zhang et al., 2000). Such gridded data can serve for evaluation of 20th century simulations of GCMs and ESMs. This dataset, however, does not contain soil temperature or moisture conditions.

Any gridded data sets bear some uncertainty from various sources. First the data stem from routine monitoring that typically has less accuracy than specialized field campaigns. Furthermore, these data have been collected for other reasons than evaluation of GCMs or ESMs. Thus, the monitoring networks may not be representative for the landscape that a GCMs or ESMs is to cover. For evaluation of CCSM3 simulated soil-temperature climatologies in Siberia PaiMazumder et al. (2008) used gridded data based on over 400 agricultural monitoring sites. They found December-biases in soil-temperature climatology for CCSM3 of up to 6 K at 0.2 m depth of which they could explain about 2.5 K by incorrect simulated atmospheric forcing. It is obvious that the soil conditions represented by the gridded data derived there from are biased with respect to the conditions in well drained, fertile soils with other density than non-plowed soils. Moreover, agriculture is typically made on soils that have a relative deep active layer depth. Thus, great care is needed in interpreting simulated soil conditions when using these kinds of gridded data. Investigations by PaiMazumder and Mölders (2008) showed that such bias in representing the soil distribution can lead to overestimation of soil-temperature amplitudes of more than 1 K and difficulties in capturing the phase. These findings also suggest that some of the discrepancies found for GCM or ESM soil temperature simulations may be explained the networks on which the gridded climatologies are based. Taking the errors resulting from incorrect forcing and the gridded data into account, only about 1.5 K of the bias found by PaiMazumder et al. (2008) for the CCSM3 soil-temperature simulations may stem from model deficits or other error sources.

FUTURE DIRECTIONS

As discussed in the previous sections, permafrost modeling in NWPMs, CTMs, GCMs and ESMs still has several short-comings. Some of them may be addressed easily as available computer resources increase with the next generations of supercomputers, while others require serious research and data collection efforts.

Increased computational power will permit us to consider more layers and locate the lower boundary of NWPM, CTM, GCM and ESM soil models at deeper levels, i.e. reduce uncertainty related to the choice of the lower boundary condition. This way a greater depth of the soil model does not compromise the required fine resolution in the upper soil that is required to capture the diurnal and seasonal cycle of active layer depth.

As has been shown by Narapusetty and Mölders (2006) finite element schemes permit us to better capture the phase and amplitude of soil temperature variations and hence the active layer depth. Currently the computational burden is too high to run GCMs and ESMs for several decades using such methods. Therefore this improvement has to be postponed until the next generations of supercomputer will become available.

The difficulties related to initialization of soil moisture and temperature in NWPMs and CTMs could be addressed by developing a kind of analysis procedure like applied to initialize the atmosphere in NWPMs. Such an analysis method would require to measure worldwide soil temperature and moisture at the same universal coordinated time (UTC) several times a day like is common practice for meteorological data. These soil data would have to be reported and collected at a central place in an agreed upon format like GRIB that is used by the World Meteorological Organization (WMO) for reporting the huge amount of meteorological data. Some kind of interpolation procedure would have to be run to produce a hydro-thermodynamically consistent gridded global soil temperature and moisture dataset based on the latest observations.

Data of soil type distribution exist in various different data sources and must be gathered in a data center to derive a quality assessed and quality assured global gridded dataset.

ESMs that consider random aerosol release from wildfires and wildfire related land-cover changes in the biogeochemical cycles and CTMs that serve for wildfire smoke forecasts in boreal regions could be enlarged to also consider the impact of wildfires on permafrost. Doing so would require assuming a wildfire-related heat source at the top of the soil where currently the wildfire related aerosols are released and later land-cover is changed.

Non-representative network design, low site density, shut-down and/or adding sites to long-term monitoring networks can introduce substantial uncertainty in gridded data (e.g. PaiMazumder and Mölders 2009). Therefore it is an urgent need (1) to assess the potential influence of networks on gridded data derived there from, (2) to develop evaluation strategies for application of gridded data from "imperfect" existing long-term networks, and (3) to develop recommendation to improve existing networks and/or design better networks in the future.

To avoid errors from non-representative networks in gridded soil-temperature data some kind of data assimilation could be used. Similar to reanalysis in atmospheric sciences all available soil data plus meteorological forcing data as upper boundary condition in conjunction with physical soil modeling could be performed to provide some kind of reanalysis (e.g., Kalany et al., 1996, Uppala et al., 2005) for soil temperature. This method could consider the various soil types within a grid and the gridded dataset would provide a weighted soil temperature. The weighting would be with respect to the fractional coverage of a given soil-type within the grid area like soil temperatures are typically simulated in GCMs or ESMs.

References

Abramopoulos, F., Rosenzweig, C., & Choudhury, B. (1988) Improved ground hydrology calculations for global models (GCMs): soil water movement and evapotranspiration. *J. Climate*, 1, 921-941.

Ács, F., Hantel, M., & Unegg, J.W. (2000) Climate diagnostic with the Budapest-Vienna land-surface model SURFMOD. Austrian contributions to the IGBP, Vol. 3.

Bastidas, L.A., Gupta, H.V., & Sorooshian, S. (2003) Parameter, Structure and Performance Evaluation for Land Surface Models. In: Q. Duan, H.V. Gupta, S. Sorooshian, A. Rousseau, & R. Turcotte (Eds.), *Advances in the Calibration of Watershed Models*. Water Science and Application Series, 6, Washington, D.C., American Geophysical Union., pp .229-238.

Beringer, J., Lynch, A.H., Chapin III, F.S., & Mack, M. (2001) The representation of Arctic soils in the land surface model: The importance of mosses. *J. Climate,* 14, 3324-3335.

Blümel, K. (1998) Estimation of sensible heat fluxes from the surface temperature wave and one-time-of-day air temperature observation. *Boundary-Layer Meteorol.*, 86, 193-232.

Blümel, K. (1999) A simple formula for estimation of the roughness length for heat transfer over partly vegetated surfaces. *J. Appl. Meteor.,* 38, 814-829.

Blümel, K. (2001) Estimation of sensible heat fluxes from the surface temperature wave and one-time-of-day air temperature observation, part II – heterogeneous surfaces. *Boundary-Layer Meteorol.*, 98, 183-218.

Boone, A., Masson, V, Meyers, T., & Noilhan, J. (2000) The influence of the inclusion of soil freezing on simulations by a soil-vegetation-atmosphere transfer scheme. *J. Appl. Meteor.*, 39, 1544-1569.

Braden, H. (1996) The model AMBETI – A detailed description of a soil-plant-atmosphere model. *Berichte des Deutschen Wetterdienstes*, 195, Offenbach am Main, Germany, Deutscher Wetterdienst, 117pp.

Beven, K.J., & Kirby, M.J. (1979). A physically based, variable contributing area model of basin hydrology. *Hydrol. Sci. Bull.*, 24, 43-69.

Bonan, G.B., Oleson, K.W., Vertenstein, M., Levis, S., Zeng, X., Dai, Y., Dickinson, R.E., & Yang, Z.-L. (2002) The land surface climatology of the community land model coupled to the NCAR community climate model. *J. Climate*, 15, 1115-1130.

Bronstert, A., Leavesley, G., & Mölders, N. (2005). Scale issues. In: Bronstert, A., Carrera, J., Kabat, P., Lütkemeier, S. (eds.), *Coupled Models for the Hydrological Cycle - Integrating Atmosphere, Biosphere, and Pedosphere*, Springer, Berlin/Heidelberg, pp. 21-39.

Calhoun, F.G., Smeck, N.E., Slater, B.L., Bigham, J.M., & Hall, G.F. (2001) Predicting bulk density of Ohio soils from morphology, genetic principles, and laboratory characterization data. *Soil Sci. Soc. Am. J.*, 65, 811-819.

Callies, U., Rhodin, A., & Eppel, D.P. (1998) A case study on variational soil moisture analysis from atmospheric observations. *J. Hydrol.,* 212-213, 95-108.

Carey, S.K., & Woo, M. (1999) Hydrology of two slopes in subarctic Yukon, Canada, *Hydrol. Process.*, 13, 2549-2562.

Chen, F., Mitchell, K., Schaake, J., Xue, Y., Pan, H.-L., Koren, V., Duan, Q.Y., Ek, M., & Betts, A. (1996) Modeling of land surface evaporation by four schemes and comparison with FIFE observations. *J. Geophys. Res.*, 101 (D), 7251-7268.

Chen, T.H., Henderson-Sellers, A., Milly, P.C.D., Pitman, A.J., Beljaars, A.C.M., Polcher, J., Abramopoulos, F., Boone, A., Chang, S., Chen, F., Dai, Y., Desboroough, C.E., Dickinson, R.E., Dümenil, L., Ek, M., Lettenmaier, J.R., Liang, X. Mahouf, J.-F., Mengelkamp, H.T., Mitchell, K., Nasonova, O.N., Noilhan, J., Robock, A., Verseghy, D.L., Wetzel, P., Wood, E.F., Xue, Y., Yang, Z.-L., & Zeng, Q. (1997) Cabauw experimental results from the project for intercomparison of land-surface parameterization schemes. *J. Climate*, 10, 1194-1215.

Chen, F., & Dudhia, J. (2001a) Coupling an advanced land surface hydrology model with the Penn State/NCAR MM5 modeling system. Part I: Model description and implementation. *Mon. Wea. Rev.*, 129, 569-585.

Chen, F., & Dudhia, J. (2001b) Coupling an advanced land surface hydrology model with the Penn State/NCAR MM5 modeling system. Part II: Model description and implementation. *Mon. Wea. Rev.*, 129, 587-604.

Cherkauer, K.A., & Lettenmaier, D.P. (1999) Hydrologic effects of frozen soils in the upper Mississippi river basis. *J. Geophys. Res.*, 104 (D), 19611-19621.

Clapp, R.B., & Hornberger, G.M. (1978) Empirical equations for some soil hydraulic properties. *Water Resour. Res.*, 14, 601-604.

Claussen, M. (1988) On the surface energy budget of coastal zones with tidal flats. *Contrib. Atmos. Phys.*, 61, 39-49.

Cosby, B.J., Hornberger, G.M., Clapp, R.B., & Ginn, T.R. (1984) A statistical exploration of the relationships of soil moisture characteristics to the physical properties of soils. *Water Resour. Res.,* 20, 682-690.

Cogley, J.G., Pitman, A.J., & Henderson-Sellers, A. (1990) A land-surface for large scale climate models. *Tech. Note* 90-1, Trent Univ., Peterborough, Ont. 129pp.

Cuenca, R., Ek, M., & Mahrt, L. (1996) Impact of soil water property parameterization on atmospheric boundary layer simulation. *J. Geophys. Res.*, 101 (D), 7269-7277.

Deardorff, J.W. (1978) Efficient prediction of ground surface temperature and moisture, with inclusion of a layer of vegetation. *J. Geophys. Res.*, 84 (C), 1889-1903.

de Groot, S.R. (1951) *Thermodynamics of Irreversible Processes*. New York, Interscience Publishers Inc., 242 pp.

Desborough, C.E. (1997) The impact of root-weighting on the response of transpiration stress in a land surface scheme. *Mon. Wea. Rev.*, 125, 1920-1930.

De Rosnay, P., & Polcher, J., (1999) Modeling root water uptake in a complex land surface scheme coupled to a GCM. *Hydrol. Earth Sys. Sci.*, 2, 239-255.

DéQué, M., Dreveton, C., Braun, A., & Cariolle, D. (1994) The ARPEGE/IFS atmosphere model – A contribution to the French community climate modelling. *Climate Dyn.*, 10, 249-266.

Déry, S.J., Taylor, P.A., & Xiao, J. (1998) The thermodynamic effects of sublimating, blowing snow in the atmospheric boundary layer. *Boundary-Layer Meterol.*, 89, 251-283.

Déry, S.J., & Yau, M.K. (1999) A climatology of adverse winter-type weather events. *J. Geophys. Res.*, 104 (D), 16657-16672.

Déry, S.J., & Yau, M.K. (2002) The thermodynamic effects of sublimating, blowing snow and surface sublimation. *J.Geophys. Res.*, 107 (D), 4679, doi:10.1029/2001JD001251, ACL 8-1 to 8-17.

de Vries, D.A. (1958) Simultaneous transfer of heat and moisture in porous media. *Transactions Am. Geophys. Union*, 39, 909-916.

Dickinson, R.E., Henderson-Sellers, A., Kennedy, P.J., & Wilson, M.F. (1986) Biosphere atmosphere transfer scheme (BATS) for the NCAR community climate model. *NCAR Tech. Note NCAR/TN-275+STR*, 69 pp.

Dickinson, R.E., Henderson-Sellers, A., & Kennedy, P.J. (1993) Biosphere Atmosphere Transfer Scheme (BATS) version 1e as coupled to the NCAR community climate model. *NCAR Technical Note, NCAR/TN-378+STR.*

Dingman, S.L. (1994) *Physical Hydrology*. New York/Oxford/Singapore/Sydney, Macmillan Publishing Company, 575 pp.

Douville, H., Royer, J.F., & Mahouf, J.F. (1995) A new snow parameterization for the meteo-france climate model. 1. Validation in stand-alone experiments. *Climate Dyn.*, 12, 21-35.

Douville, H., & Chauvin, F. (2000) Relevance of soil moisture for seasonal climate predictions: a preliminary study. *Climate Dyn.*, 16, 719-736.

Ducoudre, N., Laval, K., & Perrir, A., (1993) SECHIBA, a new set of parameterizations of the hydrologic exchanges at the land-atmosphere interface within the LMD atmospheric general circulation model. *J. Climate*, 6, 248-273.

Dudhia, J. (1993) A non-hydrostatic version of the Penn State/NCAR mesoscale model: Validation tests and simulation of an Atlantic cyclone and cold front. *Mon. Wea. Rev.*, 121, 1493-1513.

Dümenil, L., & Todini, E. (1992) A rainfall-runoff scheme for use in the Hamburg climate model. In: J.P. O'Kane (Ed.), *Advances in Theoretical Hydrology*, European Geophysical Society Series on Hydrological Sciences, **1**. Amsterdam, Elsevier Science Publishers, pp. 129-157.

Eppel, D.P., Kapitza, H., Claussen, M., Jacob, D., Koch, W., Levkov, L., Mengelkamp, H.-T., & Werrmann, N. (1995) The non-hydrostatic mesoscale model GESIMA. Part II: Parameterizations and applications. *Contrib. Atmos. Phys.*, 68, 15-41.

Esch, D.C., & Osterkamp, T.E. (1990) Cold region engineering: Climatic warming concerns for Alaska. *J. Cold Regions Engineering,* 4, 6-14.

Essery, R.L.H., Best, M.J., Betts, R.A., Cox, P.M., & Taylor, M.C. (2003) Explicit representation of subgrid heterogeneity in a GCM land-surface scheme. *J. Hydrometeorol.*, 4, 530-545.

Farouki, O. (1981) *Thermal properties of soils.* CRREL Monograph 81-1, Hanover, New Hampshire, U.S. Army Cold Regions Research and Engineering Laboratories.

Flerchinger, G.N., & Saxton, K.E. (1989) Simultaneous heat and water model of a freezing snow-residue-soil system I. Theory and development. *Transactions ASAE*, 32, 565-571.

Francaviglia, R., & Capri, E. (2003) Lysimeter experiments with metolachor in Tor Mancina (Italy) 1.389E-6.

Fuchs, M., Campbell, G.S., & Papendick, R.I. (1978) An analysis of sensible and latent heat flow in partially frozen unsaturated soil. *Soil Sci. Soc. Am. J.*, 42, 379-385.

Grell, G., Dudhia, J., & Stauffer, D. (1994) A Description of the Fifth-Generation Penn State/NCAR Mesoscale Model (MM5), *NCAR Technical Note, NCAR/TN-398+STR*, 122 pp..

Grell, G.A., Peckham, S.E., Schmitz, R., McKeen, S.A., Frost, G., Skamarock, W.C., Eder, B. (2005) Fully coupled "online" chemistry within the WRF model. *Atmos. Environ.*, 39, 6957-6975.

Groß, G. (1988) A numerical estimation of the deforestation effects on local climate in the area of the Frankfurt International Airport. *Contrib. Atmos. Phys.*, 61, 219-231.

Grunwald, S., Rooney, D.J., McSweeney, K., & Lowery, B. (2001) Development of pedotransfer functions for a profile cone penetrometer. *Geoderma*, 100, 25-47.

Hagemann, S., & Dümenil, L. (1998) A parameterization of the lateral waterflow for the global scale. *Climate Dyn.*, 14, 17-31.

Hantel, M. (1997) Klimatologie. In: Raith, W. (ed.) Bergmann Schaefer, *Lehrbuch der Experiementalphysik, Band 7, Erde und Planeten*, Walter de Gruyter, Berlin, New York, pp. 311-426.

Henderson-Sellers, A. (1993) A factorial assessment of the sensitivity of the BATS land-surface parameterization. *J. Climate*, 6, 227-247.

Henderson-Sellers, A., Pitman, A.J., Love, P.K., Irannejad, P., & Chen, T.H. (1995) The project for intercomparison of land surface parameterization schemes (PILPS): Phase 2 and 3. *Bull. Am. Meteorol. Soc.*, 76, 489-503.

Hinkel, K.M., Paetzold, R.F., Nelson, F.E., & Bockheim, J.G. (2001) Patterns of soil temperature and moisture in the active layer and upper permafrost at Barrow, Alaska: 1993-1999. *Global and Planetary Change*, 29, 293-309.

Hinkel, K.M., Nelson, F.E., Klene, A.E., & Bell, J.H. (2003) The urban heat island in winter at Barrow, Alaska. *Int. J. Climatol.*, 23, 1889-1905.

Jacobson, I., & Heise, E. (1982) A new economic method for the computation of the surface temperature in numerical models. *Contrib. Atmos. Phys.*, 55, 128-141.

Kalnay, E., Kanamitsu, M., Kistler, R., Collins, W., Deaven, D., Gandin, L., Iredell, M., Saha, S., White, G., Woollen, J., Zhu, Y., Leetmaa, A., Reynolds, B., Chelliah, M., Ebisuzaki, W., Higgins, W., Janowiak, J., Mo, K.C., Ropelewski, C., Wang, J., Jenne, R., & Josep, D. (1996) The NCEP/NCAR 40-year reanalysis project. *Bull. Amer. Meteor. Soc.*, 77, 437-470.

Kane, D.L., Hinzman, L.D., & Zarling, J.P. (1991) Thermal response of the active layer in a permafrost environment to climatic warming. *Cold Regions Science and Technology,* 19 (2), pp. 111-122.

Koren, V., Schaake, J., Mitchell, K., Duan, Q.-Y., Chen, F., & Baker, J.M. (1999) A parameterization of snowpack and frozen ground intended for NCEP weather and climate models. *J. Geophys. Res.*, 104 (D), 19569-19585.

Kramm, G., Dlugi, R., Mölders, N., & Müller, H. (1994) Numerical investigations of the dry deposition of reactive trace gases. In: J.M. Baldasano, C.A. Brebbia, H. Power, & P. Zannetti (Eds.), *Air Pollution II Vol. 1: Computer Simulation*, Southampton, Boston, Computational Mechanics Publications, pp. 285-307.

Kramm, G. (1995) *Zum Austausch von Ozon und reaktiven Stickstoffverbindungen zwischen Atmosphäre und Biosphäre*. Frankfurt am Main, Germany, Maraun-Verlag, 268 pp.

Kramm, G., Beier, N., Foken, T., Müller, H., Schröder, P., & Seiler, W. (1996) A SVAT scheme for NO, NO_2, and O_3 - model description. *Meteorol. Atmos. Phys.*, 61, 89-106.

Kreyszig, E. (1970) *Statistische Methoden und ihre Anwendung*. Göttingen, Vanden Hoeck & Ruprecht, 422 pp.

Kuhl, S.C., & Miller, J.R. (1992) Seasonal river runoff calculated from a global atmospheric model. *Water Resour. Res.*, 28, 2029-2039.

Kvaerno, S.H., & Deelstra, J. (2002) Spatial variability in soil hydraulic properties of a silty clay loam in south-eastern Norway. 17th WCSS, 14-21 August 2002, Thailand. Paper No. 2266.

Landsberg, J.D., Miller, R.E., Anderson, H.W., & Tepp, J.S. (2003) Bulk density and soil resistance to penetration as affected by commercial thinning in northeastern Washington. Research paper PNW-RP-551, Portland, OR, U.S. Department of Agriculture, Forest Service, Pacific Northwest Research Station.

Laurén, A., & Heiskannen, J. (1997) Physical properties of the moor layer in a Scots pine stand I. Hydraulic conductivity. *Can. J. Soil Sci.*, 77, 627-634.

Ling, F., & Zhang, T. (2003) Impact of the timing and duration of seasonal snow cover on the active layer and permafrost in the Alaskan Arctic. *Permafrost Periglac. Process.*, 14, 141-150.

Lohmann, D., Lettenmaier, D.P., Liang, X., Wood, E.F., Boone, A., Chang, S., Chen, F., Dai, Y., Desborough, V., Dickinson, R.E., Duan, Q., Ek, M., Gusev, Y.M., Habets, F., Irannejad, P., Koster, R., Mitchell, K.E., Nasonova, O.N., Noilhan, J., Schaake, J., Schlosser, A., Shao, Y., Shmakin, A.B., Verseghy, D., Warrach, K., Wetzel, P., Xue, Y., Yang, Z.-L. & Zeng, Q.C. (1998) The Project for Intercomparison of Land Surface Parameterization Schemes Phase (PILPS) phase 2(c) Red-Arkansas river basin experiement: 3. Spatial and temporal analysis of water fluxes. *Global and Planetary Change*, 19, 161-179.

Luo, L., Robock, A., Vinnikov, K.Y., Schlosser, C.A., Slater, A.G., Boone, A., Braden, H., Cox, P., de Rosany, P., Dickinson, R.E., Dai, Y., Duan, Q., Etchevers, P., Henderson-Sellers, A., Gedney, N., Gusev, Y.M., Habets, F., Kim, J., Kowalczyk, E., Mitchell, K., Nasonova, O.N., Noilhan, J., Pitman, A.J., Schaake, J., Shmakin, A.B., Smirnova, T.G., Wetzel, P., Xue, Y., Yang, Z.-L., & Zeng, Q.-C. (2003) Effects of frozen soil on soil temperature, spring infiltration, and runoff: Results from the PILPS2(d) experiment at Valdai, Russia. *J. Hydrometeor.*, 4, 334-351.

Mahouf, J.F., & Noilhan, J. (1991) Comparative study of various formulations of evaporation from bare soil using in situ data. *J. Appl. Meteor.*, 30, 1354-1365.

Mahfouf, J.-F., Manzi, A.O., Noilhan, J., Giordani, H., & DéQué, M. (1995) The land surface scheme ISBA within the Météo-France climate model ARPEGE. Part I. Implementation and preliminary results. *J. Climate*, 8, 2039–2057.

Manabe, S. (1969) Climate and ocean circulation 1. The atmospheric circulation and the hydrology of the earth's surface. *Mon. Wea. Rev*., 97, 739-774.

Marengo, J.A., Miller, J.R., Russel, G.L., Rosenzweig, C.E., & Abramapoulos, F. (1994) Calculations of river-runoff in the GISS GCM: impact of a new land-surface parameterization and runoff routing model on the hydrology of the Amazon river. *Climate Dyn.*, 10, 349-361.

McCumber, M.C. (1980) *A numerical simulation of the influences of heat and moisture fluxes upon mesoscale circulation*. Ph.D. thesis, Dept. Environ. Sci., University of Virginia, Charlottesville, VA.

McCumber, M., & Pielke, R.A. (1981) Simulation of the effects of surface fluxes of heat and moisture in a mesoscale numerical model 1. Soil layer. *J. Geophys. Res*., 86 (C), 9929-9938.

Mendoza, G., & Steenhuis, T.S. (2003) www.bee.cornell.deu/research/SoilandWater/HTML/~hill-lysimeter.html

Mengelkamp, H.-T., Warrach, K., & Raschke, E. (1999) SEWAB – a parameterization of the surface energy and water balance for atmospheric and hydrologic models. *Adv. Wat. Resour*., 23, 165-175.

Mengelkamp, H.-T., Warrach, K., Ruhe, C., & Raschke, E. (2001) Simulation of runoff and streamflow on local and regional scales. *Meteor. Atmos. Phys*., 76, 107-117.

Meyer, P.D., Rockhold, G.W., & Gee, G.W. (1997) Uncertainty analyses of infiltration and subsurface flow and transport for SDMP sites. Pacific Northwest National Laboratory, Richland, WA.

Mohanty, B.P., & Mousli, Z. (2000) Saturated hydraulic conductivity and soil water retention properties across a soil slope transition. *Water Resour. Res.*, 36, 3311-3324.

Montaldo, N., & Albertson, J.D. (2001) On the use of the force-restore SVAT model formulation for stratified soils. *J. Hydrometeor.*, 2, 571-578.

Miller, D.A., & White, R.A. (1998) A conterminous United States Multi-layer soil characteristics data set for regional climate and hydrological modeling. *Earth Interactions*, 2. (Available on-line at http://EarthInteractions.org).

Miller, J.R., Russel, G.L., & Caliri, G. (1994) Continental-scale river flow in climate models. *J. Climate,* 7, 914-928.

Mölders, N., Strasser, U., Schneider, K., Mauser, W., & Raabe, A. (1997) A sensitivity study on the initialization of surface characteristics in meso-β/γ-modeling using digitized vs. satellite derived landuse data. *Contrib. Atmos. Phys*., 70, 173-187.

Mölders, N., & Raabe, A. (1997) Testing the effect of a two-way-coupling of a meteorological and a hydrologic model on the predicted local weather. *Atmos. Res.*, 45, 81-108.

Mölders, N., Raabe, A., & Beckmann, T. (1999) A technique to downscale meteorological quantities for the use in hydrologic models - Description and first results. *IAHS Publ.*, 254, 89-98.

Mölders, N. (2001) On the uncertainty in mesoscale modeling caused by surface parameters. *Meteor. Atmos. Phys.*, 76, 119-141.

Mölders, N. (2001) Concepts for coupling hydrological and meteorological models. In: K. Arnold & A. Raabe (Eds.), *Meteorologische Arbeiten aus Leipzig (VI)*. Wiss. Mitt. Univ. Leipzig, 22, 1-15.

Mölders, N., & W. Rühaak W (2002) On the impact of explicitly predicted runoff on the simulated atmospheric response to small-scale land-use changes - An integrative modeling approach. *Atmos. Res.*, 63, 3-38.

Mölders, N., Haferkorn, U., Döring, J., & Kramm, G. (2003a) Long-term numerical investigations on the water budget quantities predicted by the hydro-thermodynamic soil vegetation scheme (HTSVS) – Part I: Description of the model and impact of long-wave radiation, roots, snow, and soil frost. *Meteorol. Atmos. Phys.*, 84, 115-135.

Mölders, N., Haferkorn, U., Döring, J., & Kramm, G. (2003b) Long-term numerical investigations on the water budget quantities predicted by the hydro-thermodynamic soil vegetation scheme (HTSVS) – Part II: Evaluation, sensitivity, and uncertainty. *Meteorol. Atmos. Phys.*, 84, 137-156.

Mölders, N., & Walsh, J.E. (2004) Atmospheric response to soil-frost and snow in Alaska in March. *Theor. Appl. Climatol.*, 77, 77-105.

Mölders, N., Jankov, M., & Kramm, G. (2005) Application of Gaussian error propagation principles for theoretical assessment of model uncertainty in simulated soil processes caused by thermal and hydraulic parameters. *J. Hydrometeorol.*, 6, 1045-1062.

Müller, E., Frühwald, D., Jacobsen, I., Link, A., Majewski, D., Schwirner, J.-U., & Wacker, U. (1986) Results and prospects of mesoscale modeling at the Deutscher Wetterdienst. In: *Collection of papers presented at the WMO/IUGG NWP Symposium*, Tokyo, 4-8 August 1986. pp. 533-546.

Nicolsky, D.J, Romanovsky, V.E., Alexeev, V.A., & Lawrence, D.M. (2007) Improved modeling of permafrost dynamics in a GCM land-surface scheme. *Geophys. Res. Lett.*, 34, L08501, doi:10.1029/2007GL029525.

Niu, G.-Y., Yang, Z.-L., Dickinson, R.E., & Gulden, L.E. (2005) A simple TOPMODEL-based runoff parameterization (SIMTOP) for use in global climate models. *J. Geophys. Res.*, 110, D21106, doi:10.1029/2005JD006111.

Noilhan, J., & Planton, S. (1989) A simple parameterization of land surface processes for meteorological models. *Mon. Wea. Rev.*, 117, 536-549.

Oechel, W.C., Vourlitis, G.L., Hastings, S.J., Zulueta, R.C., Hinzman, L., & Kane, D. (2000) Acclimation of ecosystem CO_2 exchange in the Alaskan Arctic in response to decadal climate warming. *Nature*, 406, 978-981.

Oke, T.R. (1978) *BoundaryLayer Climates*. London/New York, Methuen, 435 pp.

Oleson, K.W., Dai, Y., Bonan, G., Bosilovich, M., Dickinson, R.E., Dirmeyer, P., Hoffman, F., Houser, P., Levis, S., Niu, G.-Y. Thornton, P., Vertenstein, M., Yang, Z.-L., & Zeng, X. (2004) Technical description of the Community Land Model (CLM). *NCAR Tech. Note NCARTN-461+STR*, 174 pp.

PaiMazumder, D., Miller, J., Li, Z., Walsh, J. E., Etringer, A., McCreight, J., Zhang, T., & Mölders, N. (2008) Evaluation of Community Climate System Model soil-temperatures using observations from Russia. *Theor. Appl. Climatol.*, 94, 187-213.

PaiMazumder, D., & Mölders, N. (2008) Sources of discrepancy between CCSM simulated and gridded observation-based soil-temperature over Siberia: The influence of site density and distribution. *Proceedings of the Ninth International Conference On Permafrost, NICOP,* Fairbanks, AK, 1351-1356.

PaiMazumder, D., & Mölders, N., 2009. Theoretical assessment of uncertainty in regional averages due to network density and design. *J. Appl. Meteor. Clim.* (in press)

Pauwels, V.R.N., & Wood, E.F. (1999a) A soil-vegetation-atmosphere transfer scheme for modeling of water and energy balance processes in high latitudes, 1, Model improvements. *J. Geophys. Res.*, 104 (D), 27811-27822.

Pauwels, V.R.N., & Wood, E.F. (1999b) A soil-vegetation-atmosphere transfer scheme for modeling of water and energy balance processes in high latitudes, 2, Application and validation. *J. Geophys. Res.*, 104 (D), 27823-27839.

Perfect, E., Sukop, M.C., & Haszler, G.R. (2002) Prediction of dispersivity for undistributed soil columns from water retention parameters. *Soil Sci. Soc. Am. J.*, 66, 696-701.

Peters-Lidard, C.D., Blackburn, E., Liang, X., & Wood, E.F. (1998) The effect of soil thermal conductivity parameterization on surface energy fluxes and temperatures. *J. Atmos. Sci.*, 55, 1209-1224.

Philip, J.R., & de Vries, D.A. (1957) Moisture in porous materials under temperature gradients. *Transactions Am. Geophys. Soc.*, 18, 222-232.

Pielke, R.A. (1984) Mesoscale Meteorological Modelling. London, Academic Press Inc., 612 pp.

Pitman, A.J., Yang, Z.L., Cogley, J.G., & Henderson-Sellers, A. (1991) Description of bare essentials of surface transfer for the Bureau of Meteorology Research Centre AGCM. *BMRC Research Rep.*, 32, 117 pp.

Polcher, J., Laval, K., Dümenil, L., Lean, J., & Rowntree, P.R. (1996) Comparing three land surface schemes used in general circulation models. *J. Hydrol.*, 180, 373-394.

Prigogine, I. (1961) *Introduction to Thermodynamics of Irreversible Processes.* New York, London, Interscience Publ. Inc., 119 pp..

Riseborough, D.W. (2002) The mean annual temperature at the top of permafrost, the TTOP model, and the effect of unfrozen water. *Permafrost Periglac. Process.*, 13, 137-143.

Robock, A., Vinnikov, K.Y., Schlosser, C.A., Speranskaya, N.A., & Xue, Y. (1995) Use of midlatitude soil moisture and meteorological observations to validate soil moisture simulations with biosphere and bucket models. *J. Climate*, 8, 15-35.

Romanovsky, V.E., & Osterkamp, T.E. (1997) Thawing of the active layer on the coastal plain of the Alaskan Arctic. *Permafrost and Periglac. Process.*, 8, 1-22, 1997.

Romanovsky, V.E., Osterkamp, T.E., & Duxbury, N. (1997) An evaluation of three numerical models used in simulations of the active layer and permafrost temperature regimes. *Cold Regions Science and Technology*, 26, 195-203.

Romanovsky, V.E., & Osterkamp, T.E. (2000) Effects of unfrozen water on heat and mass transport processes in the active layer and permafrost. *Permafrost and Periglac. Process.*, 11, 219-239.

Romanovsky, V.E., & Osterkamp, T. E. (2001) Permafrost: Changes and impacts. In: R. Paepe & V. Melnikov (Eds.), *Permafrost Response on Economic Development, Environmental Security and Natural Resources*. Dordrecht, The Netherlands, Kluwer Academic Publishers, pp. 297-315.

Romanovsky, V., Burgess, M., Smith, S., Yoshikawa, K., & Brown, J. (2002) Permafrost Temperature Records: Indicators of Climate Change. *EOS*, *AGU Transactions*, 83 (50), 589-594.

Sato, N., Sellers, P.J., Randall, D.A., Schneider, E.K., Shukla, J., Kinter II. J.L., Hou, Y.-T., & Albertazzi, E (1989) Effects of implementing the Simple Biosphere Model in a general circulation model. *J. Atmos. Sci.*, 46, 2757-2782.

Sausen, R., Schubert, S., & Dümenil, L. (1994) A model of the river runoff for use in coupled atmosphere-ocean models. *J. Hydrol.*, 155, 337-352.

Schaake, J.C., Koren, V.I., Duan, Q.Y., Mitchell, K., & Chen, F. (1995) A simple water balance model (SWB) for estimating runoff at different spatial and temporal scales. *J. Geophys. Res.*, 101, 7461-7475.

Schädler, G. (1990) Numerische Simulationen zur Wechselwirkung zwischen Landoberflächen und atmosphärischer Grenzschicht. Wiss. Berichte Inst. Meteor. u. Klimaf., Univ. Karlsruhe.

Schlosser, C.A., Slater, A.G., Robock, A., Pitman, A.J., Vinnikov, K.Y., Henderson-Sellers, A., Speranskaya, N.A., Mitchell, K., & the PILPS2(D) Contributors (2000) Simulations of a boreal grassland hydrology at Valdai, Russia: PILPS phase 2(d). *Mon. Wea. Rev.*, 128, 301-321.

Schlotzhauer, S.M., & Price, J.S. (1999) Soil water flow dynamics in a managed cutover peat field, Quebec: Field and laboratory investigations, *Water Resour. Res.*, 35, 3675-3683.

Schwartz, R.C., Unger, P.W., & Evett, S.R. (2002) Land use effects on soil hydraulic properties. ISTRO 2000; 1-10.

Sellers, P.J., Mintz, Y., Sud, Y.C., & Dalcher, A. (1986) A simple biosphere atmosphere model (SiB) for use within general circulation models. *J. Atmos. Sci.*, 43, 505-531.

Shao, Y., & Henderson-Sellers, A. (1996) Modeling soil moisture: A project for intercomparison of land surface parameterization schemes phase 2 (b). *J. Geophys. Res.*, 101 (D), 7227-7250.

Shao, Y., & Irannejad, P. (1999) On the choice of soil hydraulic models in land-surface schemes. *Boundary-Layer Meteorol.*, 90, 83-115.

Shao, Y., Sogalla, M., Kerschgens, M., & Brüchner, W. (2001) Effects of land-surface heterogeneity upon surface fluxes and turbulent conditions. *Meteor. Appl. Phys.*, 78, 157-181.

Sievers, U. Forkel, R., Wilson, M.F., Henderson-Sellers, A., Dickinson, R.E., & Kennedy, P.J. (1987) Sensitivity of the biosphere-atmosphere transfer scheme (BATS) to the inclusion of variable soil characteristics, *J. Clim. Appl. Met.*, 26, 341-362.

Skamarock, W.C., Klemp, J.B., Dudhia, J., Gill, D.O., Baker, D.M., Wang, W. & Powers, J.G. (2005) A description of the advanced research WRF version 2. *NCAR Technical Note, NCAR/TN-468+STR*, pp. 88.

Slater, A.G., Pitman, A.J., & Desborough, C.E. (1998) Simulation of freez-thaw cycles in a general circulation model land surface scheme. *J. Geophys. Res.*, 103 (D), 11303-11312.

Slater, A.G., Schlosser, C.A., Desborough, C.E., Pitman, A.J., Henderson-Sellers, A., Robock, A., Vinnikov, K.Y., Speranskaya, N.A., Michell, K., Boone, A., Braden, H., Chen, F., Cox, P., de Rosnay, P., Dickinson, R.E., Dai, Y.-J., Duan, Q., Entin, J., Etchevers, P., Gedney, N., Gusev, Y.M., Habbets, F., Kim, J., Koren, V., Kowalczyk, E., Nasonova, O., Noilhan, J., Schaake, J., Shmakin, A.B., Smironova, T., Verseghy, D. Wetzel, P., Xue, Y., & Yang, Z.-L. (2001) The representation of snow in land-surface schemes: results from PILPS 2(d). *J. Hydrometeor.*, 1, 7-25.

Smirnova, T.G., Brown, J.M., & Benjamin, S.G. (1997) Performance of different soil model condfigurations in simulating ground temperature and surface fluxes. *Mon. Wea. Rev.*, 125, 1870-1884.

Smironova, T.G., Brown, J.M., Benjamin, S.G., & Kim, D. (2000) Parameterization of cold season processes in the MAPS land-surface scheme. *J. Geophys. Res.*, 105D, 4077-4086.

Smith, M.W. & Riseborough, D.W. (2001) Climate and the limits of permafrost: a zonal analysis. *Permafrost and Periglac. Process.,* 13, 1-15.

Smith, J., Bhuiyan, M., Kirkby, C., & Humphreys, L. (2003) Determination of rice soil hydraulic properties by Guelph and disc permeametry, RIRDC web publication No 03/004.

Stendel, M., & Christensen, J. H. (2002) Impact of global warming on permafrost conditions in a coupled GCM. *Geophys. Res. Lett.,* 29, 10.1029/2001GLO14345.

Uppala, S.M., Kållberg, P.W., Simmons, A.J., Andrae, U., Bechtold, V.D., Fiorino, M., Gibson, J.K., Haseler, J., Hernandez, A., Kelly, G.A., Li, X., Onogi, K., Saarinen, S., Sokka, N., Allan, R.P., Andersson, E., Arpe, K., Balmaseda, M.A., Beljaars, A.C.M., Berg, L., Bidlot, J., Bormann, N., Caires, S., Chevallier, F., Dethof, A., Dragosavac, M., Fisher, M., Fuentes, M., Hagemann, S., Hólm, E., Hosking, B.J., Isaksen, L., Janssen, P.A.E.M., Jenne, R., McNally, A,P,, Mahfouf, J.F., Morcrette, J.J., Rayner, N.A., Saunders, R.W., Simon, P., Sterl, A., Trenberth, K.E., Untch, A., Vasiljevic, D., Viterbo, P., & Woollen, J. (2005) The ERA-40 reanalysis. *Q.J.R. Meteorol. Soc.*, 131, 2961-3012.

van den Hurk, J.J.M., Bastiaanssen, W.G.M., Pelgrum, H., & van Meijgaard, E. (1997) A new methodology for assimilation of initial soil moisture fields in weather prediction models using Meteosat and NOAA data. *J. Appl. Meteor.*, 36, 1271-1283.

Verseghy, D.L. (1991) CLASS – A Canadian land surface scheme for GCMs, 1, soil model. *Int. J. Climatol.*, 11, 111-133.

Verseghy, D.L., McFarlane, N.A., & Lazare, M. (1993) CLASS – A Canadian land surface scheme for GCMs. II: Vegetation model and coupled runs. *Int. J. Climatol.*, 13, 347-370.

Viterbo, P., & Beljaars, A., (1995) An improved land surface parameterization scheme in the ECMWF model and its validation. *J. Climate*, 8, 2716-2748.

Viterbo, P., Beljaars, A., Mahouf, J.-F., & Teixeira, J., (1999) The representation of soil moisture freeying and its impact on the stable boundary lazer. *Q. J. Roy. Met. Soc.*, 125, 2401-2426.

Walko, R.L., Band, L.E., Baron, J., Kittel, T.G.F., Lammers, R., Lcc, T.J., Ojima, D., Pielke, R.A. Sr., Taylor, C., Tague, C., Tremback, C.J., & Vidale, P.L. (2000) Coupled atmosphere-biophysics-hydrology models for environmental modeling. *J. Appl. Meteor.*, 39, 931-944.

Wallace Laboratories (2003) www.nuscience.com/wallace3.htm

Wang, W., & Kumar, A., (1998) A GCM assessement of atmospheric seasonal predictability associated with soil moisture anomalies over North America. *J. Geophys. Res.*, 103, 28637-28646.

Warrach, K., Stieglitz, M., Mengelkamp, H.-T., & Raschke, E. (2002) Advantages of a topographically controlled runoff simulation in a soil-vegetation-atmosphere transfer model. *J. Hydrometeor.*, 3, 131-148.

Wetzel, P.J., & Chang, J.T. (1988) Evapotranspiration from nonuniform surfaces: A first approach for short-term numerical weather prediction. *Mon. Wea. Rev.*, 116, 600-621.

Wetzel, P.J., & Boone, A. (1995) A parameterization for land-cloud-atmosphere exchange (PLACE): Documentation and testing of a detailed process model of the partly cloudy boundary layer over heterogeneous land. *J. Climate*, 8, 1810-1837.

Williams, P.J., & Smith, M.W. (1989) *The frozen earth – Fundamentals in Geocryology*. Cambridge, New York, Port Chester, Melbourne, Sydney, Cambridge University Press, 306 pp.

Wilson, M.F., Henderson-Sellers, A., Dickinson, R.E., & Kennedy, P.J. (1987) Sensitivity of the biosphere-atmosphere transfer scheme (BATS) to the inclusion of variable soil characteristics. *J. Clim. Appl. Met.*, 26, 341-362.

Woo, M. (1986) Permafrost hydrology in North America. *Atmosphere-Ocean*, 24, 201-234.

Xue, Y., Sellers, P.J., Kinter, J.L., & Shukla, J. (1991) A simplified biosphere model for global climate studies. *J. Climate*, 4, 345-364.

Yang, K., Koike, T., Ye, B., & Bastidas, L. (2005) Inverse analysis of the role of soil vertical heterogeneity in controlling surface soil state and energy partition. *J. Geophys. Res.*, 110, D08101, doi:10.1029/2004JD005500.

Yang, Z.L., Dickinson, R.E., Henderson-Sellers, A., & Pitman, A.J. (1995) Preliminary study of spin-up processes in land surface models with the first stage data of Project for Intercomparison of Land Surface Parameterization Schemes Phase 1(a). *J. Geophys. Res.*, 100 (D), 16553-16578.

Yang, Z.-L., Dickinson, R.E., Robock, A. & Vinnikov, K.Y., (1997) Validation of the snow submodel of the biosphere-atmosphere transfer scheme with Russian snow cover and meteorological observation data. *J. Climate*, 10, 353-373.

Yang, Z.L., Pitman, A.J., McAvaney, B., & Henderson-Sellers, A. (1995) The impact of implementing the bare essentials of surface transfer land surface scheme into the BMRC GCM. *Climate Dyn.*, 11, 279-297

Zdunkowski, W. (1983) Transport equations for heat and moisture in the soil and their application to boundary layer problems. *Contr. Phys. Atmos.*, 56, 58-83.

Zdunkowski, W.G., Paegle, J., & Reilly, J.P. (1975) The effect of soil moisture upon the atmospheric and soil temperature near the air-soil interface. *Arch. Met. Geoph. Biokl.*, 24A, 245-268.

Zeng X, Shaikh M, Dai Y, Dickinson RE, & Myneni R (2002) Coupling of the Common Land Model to the NCAR Community Climate Model. *J. Climate*, 15, 1832-1854

Zhang, T., Barry, R.G., Knowles, K., Heginbottom, J.A., & Brown, J. (2000) Statistics and characteristics of permafrost and ground ice distribution in the Northern Hemisphere. *Polar Geography*, 23, 132-154.

Zhang, T., Osterkamp, T.E., & Stamnes, K. (1996) Influence of the depth hoar layer of the seasonal snow cover on the ground thermal regime. *Water Resour. Res.*, 32, 2075-2086.

Zhang, T., Osterkamp, T.E., & Stamnes, K. (1997) Effect of climate on the active layer and permafrost on the North Slope of Alaska, U.S.A. *Permafrost and Periglac. Process.*, 8, 45-67.

Zhuang, Q., Romanovsky, V.E., & McGuire, A.D. (2001) Incorporation of a permafrost model into a large-scale ecosystem model: Evaluation of temporal and spatial scaling issues in simulating soil thermal dynamics. *J. Geophys. Res.*, 106, 33649-33670.

In: New Permafrost and Glacier Research
Editors: Max I. Krugger and Harry P.Stern
ISBN: 978-1-60692-616-1

Chapter 3

MULTIDRUG-RESISTANT BACTERIA IN PERMAFROST: ISOLATION, BIODIVERSITY, PHENOTYPIC AND GENOTYPIC ANALYSIS

Sofia Mindlin[1], Mayya Petrova[1], Zhosefine Gorlenko[1], Vera Soina[2], Natalia Khachikian[1] and Ekaterina Karaevskaya[2]

[1]Institute of Molecular Genetics, Russian Academy of Sciences, Moscow, Russia
[2]Department of Soil Biology, Moscow State University, Moscow, Russia

ABSTRACT

Bacterial strains resistant to beta-lactams, aminoglycosides, tetracycline, chloramphenicol, sulphathiazole and trimethoprim were isolated from more than 60 samples of Arctic and Antarctic permafrost subsoil sediments dated from 5 thousand to 3 million years of age. About 30% of the isolated strains were cross-resistant to two and more antibiotics of different classes. The diversity of multidrug-resistant ancient bacteria, the genetic structure of resistance determinants and their association with different mobile elements were studied.

Principal attention was given to multidrug-resistant strains of Gram-negative bacteria belonging to genera *Acinetobacte*r, *Pseudomonas*, *Psychrobacter, Stenotrophomonas*, and *Xanthomonas*. It was shown that multidrug resistance of some strains of *Acinetobacter* sp. can be transferred by transformation of chromosomal genes and most probably results from expression of efflux pumps. We also revealed that many of the strains contained antibiotic resistance genes closely related to those of modern bacteria. In particular, among different strains resistant to streptomycin, we identified strains with *strA-strB* genes, strains with *aadA* genes and strains containing both types of genes. Genes closely related to *tetR-tet(H)* genes were detected in a strain of *Psychrobacter psyhrophilus* resistant to tetracycline and streptomycin. Finally, we demonstrated that many of the resistance determinants are associated with mobile elements such as plasmids and transposons.

The results of the study strengthen the hypothesis that antibiotic resistance genes were present in natural bacterial populations long before the 'antibiotic era'. Also, the association of the resistance determinants with different mobile elements confirm an important role of horizontal transfer in distribution of these genes among environmental bacteria.

INTRODUCTION

The problem of multidrug resistance in pathogenic bacteria causes constant anxiety in the medical community and the pharmaceutical industry [Poole, 1994; Fluit et al., 2001]. Historically, the development of multidrug resistance in bacteria has been attributed to the presence of various mobile elements such as R plasmids, transposable elements and integrons [Davies, 1994; Tenover, 2006]. According to the current view, resistance genes first appeared in antibiotic producers, mainly streptomycetes, as an indispensable mechanism for their self-protection and afterwards, via sequential rounds of horizontal transfer mediated by mobile elements distributed among other microbial genera, and from those have moved into clinical strains of Gram-positive and Gram-negative bacteria [Davies, 1994].

The most straightforward way to verify this hypothesis is to investigate antibiotic resistance of environmental bacterial strains isolated from soil and water. Recently, bacteria resistant to the most commonly prescribed antibiotics were found among soil bacterial isolates and some of them were resistant to even more than two antibiotics [Esiobu et al., 2002, Riesenfeld et al., 2004, D'Costa et al., 2006]. It was concluded that environmental bacteria provide a natural reservoir of antibiotic resistance genes, which can then be transferred to clinically relevant bacteria [D'Costa et al., 2006]. However, one cannot rule out a possibility that the antibiotic resistance determinants in modern environmental bacteria have come in fact from external sources such as commensal bacteria or human pathogens.

Earth permafrost from the polar region is the most static and balanced environment, where microbial communities survive for thousands and millions of years [Vorobyova et al., 1997; Soina & Vorobyova, 2004]. Thus, study of permafrost sediments of a different age can provide a unique opportunity for analysis of biotopes of the preantibiotic era and allow for direct molecular comparison of the antibiotic resistance determinants in ancient permafrost and modern bacteria. However, only a few published works have analyzed native antibiotic resistance of bacterial strains isolated from permafrost [Tiedje et al., 1994; Vishnivetskaya et al., 2006] and neither of them has tried to elucidate the mechanisms of this resistance.

Previously, we described permafrost bacterial strains resistant to mercury compounds which harboured resistance determinants exhibiting a high level of homology with *mer*-operons of present-day bacteria [Petrova et al., 2002; Kholodii et al., 2003; Mindlin et al., 2005]. Recently, we succeeded in isolation of permafrost bacterial strains resistant to different antibiotics [Mindlin et al., 2008] and demonstrated that some of the resistance determinants were associated with mobile elements such as plasmids and transposons [Petrova et al., 2008].

The present work was designed to investigate the phenotypic and genotypic properties of multidrug-resistant strains of permafrost bacteria and determine the contribution of acquired resistance in their origin.

CHARACTERIZATION OF PERMAFROST SEDIMENTS

It is well established now that Earth permafrost from Arctic and Antarctic regions, the age of which is defined as the time of stay at subzero temperatures and can vary from several thousands up to a few million years, represents a unique opportunity to investigate ancient forms of life. Detection of considerable numbers of viable microorganisms in frozen Arctic and Antarctic sediments [Vorobyova et al., 1997, Soina &Vorobyova, 2004, Vishnivetskaya et al., 2000, 2006, Gilichinsky et al., 2007] allows studying of ancient bacterial communities in various aspects—from their age and period of preservation in permafrost and population diversity, to more special characteristics, including the origin and evolution of antibiotic resistance.

For isolation of ancient antibiotic resistance strains, we used permafrost samples of different ages and genesis (Table 1).

Arctic permafrost samples were collected at the Polar Regions of the Kolyma lowland, Coast of Laptev Sea and Coast of East-Siberian Sea. The region is characterized by the presence of a thick depth layer of polygonal ice (Fig 1).

Figure 1. Polygonal ice sheets on the cost of East-Siberian Sea

Antarctic permafrost samples were recovered from the Dry Valleys polar desert, which is the largest of the ice-free Antarctic regions. In earlier works it was shown that Dry Valley permafrost had several orders of magnitude lower number of viable cells and had less microbial diversity than arctic permafrost, but the spectrum of microorganisms isolated from Antarctic permafrost on the whole was similar to that of Arctic surface ecosystems and other cold habitats [Vorobyova et al., 2005, Gilichinsky et al., 2007].

The preparation of the samples for microbiological analysis was performed as described previously [Vishnivetskaya et al., 2000, 2006]. The sampling from permafrost cores, transportation and storage of the samples, control and specialized tests, have ensured that microorganisms recovered from permafrost samples were not contaminants but were indigenous to the samples [Vishnivetskaya et al., 2000, 2006, Gilichinsky et al., 2007].

Table 1. Characteristics of Arctic and Antarctic permafrost samples

Region of sampling	Borehole number	Age (YBP)*	Depth (m)	Bacterial cell counts, (CFU/g)**
Arctic region Coast of Laptev Sea	1/01-DAV	Late-Pleistocene-Ice-Complex 15- 40 K	28.0	$1.5x 10^2$
	03/03-Tiksi		2.0	$1.0x 10^7$
	12/03-Tiksi		5.0	$7.2x 10^6$
	3/01-DC	Middle-Pleistocene 220-390 K	22.5	$6.0x 10^5$
Arctic region Kolyma Lowland				
Bank of river Grand Chukochia	1/91	Late Pliocene- Early Pleistocene 2-3 M	34.0	$4.2x10^4$
Bank of river Alazeya	15/91	Late Pliocene 3 M	28.0	$3.2x10^4$
Bank of river Homus-Yuryiah	11/89	Late Pliocene- Early Pleistocene 2-3 M	24.0	$6.8x10^3$
			25.5	$1.4x10^3$
Bank of river Homus-Yuryiah	2/94	Late Pliocene- Early Pleistocene 2-3 M	39.9	$4.4x10^3$
Bank of river Homus-Yuryiah	4/93	Middle-Pleistocene 200-220 K	4.0	$8.7x10^3$
Bank of river Homus-Yuryiah	3/93	Middle-Pleistocene 200-600 K	6.0	$6.2x10^3$
Shore of Grand Oler Lake	1/95	Holocene 3-5 K	1.75	$4.8x10^4$
			8.75	$7.8x10^4$
Arctic region Coast of East-Siberian Sea	14/99	Late-Pleistocene-Ice-Complex 15- 35 K	13.6	$2.0x10^6$
			14.6	$5.0x10^5$
Antarctitida Beacon Dry Valley	A99	50-300K [Ng et al., 2005]	0.00	$2.3x10^4$
	6/99		1.3	$4.1x10^3$
	6/99		4.85	$5.2x10^2$

*Age of frozen subsoil layers (corresponds to the period of freezing before the present) in years (YBP): $K=10^3$ years, $M=10^6$ years; **The total number of viable microorganisms was determined after 3 days of incubation at 25°C on solid nutrient media by direct counting of colony forming units (CFU/g).

The number of platable cells in different Arctic and Antarctic samples ranged from 10^2 to 10^7 CFU/g (Table 1) and was comparable with earlier published data on permafrost sediments [Vorobyova et al., 1997; Vishnivetskaya et al., 2006]. The majority of the strains (70-80%) initially isolated from permafrost samples of different age were non spore-forming Gram-positive bacteria. Spore-forming Gram-positive bacteria composed from 10 to 20% of all bacteria in the samples. Only 2-5% of isolated strains could be referred to Gram-negative bacteria.

Isolation and Primary Characterization of Antibiotic Resistant Strains from Permafrost

To study the antibiotic resistance in ancient bacteria, five antibiotics including three aminoglycosides (gentamicin, kanamycin, streptomycin), tetracycline and chloramphenicol were chosen. For isolation of antibiotic resistance strains, bacteria from water suspensions of permafrost samples were plated on antibiotic supplemented solid media. Single colonies, formed on antibiotic containing plates were replated three times on the same media and incubated at 25°C. The concentrations of antibiotics in nutrient media were as follows (μg/ml): chloramphenicol (Cm), - 20; gentamicin (Gm), - 5-10; kanamycin (Km), - 25-50; streptomycin (Sm), - 50-100; tetracycline (Tc), - 10-20. At these concentrations, the fraction of bacteria, able to grow on selective nutrient media, to those unable to grow, was usually varied from between $1/10^{-2}$ and $1/10^{-3}$ CFU/g.

We succeeded in isolation of antibiotic resistant bacteria in all analyzed samples from both Arctic and Antarctic sediments. In general, the number of viable cells that were able to grow on media supplied with antibiotics was more than 10^3 CFU/g. It was noted that the number of antibiotic resistant bacteria did not decrease with the age of permafrost (data not shown).

In addition to direct selection of antibiotic resistant bacteria we isolated about 300 strains grown on nutrient media in the absence of antibiotics, and further studied their resistance to different antibiotics.

Study of cultural, morphological and biochemical properties of antibiotic resistant bacteria, as well as method of partial sequencing of 16S rRNA genes, allowed to identify the strains as representatives of genera *Acinetobacter, Pseudomonas, Xanthomonas, Bacillus, Arthrobacter, Micrococcus, Flavobacterium, Psychrobacter, Stenotrophomonas, Sphingomonas, Brevundimonas, Paenibacillus.* It was revealed, that the majority of the antibiotic resistant strains were Gram-positive bacteria. It is noteworthy that the bacterial diversity in our collection of antibiotic resistant permafrost strains was similar to that of strains isolated earlier from other cold Arctic and Antarctic environments [Vishnivetskaya et al., 2006, Gilichinsky et al., 2007]. In particular, antibiotic resistant strains included psychrophilicbacteria.

Study of Antibiotic Resistance Spectrum of Permafrost Strains

About half of the antibiotic resistant strains obtained after initial selection, were chosen for further work. Among them we identified 53 strains of Gram-positive and 30 strains of Gram-negative bacteria. For each of these strains we determined the spectrum of resistance to all antibiotics used previously (Km, Gm, Sm, Tc, Cm). This analysis revealed not only strains resistant to a single antibiotic, but also strains resistant to two and three different antibiotics. The frequency of strains with double antibiotic resistance exceeded 20% both in Gram-positive and Gram-negative bacteria, while the strains with triple resistance were found only among Gram-negative bacteria with 20% frequency [Mindlin et. al., 2008].

Twenty five strains of Gram-negative bacteria and four strains of Gram-positive bacteria, most of them resistant to at least two antibiotics, were chosen for detailed analysis of resistance determinants (Table 2). Phylogenetic affiliation of all isolates using 16S rRNA gene sequencing allowed to identify ten strains belonging to genus *Pseudomonas*, eight *Acinetobacter*, two *Stenotrophomonas*, two *Sphingomonas* strains, and four strains belonging to Gram-positive genus *Paenibacillus*. The remaining three strains were found to belong to *Xanthomonas*, *Psychrobacter* and *Brevundimonas* genera, respectively (Table 2).

For each of these strains we determined the spectrum of resistance to wide range of antibiotics (beta-lactams, aminoglycosides, tetracyclines, sulphonamide drugs, trimethoprim and quinolones), including those used during the first step of selection. Besides, we determined the mercury resistance of these strains. Susceptibility of the resistant strains to different antimicrobial agents was determined by the agar diffusion and the agar dilution methods using the Mueller-Hinton agar [Hirai et al., 1986; Andrews, 2007]. This analysis revealed that the most permafrost strains studied were resistant to antibiotics of different classes (Table 2). For instance, strain MR29-12 of *Psychrobacter psychrophilus* was simultaneously resistant to streptomycin, tetracycline and trimethoprim; strain MR5-8 of *Acinetobacter* sp. was resistant to five antibiotics (Ap, Cm, Gm, Km, Tb); and strain ED23-35 of *Acinetobacter* sp. was simultaneously resistant to $HgCl_2$ and streptomycin. It is interesting to note that, independently of the isolation procedure (the presence or absence of particular antibiotics during isolation), many strains were simultaneously resistant to two or more structurally unrelated antibiotics. As an example, strain Tik1 of *Pseudomonas putida* that was isolated in the absence of antimicrobial agents was simultaneously resistant to $HgCl_2$, carbenicillin, chloramphenicol, streptomycin and trimethoprim, which belong to different classes of antibacterial agents (Table 2).

Table 2. Drug resistance of permafrost strains

Strain designation	Region[1]	Age (YBP)[2]	Taxonomic position	Resistance pattern
ED23-35	A	15-40K	*Acinetobacter* sp.	Hg, Sm
ED45-25	A	15-40K	*Acinetobacter* sp.	Hg, Sm
M2-7	A	15-40K	*Acinetobacter jonsonii*	Ap, Hg, Tp
MR5-8	A	15-40K	*Acinetobacter* sp.	Ap, Cm, Gm, Km, Tb
MR5-11	A	15-40K	*Acinetobacter* sp.	Ap, Cm, Sm, Sp
VS15	A	2-3M	*Acinetobacter* sp.	Ap, Cm, Sm, Sp
EK30A	A	1M	*Acinetobacter* sp.	Ap, Sm, Sp
EK67	A	2-3M	*Acinetobacter* sp.	Ap, Sm, Sp
MR29-12	A	15-35K	*Psychrobacter psychrophilus*	Sm, Tc, Tp
MR7-1	A	200-390K	*Stenothrophomonas* sp.	Cm, Gm, Cb, Km, Nm, Sp, Tc, Tp
MR7-3	A	200-390K	*Stenothrophomonas maltophilia*	Ap, Cm,Gm,Km, Nm, Sp, Tp

Table 2. (Continued)

MR9-2	A	15-40K	*Xanthomonas retroflexus*	Am, Ap, Cm, Gm, Km, Nm, Sp, Tp
ED23-9	A	15-40K	*Sphingomonas*[3] sp.	Gm, Km, Nm, Sm,Tb
EK42	B	50-300K	*Sphingomonas mucosissima*	(Ap), Sm
ED23-10	A	15-40K	*Pseudomonas libanensis*[4]	Cb, Cm, Hg, Sp, Tp
ED23-26	A	15-40K	*Pseudomonas fluorescens*	Cb, Cm, Hg, Km, Sp, Tp
ED94-71	A	15-40K	*Pseudomonas* (non-fluorescent)	Cb, Cm, Hg, Km, Sp,Tp,
EDM6-1	A	15-40K	*Pseudomonas* sp. (fluorescent)	Cb, Hg, Tp
VS27	A	3-5K	*Pseudomonas* sp.	Cm, Cb, Sm, Tp
VS38	A	3-5K	*Pseudomonas* sp.	Cm, Cb, Sm, Sp, Su, Tc,Tp, Nal
VS50	A	3-5K	*Pseudomonas* sp.	Cm, Cb, Sm, Sp, Su, Tp
VS51	A	3-5K	*Pseudomonas* sp.	Cm, Cb, Sm, Sp, Tp
Tik1	A	200-600K	*Pseudomonas putida*	Hg, Cm, Cb, Sm, Sp, Tp
Tik3	A	200-600K	*Pseudomonas* sp.	Cm, Sm, Sp, Su, Tp
EK41	B	50-300K	*Brevundimonas*[5] *vesicularis*	Ap, Sm
MRI-1	A	3-5K	*Paenibacillus amyloliticus*	Sm, (Sp)
VSH72	A	200-600K	*Paenibacillus amyloliticus*	(Ap), Sm
VSH76	A	200-600K	*Paenibacillus amyloliticus*	Sm, (Sp)
EK63	B	50-300K	*Paenibacillus* sp.	(Ap), Sm, (Sp)

Antibiotic designations and concentrations used (μg/ml): ampicillin (Ap), -200; chloramphenicol (Cm), -20; gentamicin (Gm), -5; $HgCl_2$ (Hg), -5; carbenicillin (Cb), -400; kanamycin (Km), -25 (50-200); nalidixic acid (Nal), -20; neomycin (Nm), -25; streptomycin (Sm), -50(100); spectinomycin (Sp), -100; sulphathiazole (Su), -25; tobramycin (Tb), -40; tetracycline (Tc), -15; trimethoprim (Tp), -25. Resistance to Cb always acompany Ap-resistance; Parentheses indicate low-level resistance

[1]A-Arctic, B-Antarctic; [2]Age in years before present (YBP). $K=10^{3}$ years; $M=10^{6}$ years; [3]*Sphingomonas* were originally named *Pseudomonas paucimobilis* but were later reclassified into distinct species within a new genus *Sphingomonas* [Yabuuchi et al., 1990];[4]A new *Pseudomonas* species, *Pseudomonas libanensis* sp. nov. is part of the *Pseudomonas fluorescens* intrageneric cluster [Dabboussi et al., 1999]; [5]Bacteria belonging to *Brevundimonas* were previously classified in the genus *Pseudomonas* but later were included into a deeply branching cluster in the stem of gram-negative bacteria [Ochi, 1995].

LOCALIZATION OF RESISTANCE DETERMINANTS IN GENOMES OF PERMAFROST STRAINS

It is known that bacteria of various systematic groups are characterized by different intrinsic antibiotic resistance; moreover they may be cross-resistant to several classes of antimicrobial agents [Piddock, 2006]. Bacteria may also acquire resistance via the horizontal transfer of specific genes from other organisms [Davies, 1994; Tenover, 2006]. To study the mechanisms of antibiotic resistance in our collection of multidrug resistant permafrost bacteria we determined the localization of the resistance genes and their possible association with different mobile elements (plasmids and transposons).

We first analyzed whether the resistance determinants can have plasmid localization. Indeed, it was shown that most of the permafrost antibiotic resistant strains contained plasmids of different sizes (data not shown). For analysis of plasmid localization of the resistance genes, antibiotic resistant strains harbouring large plasmids were chosen (see Fig. 2). Plasmid DNA was extracted by an alkaline lysis method and analyzed by electrophoresis on 0.7 % (wt/vol) agarose gel in Tris-borate buffer, followed by staining with ethidium bromide and visualization under UV light [Sambrook et al., 1989]. The molecular sizes of the plasmids were evaluated by comparison with reference plasmids R388 (34 kb) [Datta & Hedges, 1972] and RP4 (60 kb) [Jacob & Grinter,1975].

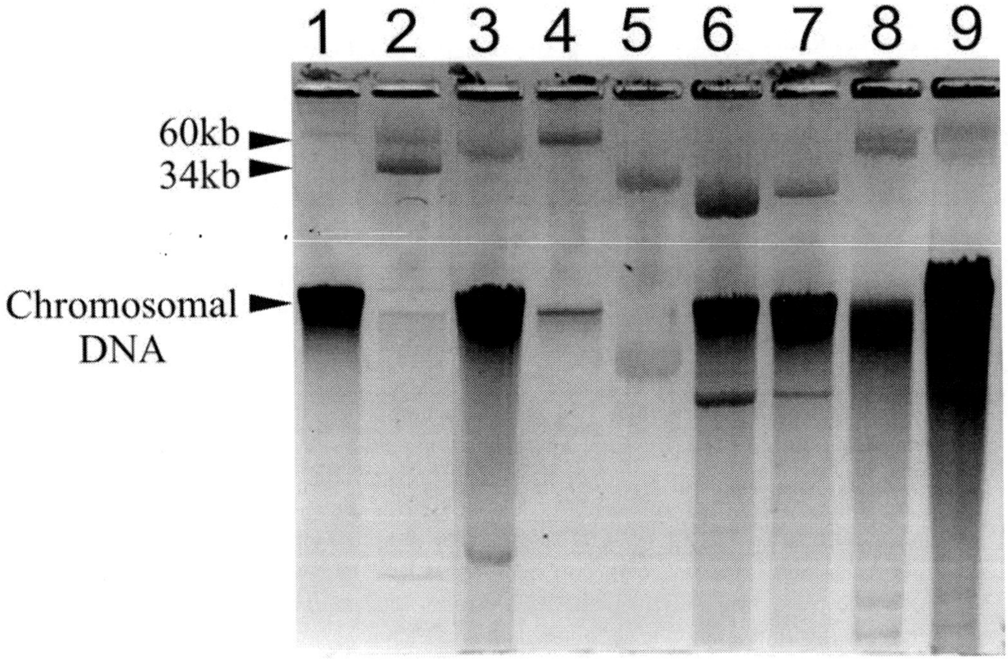

Figure 2. Plasmids of bacterial strains isolated from permafrost. Horizontal agarose gel electrophoresis (run at 50V for 18h): 1=Tik1, 2=R388, 3=ED6-1, 4=RP4, 5=MR29-12, 6=ED94-71, 7=ED23-26, 8= ED23-35, 9=ED45-25

For detailed investigations we selected five strains of *Pseudomonas*, three strains of *Acinetobacter* and one strain of *Psychrobacter psychrophilus*, each containing a single large plasmid (about 25-55 kb) (see Fig. 2). The association of resistance genes with R plasmids was determined in conjugation and transformation experiments with laboratory recipient bacterial strains. For *Acinetobacter* and *Psychrobacter* strains, the *Acinetobacter calcoaceticus* strain BD413rif was used as a recipient; and for *Pseudomonas* strains, the *Pseudomonas fluorescens* strain P22-1-2 was used (Table 3). It should be noted that the latter strain was resistant to four different antibiotics (Cb, Cm, Sp, Tp). For this reason association of the corresponding resistance determinants with plasmids in *Pseudomonas* strains by conjugation experiments could not be evaluated with assurance.

We determined that in four strains, only mercury resistance was transferred to appropriate recipient; in three, both mercury and streptomycin resistance determinants were transferred, and in additional two plasmid localization was revealed for streptomycin and tetracycline resistance, and streptomycin and sulphathiazole resistance, respectively (Table 3).

Table 3. Plasmid localization of resistance determinants in permafrost strains

Permafrost strains	Resistance pattern	Resistance determinants associated with R plasmids
Acinetobacter sp. ED23-35	Hg, Sm, Sp	Hg, Sm
Acinetobacter sp. ED45-25	Hg, Sm	Hg, Sm
Acinetobacter jonsonii M2-7	Ap, Hg, Tp	Hg
Psychrobacter psychrophilus MR29-12	Sm, Tc, Tp	Sm, Tc
Pseudomonas fluorescens ED23-26	Hg, Km, **Cb, Cm, Sp, Tp**	Hg
Pseudomonas sp. ED94-71	Hg, Km, **Cb, Cm, Sp, Tp**	Hg
Pseudomonas sp. (fluorescent) EDM6-1	Hg,**Cb, Tp**	Hg
Pseudomonas putida Tik1	Hg, Km, **Cb, Cm, Sp, Tp**	Hg, Sm
Pseudomonas sp. Tik3	Sm, Su, **Cm, Sp, Tp**	Sm, Su
Recipient strains:		
Acinetobacter calcoaceticus BD413rif [1]	[rif-r]	
Pseudomonas fluorescens P22-1-2[2]	**Cb, Cm, Sp, Tp** [str-r, rif-r]	

Resistance determinants that were also present in *Pseudomonas* recipient strain, and therefore could not be analyzed by plasmid transfer, are shown in bold

[1] used in matings with strains of *Acinetobacter* and *Psychrobacter psychrophilus*;[2] used in matings with strains of *Pseudomonas*; str-r - 30S-ribosome resistance; rif-r - mutations of RNA-polymerase gene *rpoB*

In order to investigate the possible role of R-plasmids in resistance transfer between different bacteria we determined the host range of permafrost plasmids revealed. As recipients in conjugation experiments, laboratory strains *P. fluorescens* P22-1-2, *A. calcoaceticus* BD413rif and *E. coli* K-12 C600rif were used. It was revealed that none of the plasmid had the properties of broad host range plasmids capable to disseminate the resistance markers between unrelated bacteria (Table 4). At the same time, plasmids of some pseudomonads transferred into strains belonging to other *Pseudomonas* species. For instance, plasmids of *P. putida* Tik1 and *P.* sp. ED94-71 (non-fluorescent) could be transferred between various laboratory strains of *P. putida* and *P. fluorescens* (not shown).

It is of interest that among *Pseudomonas* and *Acinetobacter* plasmids studied we revealed several ones that harboured only mercury resistance genes. At the same time, we succeeded in detection of R plasmids harbouring both mercury-resistance and antibiotic-resistance determinants. In this connection it should be noted that in the history of clinical *Pseudomonas* plasmids, resistance to $HgCl_2$ was the first marker recognized before 1969. Drug resistance plasmids of *Pseudomonas* studied thereafter conferred resistance to antibiotics of different families in addition to the mercury resistance [Kontomichalou, Papachristou & Angelatou, 1976]. Therefore, our results support the previous conclusions [Hughes & Datta, 1983;

Davies, 1994] that antibiotic resistance determinants could be "picked up" by bacterial plasmids as a consequence of introducing antibiotics in medicine and veterinary practice. The same phenomenon likely occurred in the environment but with a substantially lower rates.

Table 4. Determination of host-range of plasmids from permafrost bacterial strains

Donor strain carrying R-plasmid	Transfer ability of R-plasmids in crosses with recipient strains		
	Acinetobacter BD413rif	*Pseudomonas* P22-1-2	*E.coli* C600rif
Acinetobacter sp.ED23-35	+	-	-
Acinetobacter sp.ED45-25	+	-	-
P. psychrophilus MR29-12	+	-	-
P. fluorescens ED23-26	-	+	-
*P.*sp.(non-fluorescent) ED94-71	-	+	-
P. putida Tik1	-	+	-

+ transfer of R-plasmid is observed; - transfer is absent

We further analyzed whether the antibiotic resistance determinants with plasmid localization can be associated with transposons. We succeeded in isolation of two transposons containing antibiotic resistance genes from strains *Psychrobacter psychrophilus* MR29-12 and *Pseudomonas* sp. Tik3, respectively (see Tables 2 and 3). The first of these transposons carried determinants conferring resistance to streptomycin and tetracycline; the second – to streptomycin and sulphothiazole. [Petrova et al., 2008]. We further demonstrated that both transposons could translocate onto plasmids with broad-host-range spectrum and could be transferred into bacteria of different systematic groups including *Escherichia coli* and *Pseudomonas* spp. (data not shown). Further chracterization of R-plasmids and antibiotic resistance transposons identified in permafrost bacterial strains will be an important goal of future studies.

For a substantial fraction of the antibiotic resistant strains from our collection (in particular, various *Pseudomonas* and *Acinetobacter* strains), antibiotic resistance determinants could be transferred by neither conjugative transfer of plasmid DNA nor plasmid DNA transformation. One can suggest that in this case the resistance phenotype could result from chromosomally encoded intrinsic resistance [Gomez & Neyfakh, 2006]. To check this hypothesis, we analyzed whether the resistance determinants not associated with plasmids can be transferred between species by genomic DNA transformation.

Multidrug resistant *Acinetobacter* strains MR5-11 and VS15 which contained no plasmids were used as donor strains and model laboratory auxothrophic strain *Acinetobacter calcoaceticus* BD413ivl (Ivl-10) competent for transformation [Juni, 1972] was used as a recipient strain. For transformation of genomic DNA, a simple transformation assay procedure described for *Acinetobacter* spp was used [Juni, 1972]. Preparation of crude transforming DNA from donor strains was performed as described in [Sambrook et al., 1989]. As a marker phenotype to follow the transformation of antibiotic resistance, we used chloramphenicol resistance (Cm^R). To evaluate the efficiency of transformation, we compared the frequency of transformation of chloramphenicol resistance ($Cm^S \rightarrow Cm^R$) with that of transformation from auxotrophy to prototrophy ($Ivl^- \rightarrow Ivl^+$).

The results of experiments (Table 5) showed that frequency of homologous transformation of BD413ivl to prototrophy with DNA isolated from prototrophic BD413 strain (2.3×10^{-5}) exceeds that of heterologous transformation with DNA isolated from permafrost *Acinetobacter* strains (1.6×10^{-6} and 7.0×10^{-6}, Table 5). This agrees with published data on the efficiencies of homologous and heterologous transformation for *Acinetobacter* species [Juni, 1972]. We also observed transformation of chloramphenicol resistance (Cm-r) from the resistant permafrost strains to the sensitive BD413 ivl strain. The frequency of Cm-r transformation ($2.8–9.0 \times 10^{-7}$) was comparable to the frequency of the $Ivl^- \rightarrow Ivl^+$ transformation. Remarkably, no transformation of chloramphenicol resistance was observed when control chloramphenicol sensitive strain was used as DNA source (Table 5).

Table 5. Transformation of *Acinetobacter calcoaceticus* BD413ivl with genomic DNA

Source of DNA	Frequency of transformations (per 1μL DNA)	
	Ivl+	Cm-r
VS15 (Ivl^+,Cm-r)	1.6×10^{-6}	9.0×10^{-7}
MR5-11 (Ivl^+,Cm-r)	7.0×10^{-6}	2.8×10^{-7}
BD413 (Ivl^+,Cm-s)	2.3×10^{-5}	$< 2.0 \times 10^{-8}$

Frequency of transformation was determined as a ratio of the number of transformants to the total number of recipient strain colonies

Taking into account that a single chromosomal gene can afford intrinsic resistance to different antibiotics [Poole, 2005; Piddock, 2006], we compared the resistance pattern of the BD413 Cm-r transformants to the resistance pattern of original antibiotic resistant *Acinetobacter* strains. In particular, we analyzed resistance of the transformants to streptomycin, spectinomycin, trimethoprim and nonspecific DNA intercalating agent ethidium bromide. To assess the level of antibiotic resistance, we used a twofold serial dilution technique [Hirai et al., 1986]. The acquisition of resistance was defined as at least fourfold increase in the minimal inhibition concentration (MIC). Analysis of antibiotic susceptibility patterns showed that Cm-r transformants also displayed higher levels of resistance to streptomycin, spectinomycin and trimethoprim. Moreover, they were characterized by increased resistance to ethidium bromide (Table 6). Thus, the antibiotic resistance pattern of wild type *Acinetobacter* strains VS15 and MR5-8 could be transferred to the laboratory strain of the same genus by transformation of their chromosomal DNA, suggesting that the multidrug resistance phenotype can be explained by chromosomally encoded intrinsic resistance. Indeed, such intrinsic resistant has been previously described for some species of *Acinetobacter* [Magnet et al., 2001; Gomez & Neyfakh, 2006].

Table 6. Transfer of antibiotic resistance pattern between permafrost and laboratory strains of *Acinetobacter*

Strain	MIC (μg/ml)				
	Cm	Sm	Sp	Tp	EthBr
Permafrost resistant strains					
MR5-11	64-128	>256	256	32	128
VS15	>32	>256	>256	32	64
Laboratory sensitive strain					
BD413ivl	8	4-8	8-16	8	16-32
Transformants					
BD413ivl/VS15	64	>128	128	32	128
BD413ivl/MR5-11	64	>128	128	32	128

MICs were determined in three independent experiments performed by the agar dilution method using the Mueller-Hinton medium; Cm-chloramphenicol; Sm-streptomycin; Sp-spectinomycin; Tp-trimethoprim; EthBr – ethidium bromide.

Identification of Resistance Determinants in Permafrost Bacterial Strains

As the first step to identification of the antibiotic resistance determinants of ancient bacteria, we tested permafrost strains from our collection for the presence of known genes encoding for streptomycin resistance. Seven strains resistant to different antibiotics (Table 2) isolated from different locations of Siberian and Antarctic permafrost were chosen for comparative analysis of antibiotic resistance determinants (Table 7). Four of these strains contained plasmids and (or) transposons (Table 3 and data not shown).

These strains were screened for the presence of streptomycin, sulphathiazole and tetracycline resistance genes, which are prevalent among present-day clinically important bacteria and which are usually associated with mobile elements such as plasmids, transposons and integrons.

PCR, Southern blot analysis and partial sequencing revealed that the antibiotic resistant bacteria studied contained genes highly homologous to the known present-day antibiotic resistance genes. Thus, streptomycin-resistant permafrost strains contained streptomycin-resistance determinants of the two known types. In six of these strains streptomycin-resistance was encoded by the linked *strA-strB* genes, which encode aminoglycoside phosphotransferases and are distributed among present-day bacterial isolates from human, animals, and plants [Sundin, 2002]. In particular, we for the first time identified the *strA-strB*-like genes in ancient Gram-negative bacterial strains belonging to *Psychrobacter* and *Brevundimonas* genera and also in a strain VSH72 of a Gram-positive bacterium *Paenibacillus amyloliticus*. Other streptomycin-resistant strains of Gram-negative bacteria from permafrost contained adenylyltransferase genes related to the *aadA*-genes of present-day bacteria [Hall & Collis, 1998; Clark et al., 1999]. Intriguingly, two of the streptomycin resistance strains (Tik3 and VSH72) contained both the *strA-strB* and *aadA* genes. While the reasons for having two types of streptomycin resistance in the same cell remain unknown, similar strains harbouring both the *strA-strB* and *aadA* genes were recently detected among streptomycin resistant *Escherichia coli* of domestic animals in Norway [Sunde & Norström, 2005].

Table 7. Genes of antibiotic resistance revealed in permafrost strains

Strain designation	Region and age of permafrost	Taxonomic position	Revealed genes of resistance
ED23-35	Bank of river Homus-Yuryiah, 15-40 K	*Acinetobacter sp.*	*strA-strB*
MR29-12	Coast of East-Siberian Sea, 15-35K	*Psychrobacter psychrophilus*	*strA-strB, tetR-tet(H)*
Tik1	Coast of Laptev Sea, 200-600K	*Pseudomonas putida*	*strA-strB*
Tik3	Coast of Laptev Sea, 200-600K	*Pseudomonas sp.*	*strA-strB, aadA, sul1*
EK41	Antarctida, 50-300K	*Brevundimonas vesicularis*	*strA-strB*
VSH72	Kolyma lowland 200-600K	*Paenibacillus amyloliticus*	*strA-strB, aadA*
VSH76	Kolyma lowland 200-600K	*Paenibacillus amyloliticus*	*aadA*

A single strain that was resistant to both streptomycin and tetracycline (*Psychrobacter psychrophilus* MR29-12) carried *tetR-tet(H)* genes in addition to the *strA-strB* streptomycin resistance determinants. Previously, the *tetR-tet(H)* genes were found in clinical bacterial strains belonging to *Pasteurella*, *Mannheimia*, *Acinetobacter, Moraxella,* and *Actinobacillus* genera [Kehrenberg et al., 2001; Blanco et al., 2006].

CONCLUSION

Permafrost bacteria represent a unique opportunity to study the origin and modes of dissemination of drug-resistance determinants in environmental bacterial populations. In this work, we, for the first time, isolated strains resistant to different antibiotics from permafrost bacterial populations that originated from Arctic and Antarctic sediments of different ages and genesis. It turned out that many of the antibiotic-resistant bacteria in the permafrost collection contained genes highly homologous to the known present-day antibiotic resistance genes. Thus, our results strongly support the hypothesis that antibiotic resistance genes were spread in natural bacterial populations long before the introduction of antibiotics into clinical practice [Davies, 1994]. Unexpectedly, we also demonstrated that multidrug resistant bacterial strains can be found in permafrost bacterial communities with high frequency. We found such strains among bacteria belonging to *Pseudomonas, Acinetobacter* and *Stenotrophomonas* genera. It was revealed that multidrug resistance of some strains of *Acinetobacter* sp. can be transferred by transformation of chromosomal genes. Therefore, one can propose that multidrug resistance in many strains of permafrost bacteria is due to their intrinsic resistance to a wide range of compounds resulting from expression of efflux pump systems. Importantly, different systems of efflux pumps conferring multidrug-resistance are widely distributed among modern clinically relevant strains of *Pseudomonas aeruginosa, Acinetobacter baumannii* and *Stenotrophomonas maltophilia* [Poole, 1994; Gould &Avison 2000; Magnet et al., 2001], which are closely related to the multidrug-resistant strains isolated from permafrost.

Analysis of permafrost drug-resistant strains also provided significant insights into the gene transfer mechanisms. Thus, we succeeded in revealing several mobile elements (plasmids and transposons) associated with antibiotic resistance determinants in permafrost strains. These data demonstrate an important role of mobile elements in horizontal transfer of antibiotic resistance genes among ancient environmental bacteria.

A detailed analysis of both the molecular structure of antibiotic resistance determinants in permafrost bacteria, and mechanisms of their horizontal transfer represents a major goal for further studies and is necessary to understand the origin, evolution and spread of antibiotic resistance in nature.

ACKNOWLEDGEMENTS

The authors express their gratitude to Dr. D.A. Gilichinsky (Institute of Physicochemical and Biologicak problems in Soil Science, Russian Academy of Sciences, Russia) for scientific cooperation and assistance in providing the permafrost samples and their geological characteristic; Dr. A.L. Mulyukin and Dr. A.V. Kulbachinsky for providing useful comments and help in preparing the manuscript; E.I. Molchanova for expert technical assistance. This work was supported in part by the Russian Foundation for Basic Research, grants 05-04-49372 and 08-04-00263 and by the Russian Academy of Sciences Presidium Program in Molecular and Cellular Biology grant (to A. V. Kulbachinsky).

REFERENCES

Andrews, J. M. (2007). BSAC standardized disc susceptibility testing method (version 6). *J Antimicrob Chemother*, *60*, 20-41.

Blanco, M., Gutiérrez-Martin, C. B., Rodríguez-Ferri, E. F., Roberts, M. C. & Navas, J. (2006). Distribution of tetracycline resistance genes in Actinobacillus pleuropneumoniae isolates from Spain. *Antimicrob Agents Chemother, 50*, 702-708.

Clark, N. C., Olsvik, Ø., Swenson, J. M., Spiegel, C. A. & Tenover, F. C. (1999) Detection of a streptomycin/spectinomycin adenylyltransferase gene (aadA) in Enterococcus faecalis. *Antimicrob Agents Chemother, 43*, 157-160

Dabboussi, F., Hamze, M., Elomari, M., Verhille, S., Baida, N., Izard, D. & Leclerc, H. (1999) Pseudomonas libanensis sp. nov., a new species isolated from Lebanese spring waters. *Int J Syst Bacteriol, 49*, 1091-1101.

Datta, N. & Hedges, R. W. (1972) Trimethoprim resistance conferred by W plasmid in Enterobacteriaceae. *J Gen Microbiol, 72*, 349-355.

Davies, J. (1994) Inactivation of antibiotics and the dissemination of resistance genes. *Science, 264*, 375-382.

D'Costa, V. M., McGrann, K. M., Hughes, D. W. & Wright, G. D. (2006) Sampling the antibiotic resistome. *Science, 311* (5759), 374-377.

Esiobu, N., Armenta, L. & Ike, J. (2002) Antibiotic resistance in soil and water environments. *Int J Environ Health Res, 12*, 133-144.

Fluit, A. C., Schmitz, F. J. & Verhoef, J. (2001). Multi-resistance to antimicrobial agents for the ten most frequently isolated bacterial pathogens. *Int J Antimicrob Agents, 18*, 147-160.

Gilichinsky, D. A., Wilson, G. S., Friedmann, E. I., McKay, C. P., Sletten, R. S., Rivkina, E. M., Vishnivetskaya, T. A., Erokhina, L. G., Ivanushkina, N. E., Kochkina, G. A., Shcherbakova, V.A., Soina, V. S., Spirina, E. V., Vorobyova, E. A., Fyodorov-Davydov, D. G., Hallet, B., Ozerskaya, S. M., Sorokovikov, V. A., Laurinavichyus, K. S., Shatilovich, A. V., Chanton, J. P., Ostroumov, V. E. & Tiedje, J. M. (2007). Microbial populations in Antarctic permafrost: biodiversity, state, age, and implication for astrobiology. *Astrobiology, 7*, 275-311.

Gomez, M. J. & Neyfakh, A. A. (2006) Genes involved in intrinsic antibiotic resistance of Acinetobacter baylyi. *Antimicrob Agents Chemother, 50*, 3562-3567.

Gould, V. C. & Avison, M. B. (2000) SmeDEF-mediated antimicrobial drug resistance in Stenotrophomonas maltophilia clinical isolates having defined phylogenetic relationships. *J Antimicrob Chemother, 57*, 1070-1076.

Hall, R. M. & Collis, C. M. (1998) Antibiotic resistance in gram-negative bacteria: the role of gene cassettes and integrons. *Drug Resist Updates, 1*, 109-119.

Hirai, K., Aoyama, H., Suzue, S., Irikura, T., Iyobe, S. & Mitsuhashi, S. (1986). Isolation and characterization of norfloxacin-resistant mutants of Escherichia coli K-12. *Antimicrob Agents Chemother, 30*, 248-253.

Hughes, V. M. & Datta, N. (1983). Conjugative plasmids in bacteria of the 'preantibiotic era'. *Nature, 302*, 725-726.

Jacob, A. E., & Grinter, N. J. (1975). Plasmid RP4 as a vector replicon in genetic engineering. *Nature, 255*, 504–506.

Juni, E. (1972). Interspecies transformation of Acinetobacter: genetic evidence for a ubiquitous genus. *J Bacteriol, 112*, 917-931.

Kehrenberg, C., Salmon, S. A., Watts, J, L. & Schwarz, S. (2001). Tetracycline resistance genes in isolates of Pasteurella multocida, Mannheimia haemolytica, Mannheimia glucosida and Mannheimia varigena from bovine and swine respiratory disease: intergeneric spread of the tet(H) plasmid pMHT1. *J Antimicrob Chemother, 48*, 631-640.

Kholodii, G., Mindlin, S., Petrova, M. & Minakhina, S. (2003) Tn5060 from the Siberian permafrost is most closely related to the ancestor of Tn21 prior to integron acquisition. *FEMS Microbiol Lett, 226*, 251-255.

Kontomichalou, P., Papachristou, E. & Angelatou, F. (1976) Multiresistant plasmids from Pseudomonas aeruginosa highly resistant to either or both gentamicin and carbenicillin. *Antimicrob Agents Chemother. 9*, 866-873.]

Magnet, S., Courvalin, P.,& Lambert, T. (2001). Resistance-nodulation-cell division-type efflux pump involved in aminoglycoside resistance in Acinetobacter baumannii strain BM4454. *Antimicrob Agents Chemother, 45*, 3375-3380.

Mindlin, S., Minakhin, L., Petrova, M., Kholodii, G., Minakhina, S., Gorlenko, Zh. & Nikiforov, V. (2005) Present-day mercury resistance transposons are common in bacteria preserved in permafrost grounds since the Upper Pleistocene. *Res Microbiol, 156*, 994-1004.

Mindlin, S., Soina, V., Petrova, M. & Gorlenko, Z. (2008). Isolation of antibiotic resistance bacterial strains from East Siberia permafrost sediments. *Russian J Genetics, 44*, 27-34.

Ochi, K. (1995). Comparative ribosomal protein sequence analyses of a phylogenetically defined genus, Pseudomonas, and its relatives. *Int J Syst Bacteriol, 45*, 268-273.

Petrova, M. A., Mindlin,S. Z., Gorlenko, Zh. M., Kalyaeva, E. S., Soina, V. S.& Bogdanova, E. S. (2002) Mercury-resistant bacteria from permafrost sediments and prospects for their use in comparative studies of mercury resistance determinants. *Russian J Genetics, 38*, 1569-1574.

Petrova, M. A., Gorlenko, Zh. M., Soina, V. S, Mindlin,S. Z. (2008) Association of the *strA–strB* genes with plasmids and transposons in the present-day bacteria and in bacterial strains from permafrost. *Russian J Genetics, 44,* 1281-1286.

Piddock, L. J. (2006) Clinically relevant chromosomally encoded multidrug resistance efflux pumps in bacteria. *Clin Microbiol Rev, 19,* 382-402.

Poole, K. (1994) Bacterial multidrug resistance – emphasis on efflux mechanisms and Pseudomonas aeruginosa. . *J Antimicrob Chemother,34*, 453-456.

Poole, K. (2005) Efflux-mediated antimicrobial resistance. *J Antimicrob Chemother, 56,* 20-51.

Riesenfeld, S. C., Goodman, R. M. & Handelsman, J. (2004). Uncultural soil bacteria are a reservoir of new antibiotic resistance genes. *Environ Microbiol, 6*, 981-989.

Sambrook J., Fritsch, E.F. & Maniatis T. (1989) *Molecular cloning: a laboratory manual.* Cold Spring Harbor, NY. Cold Spring Harbor Laboratory Press.

Soina, V. S. & Vorobyova, E. A. (2004). Adaptation of bacteria to the terrestrial permafrost environment: a biomodel for Astrobiology. In J. Seckbach (Ed), *Origins:genesis, evolution and biodiversity of life* (pp. 427-444). Netherlands, Kluver Academic publishers.

Sunde, M. & Norström, M. (2005). The genetic background for streptomycin resistance in Escherichia coli influences the distribution of MICs. *J Antimicrob Chemother, 56*, 87-90.

Sundin, G. W. (2002). Distinct recent lineage of the strA-strB streptomycin-resistance genes in clinical and environmental bacteria. *Curr. Microbiol, 45*, 63-69.

Tenover, F.C. (2006). Mechanisms of antimicrobial resistance in bacteria. *Am J Med, 119,* Suppl. 1, S3-10.

Tiedje, J., Smith, G. B., Simkins, S., Holben, W. E., Finney, C. & Gilichinsky, D. A. (1994). Recovery of DNA, denitrifiers and patterns of antibiotic sensitivity in microorganisms from ancient permafrost soils of Eastern Siberia. In D. A. Gilichinsky (Ed.), *Viable microorganisms from permafrost*.(pp. 83-98). Puschino. (in Russian).

Vishnivetskaya, T., Kathariou S., McGrath J., Gilichinsky, D. & Tiedje, J. M. (2000) Low-temperature recovery strategies for the isolation of bacteria from ancient permafrost sediments. *Extremophiles*, *4,* 165-173.

Vishnivetskaya, T. A., Petrova, M. A., Urbance, J., Ponder, M., Moyer, C. L., Gilichinsky, D. A. & Tiedje, J. M. (2006). Bacterial community in ancient siberian permafrost as characterized by culture and culture-independent methods. *Astrobiology*, *6*, 400-414.

Vorobyova, E., Soina, V., Gorlenko, M., Minkovskaya, N., Zalinova, N., Mamukelashvili, A., Gilichinsky, D., Rivkina, E. & Vishnivetskaya, T. (1997). The deep cold biosphere: facts and hypothesis. *FEMS Microbiol Rev*, *20,* 277-290.

Vorobyova, E. A., Soina, V. S., Mamukelashvili, A. G., Bolshakova A., Yaminsky I. V.& Mulyukin A. L. (2005). Living Cells in Permafrost as Models for Astrobiology Research. (Chapter 19). In J.D Castello and S.O. Rogers (Eds.), *Life in Ancient Ice*. pp. 277-288, Princeton University Press.

Yabuuchi, E., Yano, I., Oyaizu, H., Hashimoto, Y., Ezaki, T. & Yamamoto, H. (1990). Proposals of Sphingomonas paucimobilis gen. nov. and comb. nov., Sphingomonas parapaucimobilis sp. nov., Sphingomonas yanoikuyae sp. nov., Sphingomonas adhaesiva sp. nov., Sphingomonas capsulata comb. nov., and two genospecies of the genus Sphingomonas. *Microbiol Immunol, 34*, 99–119.

Reviewed by Mulyukin (Muliukin) A. L. Institute of Microbiology, Russian Academy of Sciences, Moscow, Russia.

In: New Permafrost and Glacier Research
Editors: Max I. Krugger and Harry P.Stern

ISBN: 978-1-60692-616-1

Chapter 4

SIMILARITIES BETWEEN THE RECENT PERMAFROST IN NORTH-WESTERN CANADA AND THE PLEISTOCENE RELICT CRYOGENIC FORMS IN CENTRAL EUROPE (HUNGARY)

Ákos Szabolcs Fábián*[1]*, János Kovács*[1]*,
***Charles Tarnocai*[2] *and Gábor Varga*[1]**

[1]Institute of Geography, University of Pécs, Ifjúság u. 6, Pécs, Hungary H-7624
[2]Agriculture and Agri-Food Canada, Research Branch (ECORC), 960 Carling Avenue, Ottawa, Ontario, Canada K1A0C6

ABSTRACT

Pleistocene periglacial features were studied in different areas of Hungary in comparison with Canadian recent cryogenic features. Pleistocene periglacial activity forms an important component of the landscape of the Carpathian Basin. There is a general consensus about the study and interpretation of cryogenic deformation structures (e.g., frost fissures, cryoturbations and involutions) being helpful in paleoenvironmental reconstructions. During the glacial periods of the Pleistocene, the Carpathian Basin was ice-free and subject to a cryogenic environment that produced various periglacial features. The reason for the cold climate during these glacial periods is the Basin's unique geographic setting. The Carpathians, which surrounds this large basin, creates an almost closed climatic situation, producing climatic conditions not found elsewhere in Europe. In effect, the climate in Hungary during the glacial periods of the Pleistocene was somewhat similar to the recent climate of the dry tundra regions of North Siberia. But according to other researchers, the Carpathian Basin was mostly devoid of permafrost during the Quaternary. Previous researches described so many relict forms in different geomorphic positions, but none of these have been revised according to the most recent research methods and permafrost nomenclature. Our review addresses past periglacial processes and their coupling to paleoclimate. Permafrost distribution in Canada shows a latitudinal zonation. The main annual temperature associated with these permafrost zones are 0° to −5.5°C (SPZ), −5.5° to −8.3°C (WPZ) and −8.3° to −17°C (CPZ). Sand wedges, frost cracks and ice wedges all develop in permafrost environments and are currently actively forming in the Continuous Permafrost Zone in Canada. The well-developed sand-wedge polygons, frost cracks, ice wedge casts and cryoturbated soils found in Hungary suggest that the Carpathian Basin was a permafrost-affected area during a cold, glacial part of the Weichselian (Late Pleniglacial, 22,000–18,000 years BP). At that time this area was most likely underlain by continuous permafrost.

INTRODUCTION

This chapter describes recent cryogenic forms of North-western Canada and relict forms of the Carpathian Basin to draw comparison between them. The objectives of this chapter are to document them, discuss their paleoenvironmental significance from a process viewpoint and to demonstrate their application for reconstructing the past periglacial history of the central European area.

The theory of climatic geomorphology [4] and uniformitarianism [32] give us possibility to compare different time and location of the Earth in the same geomorphologic system. But the comparison is difficult without correct data on relict permafrost environments and the geographic locations; geomorphologic positions and climate conditions are undoubtedly different. It is essential to realize that recent permafrost areas show a high diversity in environmental settings and the former permafrost environment was multiple changed in time for the same region during the glacial periods [74]. There is a general consensus about the study and interpretation of cryogenic deformation structures (e.g. frost fissures, cryoturbation and involutions) being helpful in paleoenvironmental reconstructions as permafrost indicators [78].

According to former periglacial studies [33, 74] Central Europe was a permafrost area in the last glacial period, but the limit and the zonation of the permafrost are still debated [25, 51, 68, 75]. There are two main periods of continuous permafrost in the paleoenvironmental evolution of Central Europe during the last glacial periods: first *c.* 72,000–61,000, second *c.* 27,000–17,000 B.P. [74].

Pleistocene periglacial activity forms an important component of the landscape of the Carpathian Basin [61]. During the glacial periods of the Pleistocene, the Basin was subject to a cryogenic environment that produced various relict periglacial features [27, 48, 65]. The reason for the cold climate during these glacial periods is the Basin's unique geographic setting. The Carpathians, which surrounds this large basin, creates an almost closed climatic situation, producing climatic conditions not found elsewhere in Europe. In effect, Kriván [30] and Dylik [8] seem to imply that the climate in Hungary during the glacial periods of the Pleistocene was somewhat similar to the recent climate of the dry tundra regions of North Siberia. But according to other researchers, the Carpathian Basin was mostly devoid of continuous permafrost during the Pleistocene [6, 33, 70]. Previous researches described so many relict forms in different geomorphic positions [5, 8, 27, 46, 47, 55], but none of these have been revised according to the most recent research methods and permafrost nomenclature. Our review, however, addresses past periglacial processes and their coupling to paleoclimate.

Our interpretation of the periglacial deformations investigated is based on: (a) detailed field observations and (b) a synthesis of published data. Polygenetic forms were analyzed using criteria defined for the periglacial structures. We made classical geomorphologic sketches about locations to identify the geomorphic positions of cryogenic features and connections to their environment (Table 1). Sections were cleaned and described using sedimentological and pedological criteria, with respect of the topographical/geomorphologic location and available moisture. Samples were analyzed for physical and chemical properties and for grain-size analyses and optically stimulated luminescence dating (the latter was done only on Hungarian samples). For OSL dating, samples were collected from individual fill units and additional samples were collected for moisture content and background radiation measurement. The samples were dated at the Geological Institute of Hungary.

Table 1. Geomorphic positions of relict cryogenic features in Hungary.

Geomorphic positions	Patterned ground	Ice-wedge pseudomorphs	Cryoturbation	Sand wedges
Lower terraces	few	common	common	none
Higher terraces	few	localized	few	common
Alluvial fans	common	few	very common	few
Aeolian sandy surfaces	none	none	few	none
Gentle slopes	none	none	common	none

Distribution of Permafrost in Canada

Permafrost is defined on the basis of temperature as ground (i.e. soil and/or rock) that remains at or below 0°C for at least two consecutive years [13]. Permafrost is most commonly associated with various types of ice (ice-bonded permafrost), but if there is insufficient interstitial water the permafrost is dry (dry permafrost).

Permafrost distribution in Canada shows a latitudinal zonation (Figure 1). In the south it is discontinuous, while in the north it is continuous. Two zones are identified in the discontinuous portion, the Sporadic and the Widespread/Extensive Discontinuous Permafrost Zones [16, 18, 24]. In its most southerly occurrence the permafrost is sporadic, occurring under less than 30% of the area [18], and is found mainly in peatlands as islands in a generally unfrozen terrain (Sporadic Discontinuous Permafrost Zone, SPZ). Farther north the permafrost becomes widespread, occurring under 30–80% of the landscape [18], and is found in both organic (peatlands) and mineral terrain (Widespread Discontinuous Permafrost Zone, WPZ). In the most northerly regions all land surfaces, even under shallow water bodies, are underlain by permafrost (Continuous Permafrost Zone, CPZ). South of the southern limit of the Discontinuous Permafrost Zone permafrost can occur at high elevations in mountainous areas. These areas belong to the Alpine Permafrost Zone. The thickness of the permafrost varies greatly, depending on location. It is no more than a few meters thick in the southern part of the SPZ, increasing to 30–50 m in the WPZ, and can be as deep as 500 m in the CPZ [58].

The mean annual air temperatures (MAAT) associated with these permafrost zones in Canada range from 0° to −5.5°C in the SPZ [59], −5.5° to −8.3°C in the WPZ [45], and −8.3° to −17.0°C in the CPZ [79].

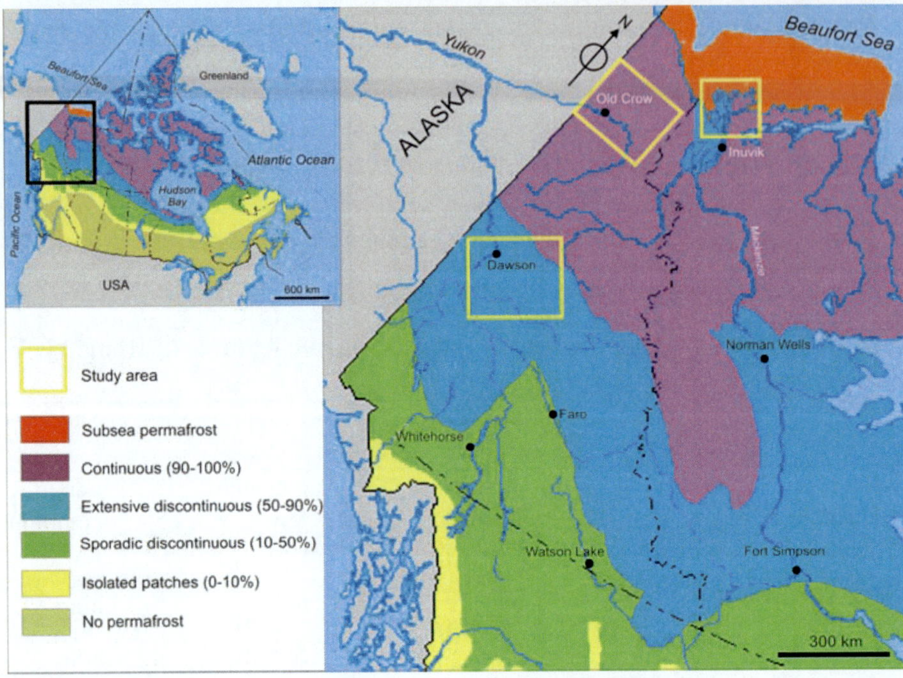

Figure 1. Permafrost zonations in Canada [44], and the study areas.

CRYOGENIC FEATURES IN CANADA

The study of paleosols in northwestern Canada (Figure 1) in central Yukon (around Dawson City), in the Old Crow Basin reviled numerous cryogenic features, which developed in the past. On the other hand those cryogenic features studied in the Mackenzie River Delta areas are represent active processes.

Cryogenic features were observed on a pre-Reid glaciofluvial terrace near Wounded Moose Dome about 70 km southeast of Dawson City and near the limit of Pleistocene glaciations in central Yukon [56]. The Wounded Moose paleosols occurs extensively and fairly continuously in the Tintina Trench between Clear Creek and Flat Creek, and on level mesa-like surfaces in the Willow Hills. It also occurs sporadically on outwash terraces on the lower Klondike River and the Yukon River between the Stewart and Fortymile rivers. Most Wounded Moose soils (and also the Stirling Bend soils) display strong cryoturbation in the form of disrupted and displaced soil horizons and oriented and shattered stones [57, 64]. Sand wedges and sand involutions of various sizes are also common. Ventifacts are commonly found at the paleosol surface.

Cryosolic paleosols were found buried within a dominantly fluvial succession overlain by Late Wisconsinan lake sediments in the Old Crow Basin. They developed in a permafrost environment and were associated with gleyed horizons, mottles, cryoturbated features, patterned ground and ice wedges. Although Cryosolic soils are common in the Old Crow Basin at the present time, the ice wedge formations found in the Old Crow paleosols (Figure 2) indicates the presence of a colder climate than now occurs. This is confirmed by the oxygen isotope record, which indicates the occurrence at that time of a cold period, which was much colder than the climate at the present time [57].

Figure 2. Ice wedge cast from the Old Crow area. (Height of wedge 2.5 m)

Figure 3. Low-centered ice-wedge polygons in the Mackenzie River Delta area, Canada.

Polygons and linear cracks are the most abundant cold-climate physical structures observed in the Mackenzie River Delta area (Figure 3). Polygons were identified on emergent bars located along the west side of Ellice Island, the south-west side of Pitt Island. Several adjoining low-relief polygons up to 5 m in diameter are defined by shallow cracks [*cf.* Fig. 7 in 23]. These features resemble the coastal polygons described by Mackay [38], which develop below high water level and are formed by thermal contraction in frozen ground [23]. Recurrent freezing of water within the cracks gradually widens the crack at the surface. Linear cracks, up to ~100 m in length, bisect the vegetated bars of Pitt Island. In cross-section these cracks are filled with fine sand, forming V-shaped wedges that extend to 0.5 m below the surface. Cryosols commonly occur in the form of high-centered lowland polygon bogs, low-centered lowland polygon fens (Figure 3) and peat mound bogs [63].

Patterned ground and cryogenic soils occur commonly in all of these permafrost regions, but their frequency and types differ somewhat from zone to zone (Table 2). Patterned ground (e.g., circles, nets and hummocks) occurs only rarely in the SPZ, but is common in the WPZ and very common in the CPZ. Ice-wedge polygons (Figures 3 and 4), a form of patterned ground, do not occur in the SPZ, but have a similar distribution to other patterned ground forms in the WPZ and CPZ. Sand wedges that have formed during the Holocene epoch (active sand wedges) occur mainly in the northern part of the CPZ. Similarly, cryoturbated soils are found only on wet, fine-textured materials in the SPZ, but they are the dominant soils in the WPZ, and the only soils in the CPZ. In the CPZ cryoturbation is even found in soils that occur on sandy materials (Figure 5). Cryosolic paleosols have well developed cryogenic features such as ice wedge casts, sand wedges, cryoturbated soil horizons and cryogenic microfabrics [62]. Sand wedges and sand involutions also occur in both contemporary soils and paleosols in northern Canada. Figures 6 and 7 show some of these sand wedges, while Figure 5 shows sand involutions. Both of these cryogenic features developed under a cold, dry climate in a permafrost environment. Because of the intense cold, thermal cracking occurred in the soil. These cracks then filled with sand that was shifted along the soil surface by high winds.

Figure 4. Contemporary ice wedge from the Old Crow area, Canada. (Height of wedge 1.2 m.)

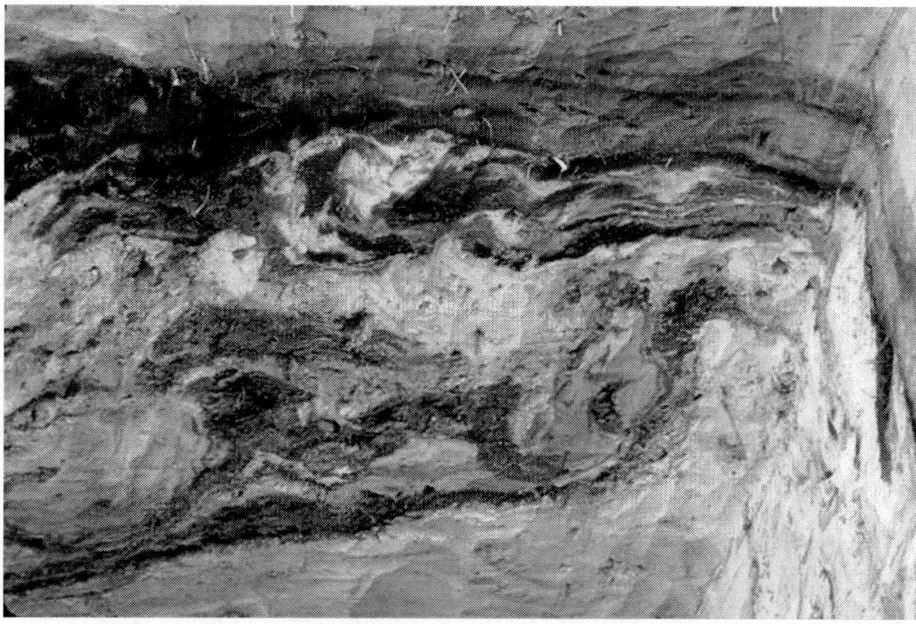

Figure 5. Strongly cryoturbated, sandy-textured, permafrost soil from the Old Crow area, northern Yukon. The strongly contorted soil horizons in the middle and lower part of the soil profile result from cryoturbation. (Depth of profile *c.* 1 m)

Figure 6. Sand wedge associated with Wounded Moose paleosol, central Yukon. (Height of wedge 1.5 m)

Figure 7. Sand involutions in the Wounded Moose paleosol, central Yukon. (Spade for scale = 1.5 m)

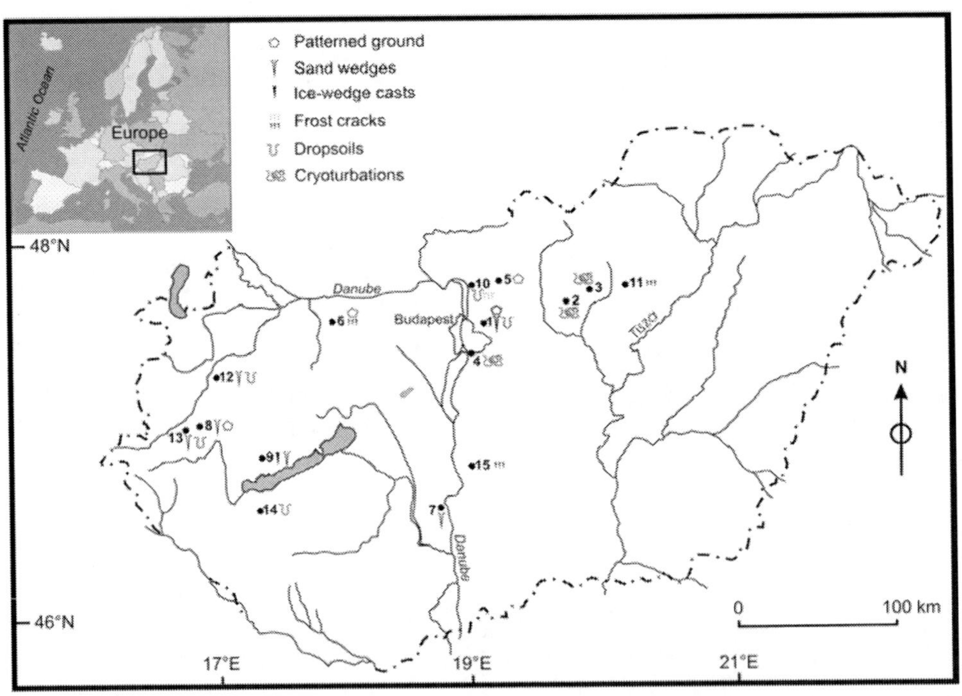

Figure 8. Locations of the recently studied (1–9) and the main published (10–15) cryogenic features in Hungary. 1 – Mogyoród, 2 –Atkár, 3 – Visonta, 4 – Bábolna, 5 – Paks, 6 – Csipkerek, 7 – Billege-erdő, 8 – Vanyarc, 9 – Soroksár, 10 – Vác 11 –Kerecsend, 12 – Ostffyasszonyfa, 13 – Vasvár, 14 – Marcali-Boronka, 15 – Tételhalom

Table 2. Cryogenic soil features in various permafrost zones in Canada.

Cryogenic features	SPZ	WPZ	CPZ
Circles, nets, and hummocks	few	common	very common
Ice-wedge polygons	none	some	very common
Sand wedges	none	none	some
Cryoturbation	localized	very common	very common

Description Of Sites in Hungary

Mogyoród Site

Large V-shaped features are to be found in a gravel pit (47°35'N, 19°13'E; 255 m a.s.l.) near Mogyoród (15 km north-east of Budapest) in the northern part of the Gödöllő Hills (Figure 8). The geological section at this site is composed of two very different deposits [12, 65]. The lower deposit is an Upper Pliocene alluvial fan of Palaeo-Danube origin overlain by aeolian sandy deposits. A thick, red paleosols developed on the lower deposit. The color is yellowish red (5YR 5/8) and the thickness of the paleosols indicates that it was most likely developed in a much warmer climate than currently exists.

The sand wedges found in the red paleosol have a polygonal system, consequently it is considered as a sand-wedge polygon. The wedges have an average vertical dimension of 1.5–2.0 m, but range up to 3 m (Figure 9). The wedge width was measured at right angle of the axial plane of the wedge [39]. Additionally these wedges are 25–30 cm and 50–60 cm or slightly more in width (Figures 9 and 10). The sand-filled wedge structures have simple V shapes with rectilinear or slightly curved sides (convex outward) and pointed toes (Figure 10). The fill of the wedges is sub-vertically laminated; the sand itself is fine- to medium-grained (1–3 φ) and moderately well sorted. The color of the sand material is light grey (2.5Y 7/2). Some wedges contain pebbles at or near the top. The host strata adjacent to sand wedges are upturned. In plan view, it forms polygonal network with cracks spaced 2–5 meters apart. The cracks are irregular and several meters long [29].

Figure 9. Relict sand wedges in Mogyoród, Hungary. (Depth of profile 3.5 m)

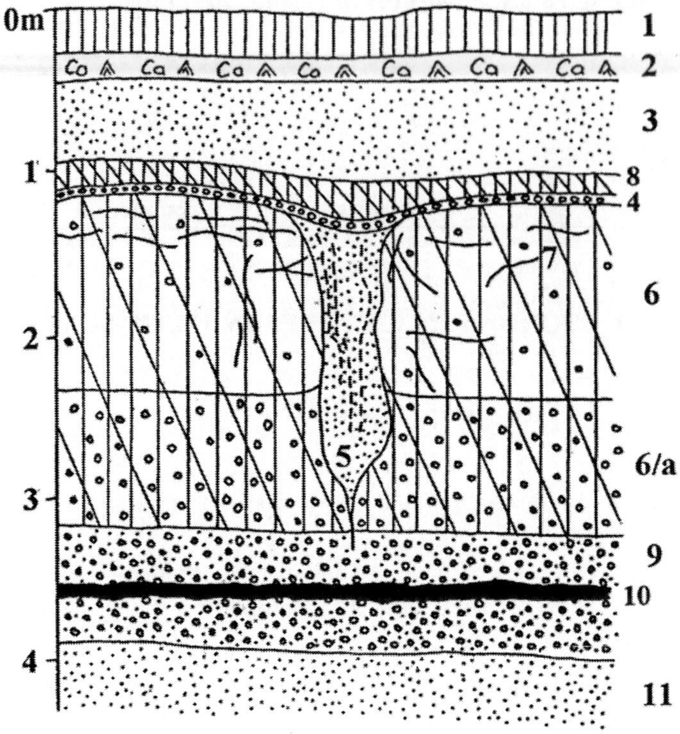

Figure 10. Detailed section through a relict sand wedge in Mogyoród, Hungary. 1 – recent soil, 2 – $CaCO_3$ accumulation, 3 – aeolian sand, 4 – sandy gravel, 5 – sand wedge, 6 – palaeosol with lower gravel content, 7 – palaeosol with higher gravel content, 8 – desert pavement, 9, 10 and 11 – Pliocene alluvial fan of Palaeo-Danube.

Visonta and Atkár Sites

According to Horváth *et al.* [20], the platy, lenticular, fitted breccia structure in the excavation in Visonta (47°45'N, 20°07'E; 135 m a.s.l.) is connected to involutions and drop soils (Figure 8 and *cf.* Plate II. Fig. 7 in [20]). The laminae of this structure follow the curve of involutions. Accepting that the platy, lenticular structure could be a good evidence for frost-action [20, 71], this observation suggests that frost played a role in the formation of the involutions.

The geology of Atkár (47°42'N, 19°52'E; 143 m a.s.l.) is similar to that of Visonta. The exposed extent of the deformation horizon is about 30 m in length and 1.5 m in height (Figure 11). Sagging load casts are very common, mainly in the upper and middle part of the deformed horizon as described by Horváth *et al.* [20]. These are 0.1 to 0.5 m sized bands in which former stratification and pedogenic horizons are discernible. These bands are more or less continuous and have a downward convex shape. Drops which are dm scaled, rounded, ellipsoidal occur in the upper and middle part of the deformed horizon but occasionally are traceable also at the lower levels (Figure 12).

Figure 11. The view of Atkár site, load casts, waves and involution are visible in this profile. (Spoon for scale = 0.2 m)

Figure 12. Well-developed drop soils in Atkár profile. (Note the scale is 0.6 m)

Soroksár and Vanyarc Sites

According to Horváth *et al.* [21], 3–5 cm wide downwards narrowing fissures were observed in the stratified sandy sediment in Soroksár (47°23'N, 19°07'E), near Budapest (archaeological excavation, trench 12), which were filled in with coarse sand (Figures 8 and 13). The parallel fissure system extends down to a very hard grey sandy level. This may mark the former dryland surface. Such wedges can appear in a sandy deposit when the otherwise loose sand behaves like a rock. This can happen when the water content freezes in the soil which then cracks like a hard rock, and the empty fissure is filled in with sediment from above. Such a phenomenon can be interpreted as a frost fissure. It is evidence to perennially frozen ground; beside the frost crack, a turbulent layer disturbance also appears in the same sandy deposit (Figure 13) probably resulted from cryoturbation. Such extremely cold weather with long-lasting frost in the soil is encountered during Pleistocene glacial periods.

Polygons were identified on the hilltop located near the village of Vanyarc (47°49'N, 19°27'E), northeast from Budapest (Figures 8 and 14). Several adjoining low-relief polygons up to 0.5–1.0 m in diameter are defined by shallow cracks [*cf.* Fig. 7 in 23]. In cross-section these cracks are filled with coarse sand, forming V-shaped wedges that extend to 0.5 m below the surface.

Figure 13. Cryoturbation in plan view from the Soroksár archaeological excavation site (courtesy of Z. Horváth).

Figure 14. Patterned ground in plan view from the Vanyarc archaeological excavation site (courtesy of Z. Horváth).

Bábolna Site

Various wedge-shaped features and sand involutions were found in a Bábolna gravel pit (47°40'N, 17°59'E; 150 m a.s.l.). The ice wedge casts in this gravel pit developed when former ice wedges melted and the wedge space filled with the surrounding gravelly materials (Figure 8). It should be pointed out that ice wedge casts are a very good indicator of a former permafrost environment since ice wedges develop only in permafrost. The vertical dimension of the wedges is 0.5–1.0 m and, in areas where the surface has been bladed by a bulldozer, the polygonal pattern is clearly visible [65].

Paks Site

Two brownish layers, most likely representing the B-horizons of two distinct paleosols, were identified in the exposed profile (46°35'N, 18°50'E 95 m a.s.l.) in the Paks sand deposits (Figure 8). A thick sand layer on which the contemporary soil has developed overlay these paleosols. This contemporary soil is associated with a thick, organic-rich, surface horizon and is quite different from those B-horizons associated with the underlying paleosols.

The B-horizons of both paleosols were contorted and disrupted, and a wedge-shaped sand body that was identified as a sand wedge also dissected the B-horizon of the lower paleosol. Closer examination of these paleosols revealed strong mottles, suggesting that a fluctuating water table affected this horizon. Thin organic layers associated with what seemed to be a former vertical crack were also observed.

Csipkerek Site

Many sand wedges were identified that display vertical foliation, form polygonal nets, and penetrate alluvial gravel deposit (Csipkerek gravel pit: 47°05'N, 16°56'E; 230 m a.s.l.). The age of the gravel is controversial. The gravel of Kemeneshát developed in Pleistocene [1], however the gravelstone of Csipkerek site in some places is overlain and cemented by red clay, therefore they are supposed to be older: Pliocene. More than ten wedges from one borrow pit wall are described here (Figure 15). Wedges are extensive with a mean width of 1 m and the mean height of 2–3 m. Wedge spacing is on average ~10 m, a similar spacing observed by Carter [7] for relict wedges in Alaska. The host strata adjacent to sand wedges are upturned. The wedge sand is fine to medium grained. The color (2.5Y 7/4) and grain size of the infilling sand is the same as the others we found nearby in Kemeneshát.

Figure 15. Sand wedge with adjacent upturned gravel strings in Csipkerek site. (Hammer for scale = 0.4 m)

Figure 16. Late Neogene alluvial fan with cryogenic features in Billege-erdő, Hungary.

Figure 17. Close up view of an ice wedge cast in Billege-erdő, Hungary.

Billege Site

Our latest research site is the Billege-erdő gravel pit (46°53'N, 17°21'E; 130 m a.s.l.) several km from the western basin of Lake Balaton [15, 47]. Large, mostly U-shaped features can be seen here which deepens into Late Neogene delta fan deposits (Figures 8 and 16). These fissures were identified that display horizontal foliation and penetrate gravelstone (Figure 17). Cracks are extensive with a mean width of 1.5 m and the mean height of 2–3 m. The fill of the fissures is horizontally laminated; the sand itself is medium-grained and moderately well sorted. The color of the sand material is light grey (2.5Y 7/2). Most ice-wedge pseudomorphs are found in gravel. This is because this coarse-grained sediment is usually ice-poor [13].

RELICT CRYOGENIC FORMS IN HUNGARY

Thermal-contraction-crack polygons are the most widespread, most visible, and most characteristic feature of permafrost terrain. Thermal-contraction cracks can be filled with ice, mineral soils (sand, loess, etc.), or a combination of both [13].

Sand-Wedge Polygons

In cold aeolian environments, the progressive primary infilling of thermal contraction cracks with sand forms distinctive sedimentary structures known as sand wedges [49]. Wedge-shaped bodies of sand can also form by secondary infilling of voids resulting from thaw of ice veins and ice wedges, producing ice-wedge casts; and by processes unrelated to thermal contraction cracking [14, 41, 43].

Thermal contraction cracking occurs when reduced frozen ground temperatures generate tensile stresses greater than the tensile strength of the ground. Such stresses are favored by both rapid cooling and low temperature [31], but the controls on cracking are complicated by such factors as snow thickness and creep of frozen ground [37]. Open cracks can be 1–100 mm wide, 2.0–5.0 m or more deep and 1 m to several tens of meters long [35, 36]. In plan view, the cracks form orthogonal or non-orthogonal networks, according to the kind of angles where cracks or their projections intersect; and complete or incomplete, depending on whether the cracks join [31]. Polygon diameter varies from a few meters to > 100 m.

The local character of the aeolian infill is indicated by the close correspondence between the grain-size parameters of the overlying sands and the crack infill material (R^2=0.96). Contraction fissures were filled from the surface with allochthonous, mostly sandy sediments. Infilling usually was (probably) carried out by wind action. Aeolian activity during the development of the sand-wedge polygons resulted in a desert pavement on the surface. The sand in these wedges shows sub-vertical lamination, suggesting that frost cracking occurred a number of times. Upturned strata of the host probably resulted from deformation caused by the addition of wedge material to permafrost and ground expansion in summer [43]. The irregular margin of the cracks (in plan view) may were formed upon thawing of the permafrost.

Numerous, mainly horizontal but some vertical, cracks 2–5 mm wide filled with $CaCO_3$ are present in the red paleosol. The origin of these cracks is most likely associated with vein ice development in permafrost environment [43].

Age Determination of Sand Wedges

It was expected that the wedges had developed during the Late Pleniglacial (Würm) glacial period, based on other western and central European information [2, 28, 42, 48, 74]. Both the samples gave ages within the last part of the Late Pleniglacial. As expected from the structure of the wedge, sample M 5/4 provided a younger age (20.75 ± 2.3 ka B.P.) and sample M 5/5 gave an older age (22.66 ± 2.86 ka B.P.) [29]. This period (22–20 ka B.P. MIS 2) could be related with North Atlantic cold Dansgaard-Oeschger stadials (DO 1–2) associated with Heinrich-events 1–2 [52, 76].

Ice-Wedge Pseudomorphs (Casts)

Ice-wedge networks are patterns in permafrost that form by filling cooling-derived tension fractures with ice. They form in a broad range of stable, aggrading or sloped surfaces, with mean annual temperatures below –4°C to –6°C, and winter temperature extremes ranging from –15°C to –35°C [51]. Individual wedges are tens to hundreds of meters long, commonly terminating orthogonally at another wedge. The overall orientation of the network usually is random, but sometimes is dictated by the trace of a river or a shoreline. Wedges usually penetrate 3–6 m where set in non-aggrading surfaces. Enclosed regions between wedges are polygonal, mostly squares to hexagons, with diameters commonly 10–30 m. Despite numerous commonalities, ice-wedge networks also differ amongst and within sites, displaying complicated behaviors that include variable frequency of fracture and growth in ice wedges, differing spacing and relative orientation of wedges in the network pattern, and varying modes of deformation of frozen ground around expanding wedges.

Casts, or pseudomorphs, represent the previous shape of a structure and can be formed in materials other than that which formed the original structure. Ice-wedge pseudomorphs are wedges of secondary mineral infilling and are thermokarst structures because they result from the thaw of excess ice [13]. Ice-wedge pseudomorphs form when the ice in the wedge slowly melts, usually as permafrost degrades. As this happens, there is a general collapse of sediment into the void that is created. These kinds of wedges are from mid-latitude regions where permafrost no longer exists. Most ice-wedge pseudomorphs are found in gravel. Ice wedges preferentially develop in ice-rich, fine-grained sediments (thaw-sensitive); their pseudomorphs are selectively preserved in ice-poor, coarse-grained sediments (thaw-stable) [13].

Patterned Ground

Most patterned-ground phenomena form within the active layer. Circles, earth hummocks, and mudboils are undoubtedly the most common. Some patterned ground may also form in seasonally frozen ground [13].

Low-centered polygons are characteristics of poorly drained tundra. They commonly possess a double raised rim, or rampart, often in excess of 50 cm in height, on either side of the ice-wedge trough. The depressed wet center contains sedges and grass. The raised rims are the results of thermal expansion within the active layer moving material from the polygon center to the periphery.

Thermokarst Involutions (Cryoturbations)

Past thermokarst activity may be recognized in the stratigraphic record by the existence of paleo-thaw layer. In present permafrost regions, paleo-thaw layers often correspond to the secondary thaw unconformities [13]. Paleo-thaw layers have been described from perennially-frozen sediments in several areas of the western North American Arctic. Late-Pleistocene paleo-thaw layers have been inferred from studies at a number of localities in the lowland of western and central Europe [9, 10, 20, 34, 65, 71, 72]. These structures occur widely in the near-surface sediments of mid-latitude areas.

Smaller deformation structures of the bird-foot or drop soil type are also caused by loading and density differences in water-saturated sediments, probably during the degradation of underlying permafrost.

Soft sediment deformations observed at Atkár site are similar but more strongly expressed than the ones at Visonta site. Traces of ice segregation on the top of the eroded Late Miocene alluvial formations suggest that frost action played a significant role in the formation of these involutions. Other observations of cryogenic features in the area [8, 40, 47] and the comparison of other cryogenic features presented by Van Vliet-Lanoë [67, 68] support the cryogenic origin. Compact aggregates of micritic carbonate and clay organized into horizontal, subhorizontal layers and occurring in association with the platy breccia structure. On the margins of these aggregates forming the platy, lenticular structure, Fe-oxide hypo-coatings are clearly discernible (*cf.* Plate III. Fig. 10. in [20]).

The platy, lenticular structure of paleosols could be correlated with segregated ice of ice lenses growing parallel to the thermal gradient in the sediment, usually roughly parallel to the soil surface. Aggregates created by ice lenses are generally very stable [67, 70].

In Visonta site, the flames are built up by powdery calcrete and are interpreted as water escape structures. The formation of water escape structures was accompanied by the development of drops, pillows or isolated patches (irregular pseudonodules) and involutions. The general presence of the platy, lenticular structure indicating frost action on the uppermost part of the Late Miocene sediments suggest that the involutions in the Post Miocene soil-sedimentary complex were formed as a result of cryoturbation.

All the soil features from Paks site appear to result from cryogenic processes. They probably developed in a cold environment, when cryogenic processes were active and permafrost conditions existed at this site.

Similar cryogenic features are found in contemporary sandy permafrost soils throughout the Canadian north. The active layers of two such soils, both having permafrost 60 to 80 cm beneath the soil surface, are shown in Figure 5. The brownish B-horizons and the dark-colored organic horizons of this soil are contorted, broken and displaced as a result of cryoturbation. The grayish sand in the middle of the photograph has been pushed up from the underlying parent material by cryoturbation. It has high organic matter content and shows similar effects of cryoturbation.

The Stirling Bend paleosol developed during the early Pleistocene epoch under a cold, probably glacial, climate [64]. As a result of cryogenic processes, it has contorted organic and mineral horizons and a hummocky microtopography.

Cryogenic Features and Their Implication For Palaeoclimate

Central Yukon has a sub-Arctic continental climate with long, bitterly cold winters, short mild summers, low relative humidity, and low to moderate precipitation (Table 3). Intrusions of mild air from the Pacific Ocean moderate the climate of the region from the Arctic climate that characterizes northern Yukon [77]. The Dawson Range shares this climate modified by higher elevations. However, higher elevations do not necessarily experience colder temperatures throughout the year. Air temperature decreases with increasing elevation during the summer months but, during the winter, cold air is frequently trapped within the Yukon River Valley and other major valleys causing a temperature inversion [77]. The Dawson Range has a periglacial climate. It is situated within a region of Yukon classified as having extensive discontinuous permafrost [19]. However, the summits of the Dawson Range lie below the regional firn line.

Table 3. Climatic data for Dawson City, Yukon Territory

Dawson City (64°N, 370m a.s.l.)	**January**	**July**	**Annual**
Daily mean temperature	-26.7	+15.6	-5
Extreme maximum (°C)	+9.7	+33.5	+34.7
Extreme minimum (°C)	-53.8	-1.5	-55.8
Mean precipitation (mm)	19.2 (snow)	48.4 (rain)	343 (snow+rain)

Source: Environment Canada, Canadian climate normals 1971–2000

Wedge-shaped sedimentary structures, interpreted as the result of thermal-contraction cracking of perennially frozen ground, are convincing evidence for the previous existence of permafrost [13]. These structures are frequently reported from mid-latitudes. Although there is general agreement as to their paleoenvironmental significance, their specific relevance to air temperature is far less clear. Cracking appears to be controlled not only by ground temperature but also by site-specific conditions such as lithology associated thermal conductivity, antecedent conditions, and the duration and thickness of the snow cover.

Palaeoclimatological information about cracking and infilling of sand wedges is limited by both a paucity of descriptions of active sand wedges and by differences between past and present environmental conditions [43]. Insufficient data exist on wedge distribution and controls to substantiate the climatic threshold values suggested in the paleoenvironmental literature [22, 26, 73]. Karte [26] suggests that primary sand-wedge formation indicates a MAAT of −12°C to −20°C and a MAP of < 100 mm. However, primary sand wedges are known to form under somewhat warmer and wetter conditions with abundant sand supply, for example locally on the Tuktoyaktuk Peninsula, Mackenzie Delta area, Canada [35], where the MAAT at Tuktoyaktuk is −10.9°C and the MAP 138 mm [11]. Sand-wedge polygons were found by Péwé [49] in the McMurdo Sound Region, Antarctica, where the mean air temperatures is about −17.2°C and annual precipitations are between 50 and 150 mm. More generally, inference of mean annual climatic parameters from cryogenic wedges is problematic because thermal contraction cracking is favored by extreme winter conditions [17].

Recent climate models suggest that the wind systems dominating the Carpathian Basin during marine oxygen isotope stage 2 (MIS 2) were from western to northwestern direction [3, 53]. The reconstructions indicate a dominant westerly to northwesterly wind in winter during the Late Pleniglacial in the central European region [53]. Prolonged periods of intensive wind intensity were recorded in loess deposits in southern Carpathian Basin. These winds were cold and dry. Significant cooling between 22,000 and 18,000 years B.P. was detected based on pollen analyses. *Picea* and *Pinus* species became dominant, and mild-climate preferring forms disappeared in this period [54]. As a cryophilous Mollusc, the *Columella columella*, typically found in tundra or tundra-like environments, was one of the dominant Mollusc elements in this phase. This indicates significant cooling, a cold peak in the Late Pleniglacial and development of the cold loess steppe environment in the Carpathian Basin [60]. The July palaeotemperature decreased to 9–12°C.

The Carpathian Basin had a unique climatic situation in the Late Pleniglacial (Dansgaard-Oeschger, between DO 1 and 2 events), as evidenced by the sand wedge formations and palaeoecological investigations.

CONCLUSION

Well-developed soft-sediment structures and wedge-shaped forms were detected in the northern, northwestern part of Hungary that we have interpreted as relict cryogenic forms. Our observations of relict permafrost features support previous inferences [14, 33, 65, 74] for the occurrence of continuous permafrost (which were based on the occurrence of ice-wedge casts and cryoturbations) in Hungary during the Pleistocene. In the Late Pleniglacial the climatic conditions may have been as cold and dry as in modern dry permafrost environments in Canada (MAAT: −8.3°C to −17°C) [66] and Antarctica [49, 80]. The climate probably was very dry and absolutely cold, favorable for the formation of desert pavement.

These characteristics indicate a very arid environment (precipitation < 100 mm, rare snow), a very poor vegetal cover, and very low temperatures (MAAT about −6 to −10°C, [74]). According to Renssen *et al.* [53], the main driving factors behind the anomalous atmospheric circulation in the Late Pleniglacial are the Laurentide Ice Sheet and a colder North Atlantic Ocean with a relatively extensive sea-ice cover, leading to an eastward relocation of the Icelandic Low and an enhanced pressure gradient over northwest Europe, which were caused the extreme cooling of the Carpathian Basin.

REFERENCES

[1] Ádám, L. In *A Kisalföld és a Nyugat-magyarországi peremvidék*; Ádám, L.; Marosi, S.; Ed.; Magyarország tájföldrajza 3; Akadémiai Kiadó: Budapest, HU, 1975; pp 251-273.

[2] Antoine, P.; Marchiol, A.; Brocandel, M.; Gros, Y. *CR Geosci* 2005, *337*, 1462-1473.

[3] Bokhorst, M. P.; Markovic, S. B.; Sümegi, P.; Prins, M. A.; Vandenberghe, J. F. (2004). Evidence for dominant wind directions in the Pannonian Basin during marine isotope stage 2. Geophysical Research Abstracts, 6, 07183.

[4] Büdel, J. *Klima-geomorphologie;* Berlin-Stuttgart: Gebrüder Borntraeger, GE, 1977; 304 p.

[5] Bulla, B. *Földrajzi Közlemények* 1939, *67*, 268-281.

[6] Bulla, B. *Magyarország természeti földrajza*; Tankönyvkiadó: Budapest, HU, 1962; pp 424.

[7] Carter, L. D. In *Permafrost: Fourth International Conference*, National Academy Press: Washington, US, 1983, pp 109-114.

[8] Dylik, J. *Földrajzi Értesítő* 1963, *12*, 453-464.

[9] Eissmann, L. *Quat Sci Rev* 2002, *21*, 1275-1346.

[10] Eissmann, L. *Z Angew Geol* 1978, *24*, 130-138.

[11] Environment Canada 2008. Canadian climate normals, 1971–2000. Temperature and precipitation, the North-Y.T. and N.W.T. Environment Canada, Atmospheric Environment Service, Ottawa, Ontario.

[12] Fábián, S. Á.; Kovács, J.; Varga, G. *Földrajzi Értesítő* 2000, *49*, 189-204.

[13] French, H. M. *The periglacial environment*, Wiley & Sons: Chichester, US, 2007, 458 p.

[14] French, H. M.; Gozdzik, J. S. *Can J Earth Sci* 1988, *25*, 2017-2027.

[15] Góczán, L. *Földrajzi Értesítő* 1960, *9*, 1-30.

[16] Harris, S. A. *Arctic* 1986, *39*, 29-38.

[17] Harry, D. G.; Gozdzik, J. S. *J Quat Sci* 1988, *3*, 39-55.

[18] Heginbottom, J. A. *Can Geographer* 1984, *28*, 78-83.

[19] Heginbottom, J. A.; Dubreuil, M. A.; Harker, P. T. 1995. Canada Permafrost. Map 2.1, National Atlas of Canada, 5th edition, Government of Canada, Ottawa, scale 1:5 000 000.

[20] Horváth, Z.; Michéli, E.; Mindszenty A.; Berényi-Üveges J. *Tectonophysics* 2005, *410*, 81-95.

[21] Horváth, Z.; Mindszenty, A.; Reményi, L.; Zádor, J. *Aquincum* 2003, *9*, 173-185.

[22] Huijzer, A. S.; Isarin R. F. B. *Quat Sci Rev* 1997, *16*, 513-533.

[23] Jenner, K. A.; Hill, P. R. *Sedimentology* 1998, *45*, 987-1004.

[24] Johnston, G. H. *Permafrost: Engineering Design and Construction*; Wiley & Sons: Toronto, 1981.

[25] Kaiser, K. *Eiszeitalter und Gegenwart* 1960, *11*, 121-141.

[26] Karte, J. *Geojournal* 1983, *7*, 329-340.

[27] Kerekes, J. *Földtani Közlöny* 1939, *69*, 131-139.
[28] Kolstrup, E.; Mejdahl, V. *Boreas* 1986, *15*, 311-321.
[29] Kovács, J; Fábián, S. Á; Schweitzer, F; Varga, G. *Permafrost Periglac Process* 2007, *18*, 379-384.
[30] Kriván, P. *Földtani Közlöny* 1958, *88*, 201-208.
[31] Lachenbruch, A. H. *Geol Soc Am Spec Pap* 1962, *70*, 69 p.
[32] Lyell, C. *Principles of Geology* John Murray: London, UK, 1830, Vol. 1, pp 1-5.
[33] Maarleveld, G. *Biuletyn Peryglacjalny* 1976, *26*, 57-78.
[34] Maarleveld, G. *Geol Mijnbouw* 1981, *60*, 347-352.
[35] Mackay, J. R. *Can J Earth Sci* 1974, *11*, 1366-1383.
[36] Mackay, J. R. *Can J Earth Sci* 1975, *12*, 1668-1674.
[37] Mackay, J. R. *Can J Earth Sci* 1993, *30*, 1720-1729.
[38] Mackay, J. R. *Geograph Branch, Mines Techn Surv, Memoir* 1963, *8*, 202 p.
[39] Mackay, J. R. *Geol Surv Canada Paper* 1977, *77-1A*, 43-44.
[40] Michéli, E.; Mindszenty, A. *Földtani Közlöny* 2002, *132*, 43-53.
[41] Murton, J. B.; French, H. M. *J Quat Sci* 1993, *8*, 185-196.
[42] Murton, J. B.; Kolstrup, E. *Prog Phys Geog* 2003, *27*, 155-170.
[43] Murton, J. B.; Worsley, P.; Gozdzik, J. *Quat Sci Rev* 2000, *19*, 899-922.
[44] National Atlas of Canada (5th edition) – Permafrost (1995) Natural Resources Canada. The Atlas of Canada – Permafrost. http://atlas.nrcan.gc.ca/sites/english/maps/environment/land/permafrost
[45] Nelson, F. E. *Phys Geogr* 1989, *10*, 233-248.
[46] Pécsi, M. *Biuletyn Peryglacjalny* 1964, *14*, 279-293.
[47] Pécsi, M. *Földrajzi Közlemények* 1961, *85*, 1-24.
[48] Pécsi, M. *Szerkezeti és Váztalajképződés Magyarországon;* Magyar Tudományos Akadémia: Budapest, HU, 1997, 296 p.
[49] Péwé, T. L. *Am J Sci* 1959, *257*, 545-552.
[50] Plug, L. J.; Werner, B. T. *Permafrost Periglac Process* 2008, *19*, 63-69.
[51] Poser, H. *Erdkunde* 1948, *2*, 53-68.
[52] Rahmstorf, S. *Geophys Res Lett* 2003, *30*, 1510.
[53] Renssen, H.; Kasse, C.; Vandenberghe, J.; Lorenz, S. J. *J Quat Sci* 2007, *22*, 281-293.
[54] Rudner, Z. E.; Sümegi, P. *Quat Int* 2001, *76/77*, 177-189.
[55] Scheuer, Gy. *Földrajzi Értesítő* 1969, *18*, 177-192.
[56] Smith, C. A. S.; Tarnocai, C.; Hughes, O. L. *Géogr Phys Quat* 1986, *40*, 29-37.
[57] Smith, C. A. S.; Veldhuis, H. In *Cryosols*; Kimble, J. E. Ed.; Springer: Berlin, GE, 2004, pp 119-138.
[58] Smith, S. L.; Burgess, M. M. *A digital database of permafrost thickness in Canada,* [52] Geological Survey of Canada, 2002, Open File 4173.
[59] Smith, S. L.; Burgess, M. M. *Ground temperature database for northern Canada,* Geological Survey of Canada, 2000, Open File Report 3954.
[60] Sümegi, P.; Krolopp, E. *Quat Int* 2002, 91, 53-63.
[61] Székely, A. *Földrajzi Értesítő* 1983, *32*, 389-398.
[62] Tarnocai, C. *Géogr Phys Quat* 1990, *44*, 363-374.
[63] Tarnocai, C. In *Cryosols*; Kimble, J. E. Ed.; Springer: Berlin, 2004, pp 95-118.
[64] Tarnocai, C.; Schweger, C. E. *Arctic* 1991, *44*, 1-11.
[65] Tarnocai, C.; Schweitzer, F. *Geogr Fis Din Quat* 1998, *21*, 84-92.

[66] Van Vliet-Lanoë, B. *Catena* 1998, *34* 157-183.
[67] Van Vliet-Lanoë, B. *Eurasian Soil Sci* 1998, *31*, 507-513.
[68] Van Vliet-Lanoë, B. *Quat Int* 1989, *3/4*, 109-113.
[69] Van Vliet-Lanoë, B.; Fox, C. A.; Gubin, S. V. In *Cryosols*; Kimble, J. E. Ed.; Springer: Berlin, 2004, pp 365-390.
[70] Van Vliet-Lanoë, B.; Magyari, A.; Meilliez F. *Global Planet Change* 2004, *43*, 103-127.
[71] Vandenberghe, J; van den Broek, P. *Boreas* 1982, *11*, 299-315.
[72] Vandenberghe, J. *Polarforschung* 1983, *53*, 97-107.
[73] Vandenberghe, J.; Isarin, R. F. B.; van Huisstedenand, J. (1998). Paleo-periglacial phenomena in Northwestern Europe. Boulder, CO: National Snow and Ice Data Center/World Data Center for Glaciology. Digital Media.
[74] Vandenberghe, J.; Pissart, A. *Permafrost Periglac Process* 1993, *4*, 121-135.
[75] Velichko, A. A. In *Paleogeography of Europe During the Last One Hundred Thousand Years*; Gerassimov, I. P.; Ed.; Nauka: Moscow, RU, 1982, 156 p.
[76] Vidal, L.; Schneider, R. R.; Marchal, O.; Bickert, T.; Stocker, T. F.; Wefer, G. *Clim Dynam* 1999, *15*, 909-919.
[77] Wahl, H. E.; Fraser, D. B.; Harvey, R. C.; Maxwell, J. B. *Climate of Yukon*. Environment Canada, Atmospheric Environment Service, Climatological Studies 40, 1987, 319 p.
[78] Washburn, A. L. *Geocryology*; E. Arnold: London, UK, 1979, 406 p.
[79] Zoltai, S. C. *Géogr Phys Quat* 1995, *49*, 45-54.
[80] The Historical Weather Data – McMurdo Base, Antarctica (2008). [website] http://www.antarticconnection.com/antarctic/weather/hist_wxdata/mcmurdo_station.shtml

In: New Permafrost and Glacier Research
Editors: Max I. Krugger and Harry P.Stern

ISBN: 978-1-60692-616-1

Chapter 5

PERIGLACIAL LANDFORMS AND PROCESSES ON DISKO ISLAND, GREENLAND

Jacob C. Yde [1,2]

[1] Department of Earth Sciences, Ny Munkegade, Bygning 1520, University of Aarhus, DK-8000 Århus C, Denmark

[2] Center for Geomicrobiology, Department of Biological Sciences, Ny Munkegade, Bygning 1540, University of Aarhus, DK-8000 Århus C, Denmark

ABSTRACT

This paper reviews our current knowledge on periglacial landforms and processes on Disko Island, central West Greenland. Disko Island is located on the southern limit of the continuous permafrost zone, and permafrost and periglacial processes are therefore sensitive to future climate warming. The landscape of Disko Island contains a wide variety of periglacial landforms such as rock glaciers, pingos, palsas, patterned ground and active layer phenomena. Also, weathering processes and fluvial and coastal landforms and processes related to cold-climate environments are widespread.

Until now, most research has concerned the incidence, morphology and palaeoclimatic significance of rock glaciers. Although this research has improved our understanding of the late-Holocene history, there are limited studies on other process-landform interrelations, making holistic geomorphological reconstructions of the landscape evolution difficult.

Disko Island contains 247 glaciers larger than 1 km^2, and 75 of these are classified as surge-type glaciers. The recession after surge events leaves proglacial areas prone to formation of periglacial landforms, providing good conditions for field studies on landform evolution on decadal and centennial timescales.

INTRODUCTION

The periglacial landscape in Greenland contains most landforms and processes associated with the periglacial environment. However, in recent years there has been limited permafrost and periglacial research in Greenland as evidenced by the lack of identified publications on lowland periglacial research, 2003–2007 (Humlum and Christiansen, 2008). This is likely a consequence of economic and logistic difficulties rather than actual research-related considerations. The exception to this is the on-going monitoring at the High-Arctic Zackenberg research facility in Northeast Greenland, where active layer thickness and snow-cover duration have been recorded since 1999 (Christiansen et al., 2008).

In Greenland, most periglacial research has focused on Disko Island in West Greenland. This reflects the presence of the Arctic Station in the coastal town of Qeqertarsuaq (Godhavn), which has provided a base for Arctic research for more than 100 years. Disko Island is located at the border between discontinuous and continuous permafrost. Hence, future climate ameliorations will most likely affect the distribution of permafrost and periglacial landforms, and accelerate thaw-related processes such as active layer detachment slides, solifluction, thermokarst development, and collapse of palsas and pingos.

This paper provides a status on our current knowledge on periglacial landforms and processes on Disko Island, West Greenland, and future research directions are suggested. These include research on glacier-permafrost interrelations, where the presence of surge-type glaciers on Disko Island provides good conditions for field investigations of landform evolution and monitoring of dynamic processes.

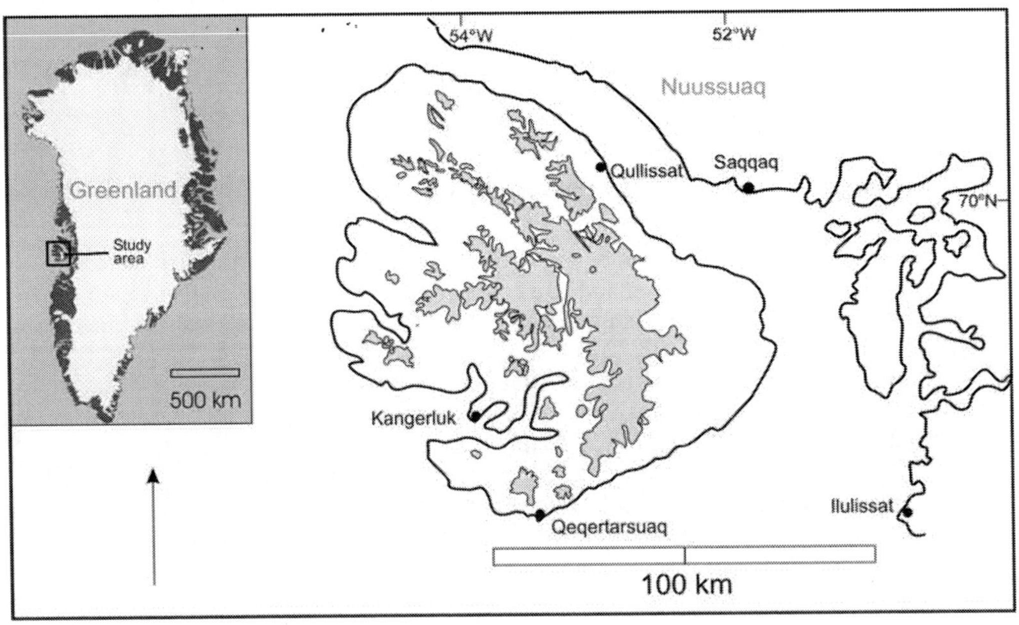

Figure 1. Location map of Disko Island, West Greenland

Study Area

Disko Island (70° N, 54° W) is located in central West Greenland (Figure 1) and is the largest island in Greenland, covering 8575 km^2 of which about 19% is glacierized (Humlum, 1987).

The climate in the south-western part of the island is polar maritime, and it gradually becomes more continental towards the northeast. The mean annual precipitation in

Qeqertarsuaq (Godhavn) is about 400 mm, whereas it was about 200 mm (1961–1972) in Qullissat, an abandoned mining village on the northeast coast (Humlum *et al.*, 1995). The termination of the Little Ice Age (1150–1920) was marked by a dramatic increase in the regional mean annual air temperature (MAAT) of about 2–4°C between 1921 and 1930. In Qeqertarsuaq (Godhavn), this was followed by a decline in temperature from -3.2°C in 1931–1960 MAAT to -4.0°C in 1961–1990 MAAT with interannual fluctuations up to 7°C (Humlum, 1999). Since 1991, the temperature has slightly increased. The temperature at sea-level is about 3°C colder in the central part of the island compared to the temperature in Qeqertarsuaq (Godhavn) (Humlum et al., 1995), and the vertical lapse rate is 6–7°C km^{-1} (Humlum, 1998).

Based on the assumption that the -6°C MAAT isotherm coincides with the position of the southern limit of continuous permafrost, it is suggested that the southern and western coastal areas of Disko Island have discontinuous permafrost, whereas the remaining part of the island has continuous permafrost. The thickness of the permafrost is not known, and there has not been conducted continuous monitoring of active layer thickness. Hence, it is difficult to assess any imbalance between the current climate and the permafrost distribution.

The non-glacierized landscape on Disko Island is characterized by an Early Tertiary basalt plateau at 600–1200 m a.s.l. intersected by several 10–50 km long, U-shaped valleys with glacier-carved cirques and talus deposits along the valley walls. The valley floors contain large areas of relict glacier ice (dead-ice), moraines, outwash plains, alluvial fans formed by tributary glacial meltwater streams, river bank terraces, and periglacial landforms. Most glacial meltwater drains through these valleys before emanating into fjords facing towards the Davis Strait to the west, or directly into the Vaigat Strait to the north or the Disko Bay to the east. At least a thousand homothermal springs are found near sea-level on Disko Island, the warmest being Puilassoq at the head of Mellemfjord with a constant temperature of about 18°C (Steenstrup, 1901; Heide-Jørgensen and Kristensen, 1999). Their presence indicates that aquifers exist within the basaltic rocks, and that the geothermal gradient is enhanced.

The lower layers of the basalt contain shale fragments or beds. In the eastern part of Disko Island, the basalt rocks overlay formations of Cretaceous sandstone and Paleocene mud- and sandstones with coal seams. The basement of Disko Island consists of gneisses, which only are exposed at sea-level along a north-south ridge through Qeqertarsuaq (Godhavn).

Disko Island contains 1070 glaciers, icefields and snowfields of which 350 are larger than 1 km^2 (Weidick et al., 1992). Of these, 247 are glaciers, while the remaining comprises ice and snowfields (Yde and Knudsen, 2007). The two major ice caps, Sermersuaq (Storbræen) and Bræpasset, cover the central part of the island, sending several outlet glaciers into the surrounding valleys, while valley glaciers, cirque glaciers and icefields are widespread all over the island. All glaciers on Disko Island terminate on land. A recent study has classified 75 glaciers on Disko Island as surge-type glaciers (Yde and Knudsen, Chapter 10). Surge-type glaciers experience a non-climatic-related cyclic behavior consisting of a short active period (a few years) of rapid frontal advance, where the glacier becomes heavily crevassed and water discharge is increased, and a long quiescent period (more than 100 years). Glacier surging may have a devastating effect on proglacial permafrost and landforms as the glacier front may advance more than 10 km down-valley.

Rock Glaciers

Nearly 1700 individual rock glaciers have been identified on Disko Island (Humlum, 1988a). The earliest descriptions of rock glaciers were made by Steenstrup (1883, 1901) and Hammer and Steenstrup (1893), who referred to rock glaciers as 'dead glaciers'. Many rock glaciers have been named by the Inuit, for instance the large Ujaragsuit rock glacier south of Qullissat on the northeast coast. Both glacier-derived rock glaciers (Figure 2) and talus-derived rock glaciers (Figure 3) are abundant all over the island. As most of Disko Island was covered by an ice cap and descending outlet glaciers during the Wisconsin glaciation (Ingólfsson et al., 1990; Weidick and Bennike, 2007), all rock glaciers can be considered as formed during the Holocene.

Figure 2. Glacier-derived rock glacier, Qivitut, Diskofjord

Figure 3. Talus-derived rock glacier, Qivitut, Diskofjord

Figure 4. Peat-covered pingo near the junction between the outwash plain in Kuannersuit Kuussuat (Kuannersuit Valley) and the delta at the head of Diskofjord

The rock glaciers on Disko Island have been intensely studied by Humlum (1982, 1988a,b, 1996, 1997, 1998, 1999, 2000). Based upon the appearance of the frontal slope, Humlum (1988a) classifies active, inactive and relict rock glaciers, and by introducing the concepts of rock glacier appearance level and rock glacier initiation line altitude he attempts to identify the initiation mechanisms of talus-derived rock glaciers as either glacial (snow accumulation-controlled) or non-glacial (temperature-controlled), in addition to the talus accumulation rate (Frich and Brandt, 1985; Humlum 2000). The size of typical rock fragments seems to be an important control on wind ventilation and thermal regime of the active layer on rock glaciers (Humlum, 1997). The topographic and altitudinal distribution of rock glaciers (Humlum, 1998), in combination with isotopic records of rock glacier ice (Humlum, 1999), may provide palaeoclimatic information on local and regional scales that may supplement data from ice cores and radiocarbon dating. However, these approaches still need to be refined in order to represent an unequivocal method.

As opposed to the glaciers on Disko Island, many rock glaciers terminate at sea-level. Donner (1978) interpreted the lack of raised marine terraces on the frontal slopes of rock glaciers reaching the shores of Diskofjord, Mellemfjord and Nordfjord as evidence of no glacioisostatic uplift since the time of the rock glaciers reaching the current sea-level. However, as many rock glaciers are active or may reactivate after a period of inactivity, they may override existing raised marine terraces, making the argumentation of Donner (1978) untenable.

PINGOS

Pingos are referred to as 'mud volcanoes' on the first topographic map of Disko Island, surveyed in 1931 – 1933 (GID, 1941). They seem to have escaped the interest of early explorers (e.g. Whymper, 1872; Steenstrup, 1901), although they are found near the junction between the outwash plains and the deltas of all the major fjords; Nordfjord (Donner, 1978), Mellemfjord (Christiansen, 1995, 1997), and Diskofjord (Figure 4). Pingos are also located in higher altitude passes such as in the northern part of Blæsedalen and in the eastern part of Blomsterdalen. They generally form as open-system pingos on high locations in well-drained, silty or clayey deposits, and often appear as 1 – 3 m high peat-covered mounds. Cold periods during the Little Ice Age may have lead to formation and growth of many pingos (Christiansen, 1995, 1997). A detailed survey on the distribution and morphology of pingos on Disko Island is much required in order to attain further knowledge on their sensitivity to physical parameters such as climate, topography, sedimentology, permafrost hydrology and vegetation.

Aerial photographs from 1953 and 1964 indicate that a pingo with a diameter of about 50 m had formed in the inner part of Kuannersuit Kuussuat (Kuannersuit Valley) at a location, which had been deglaciated around 1920 (Jost, 1940). In 1995 – 1998, the 10.5 km surge advance of Kuannersuit Glacier overrode the site of the pingo. Thus, deglaciation of the foreland of surge-type glaciers may provide a valuable field laboratory for studies on the formation and evolution of periglacial landforms such as pingos.

Figure 5. Slightly shattered basaltic stone incorporated into the surface of an outwash plain on top of buried glacier ice, Kuannersuit Kuussuat (Kuannersuit Valley). Camera etui for scale

PERIGLACIAL WEATHERING PROCESSES

In many of the major valleys large blocks of rocks have been detached from the walls and are now occupying the valley slopes (Pedersen et al., 2001). In Kuannersuit Kuussuat (Kuannersuit Valley), there are no obvious signs of glacial abrasion on the relocated blocks, although they have been eroded by fluvial processes. This leads to the preliminary interpretation that the large detachments are a paraglacial phenomenon related to pressure-release on valley walls during the deglaciation of the Wisconsin ice cap covering Disko Island. Talus deposits on valley slopes are often found in relation to ice-cored terraces, ice-cored moraines, tills or rock glaciers, although evidence of recent mass wasting events occur, i.e. on the western valley slope at the head of Kuannersuit Kuussuat. Mass wasting events in the narrow Vaigat Strait may also cause tsunamis, which lead to erosion and deposition along the northeastern shores of Disko Island.

Mechanical breakdown enhanced by frequent freeze-thaw cycles leads to rock shattering and formation of regolith deposits. Field observations on Disko Island support the notion by Mackay (1999) that rock types with schistosity facilitate water access and, hence, shatter more frequently. On an outwash plain formed by the Sorte Hak glaciers around 1940, the basaltic stones are generally unbroken or slightly shattered (Figure 5), whereas all observed shale stones have been heavily shattered (Figure 6). Observations on the outwash plain in front of Kuannersuit Glacier suggest that shale stones shatter explosively within one or a few years in this environment.

Figure 6. Explosively shattered shale stone incorporated into the surface of an outwash plain on top of buried glacier ice, Kuannersuit Kuussuat (Kuannersuit Valley). Camera etui for scale.

Figure 7. Oblique view of ice-wedge polygons associated with thermal contracting cracking, Blæsedalen.

Chemical weathering rates have not been quantified in non-glacierized catchments on Disko Island, but solute provenance estimations from Kuannersuit Glacier (Yde et al., 2005a) indicate relatively high chemical weathering yields compared to non-basaltic environments. Suspended sediment loads in rivers and streams are also relatively high due to glacial erosion and abundant unconsolidated sediments on valley floors (Rasch et al., 2003; Knudsen et al., 2007), which cause high sedimentation rates in the inner parts of fjords (Gilbert et al., 1998, 2002; Møller et al., 2001).

Patterned Ground and Active Layer Phenomena

There has not been much focus on the incidence of patterned ground and active layer phenomena on Disko Island. Ice-wedge polygons associated with thermal contracting cracking are not a widespread landform, but do occur in some areas such as at the northern end of Blæsedalen (Figure 7). However, various forms of sorted and unsorted circles (Figure 8) and stripes (Figure 9) are very common and occur from sea-level to high altitudes. Movement in the active layer is also evident in areas with high deposition rates such as on outwash plains, where up to 25 cm high stone heave mounds may form within one or a few years (Figure 10).

Figure 8. Unsorted circles on Pjetursson's moraine, east of Qeqertarsuaq (Godhavn). GPS receiver for scale.

Figure 9. Unsorted stripes on Pjetursson's moraine, east of Qeqertarsuaq (Godhavn

Figure 10. Stone heave mounds formed in glaciofluvial and glacier naled deposits, Kuannersuit Kuussuat (Kuannersuit Valley).

Solifluction is an important process of soil movement in active layers, and solifluction lobes occur on many valley slopes (Humlum et al., 1995). Active layer detachment slides are often observed in connection to solifluction lobes, increasing the rate of down-slope transportation.

Palsas seem to be rare landforms on Disko Island. The author has only observed palsas in peaty wetlands on the eastern shore of the lake in front of Pjetursson's moraine east of Qeqertarsuaq (Godhavn). Thus, it is unknown whether palsas are restricted to the more humid southwestern part of the island.

Fluvial and Coastal Landforms and Processes

Rivers and streams cause lateral and vertical erosion of ice-cored terraces and moraines by thermo-erosional undercutting, removing the debris-cover and exposing the permafrost. Most large rivers have a braided morphology due to the high sediment load and large diurnal and seasonal fluctuations in discharge common for glacio-nival flow regimes.

Taliks (unfrozen zones with groundwater movement in permafrost environments) are believed to exist below most valley rivers. When glaciers override an outwash plain, such as during a surge event, proglacial open taliks (i.e. taliks in contact with the active layer) may become closed taliks overlain by isolating glacier ice and underlain by relict permafrost. Such closed taliks have important consequences for the formation and development of subglacial drainage systems, and they may be identified by proglacial upwelling and winter discharge. They may also be important for whether a coupling of glacier ice and relict permafrost occurs.

Naled (plural naledi; also referred to as icing and aufeis) is an extrusive stratified ice accretion formed when successive water discharge inundates the valley floor during the cold season. Perennial naled assemblages are common on Disko Island and are often related to ordinary or homothermal springs (spring naled), alluvial fans, and upwelling on outwash plains (river naled) (Humlum, 1979; Humlum and Svensson, 1982). No reports of naled associated with pingos have been made to date. Naled formed as a consequence of winter discharge from glacier portals or through taliks below glacier termini (glacier naled) seems to be linked to surge-type glaciers (Yde and Knudsen, 2005). During the surge event of Kuannersuit Glacier an at least 3 m thick glacier naled was produced in front of the glacier during the winter of 1997/1998. The advancing glacier transferred longitudinal compressive stresses onto the proglacial naled, while a high hydraulic pressure below the naled caused an upward pressure, resulting in failure and stacking of large faulted naled blocks in front of the glacier (Yde et al., 2005b; Roberts et al., 2009). This landform is termed thrust-block naled. Glacier naled associated with surge events may be debris-rich relative to other types of naled, containing a significant amount of clay. Also, the observations at Kuannersuit Glacier indicate that proglacial naled may be incorporated into the basal ice of advancing glaciers.

Coastal erosion occurs as tidewater undercutting of rocks and marine terraces. Along the beaches of Disko Island a debris-rich tidal platform ice-foot forms each winter, reducing the erosional and depositional effects of waves and affecting coast morphology and sedimentology (Nielsen, 1982). Raised marine terraces and beach ridges are widespread, and the highest marine limits are between 80 – 120 m (Weidick and Bennike, 2007).

Conclusion

Periglacial landforms and processes modify the glacial landscape of Disko Island. As Disko Island is located at the southern limit of the continuous permafrost zone, a future climate warming will cause imbalance in permafrost thickness and distribution and lead to increase in the rates of degradation processes. Increase in annual precipitation and glacier ice melt will enhance the total weathering rates, especially if the number of freeze-thaw cycles is significantly increased.

As the review of periglacial research on Disko Island shows, there is still limited knowledge on many aspects of how periglacial landforms evolve and how periglacial processes work in different environments. There is a lack of systematic mapping of the incidence of most periglacial landforms, which potentially can provide important information on climatic, topographic, hydrological sedimentological and vegetational controls. The decadal and centennial evolution of previously described landforms such as rock glaciers can contribute with information on climate-related fluctuations in process dynamics. If such information becomes available it will make geomorphological reconstructions of the landscape evolution possible, which can improve our knowledge on the late-Holocene history of the region. Also, more information on the geothermal regime is vital to our understanding of the sensitivity of permafrost thaw processes.

Probably the most challenging and contributing perspectives lie in the studies on the association between glaciers and permafrost. Here, Disko Island can provide optimal field conditions for studies on the formation and evolution of periglacial landforms on previously glaciated terrain. Glacier surge events reset the overridden landscape, and as the glaciers retreat rapid formation of landforms such as pingos and patterned ground may occur. There is a lot of evidence of buried glacier ice, which now can be classified as massive ground ice (permafrost), both of recent and Early Holocene age. In Kuannersuit Kuussuat (Kuannersuit Valley), an outwash plain with a sediment thickness of approximately 1 m developed on top of detached glacier ice during the recession of the Sorte Hak glaciers after a major surge event. Subsequently, several circular thaw lakes (thermokarst lakes) formed on the outwash plain. This indicates that a continuum ranging from down-wasting and back-wasting processes to periglacial thermokarst processes controls the deglaciation of formerly ice-covered landscapes, and shows how closely related glacial and periglacial processes are.

Acknowledgements

I will like to thank the Arctic Station in Qeqertarsuaq/Godhavn for providing local logistic support for fieldwork on Disko Island. Ruth Nielsen designed the local map.

References

Christiansen, H.H. (1995). Observations of open system pingos in a marsh environment, Mellemfjord, Disko, Central West Greenland. *Danish Journal of Geography*, 95, 42-48.

Christiansen, H.H. (1997). Observations of open system pingos in a marsh environment, Mellemfjord, Disko, Central West Greenland: a reply to Gurney and Worsley. *Danish Journal of Geography*, 97, 157-159.

Christiansen, H.H., Sigsgaard, C., Humlum, O., Rasch, M. and Hansen, B.U. (2008). Permafrost and periglacial geomorphology at Zackenberg. *Advances in Ecological Research*, 40, 151-174.

Donner, J. (1978). Holocene history of the west coast of Disko, central West Greenland. *Geografiska Annaler*, 60A, 63-72.

Frich, P. and Brandt, E. (1985). Holocene talus accumulation rates and their influence on rock glacier growth: a case study from Igpik, Disko, West Greenland. *Danish Journal of Geography*, 85, 32-43.

Geodetic Institute of Denmark (GID). (1941). *Topographic map 1:250.000. Map no. 69 V.1 Godhavn*. Copenhagen, Geodetic Institute of Denmark.

Gilbert, R., Nielsen, N., Desloges, J.R. and Rasch, M. (1998). Contrasting glacimarine sedimentary environments of two Arctic fjords on Disko, West Greenland. *Marine Geology*, 147, 63-83.

Gilbert, R., Nielsen, N., Möller, H., Desloges, J.R. and Rasch, M. (2002). Glacimarine sedimentation in Kangerdluk (Disko Fjord), West Greenland, in response to a surging glacier. *Marine Geology*, 191, 1-18.

Hammer, R. and Steenstrup, K.J.V. (1893). Kaart over Nord Grönland. *Meddelelser om Grønland*, 4.

Heide-Jørgensen, H.S. and Kristensen, R.M. (1999). Puilassoq, the warmest homothermal spring of Disko Island. *Berichte zur Polarforschung*, 330, 32-43.

Humlum, O. (1979). Icing ridges: a sedimentary criterion for recognizing former occurrence of icings. *Bulletin Geological Society of Denmark*, 28, 11-16.

Humlum, O. (1982). Rock glacier types on Disko, central West Greenland. *Danish Journal of Geography*, 82, 59-66.

Humlum, O. (1987). Glacier behaviour and the influence of upper-air conditions during the Little Ice Age in Disko, central West Greenland. *Danish Journal of Geography*, 87, 1-12.

Humlum, O. (1988a). Rock glacier appearance level and rock glacier initiation line altitude: a methodological approach to the study of rock glaciers. *Arctic and Alpine Research*, 20, 160-178.

Humlum, O. (1988b). Natural cairns on rock glaciers as an indication of a solid ice core. *Danish Journal of Geography*, 88, 78-82.

Humlum, O. (1996). Origin of rock glaciers: observations from Mellemfjord, Disko Island, central West Greenland. *Permafrost and Periglacial Processes*, 7, 361-380.

Humlum, O. (1997). Active layer thermal regime at three rock glaciers in Greenland. *Permafrost and Periglacial Processes*, 8, 383-408.

Humlum, O. (1998). The climatic significance of rock glaciers. *Permafrost and Periglacial Processes*, 9, 375-395.

Humlum, O. (1999). Late-Holocene climate in central West Greenland: meteorological data and rock-glacier isotope evidence. *The Holocene*, 9, 581-594.

Humlum, O. (2000). The geomorphic significance of rock glaciers: estimates of rock glacier debris volumes and headwall recession rates in West Greenland. *Geomorphology*, 35, 41-67.

Humlum, O. and Christiansen, H.H. (2008). Lowland periglacial research: A review of published advances 2003-2007. *Permafrost and Periglacial Processes*, 19, 211-235.

Humlum, O., Christiansen, H.H., Hansen, B.U., Hasholt, B., Jakobsen, B.H., Nielsen, N. and Rasch, M. (1995). Holocene landscape evolution in the Mellemfjord area, Disko Island, central west Greenland: area presentation and preliminary results. *Danish Journal of Geography*, 95, 28-41.

Humlum, O. and Svensson, H. (1982). Naledi i Grønland. Flyfotografisk inventering af perennerende flod- og kildeis i grønlandske permafrostområder; foreløbige resultater. *Danish Journal of Geography*, 82, 51-59.

Ingólfsson, Ó., Frich, P., Funder, S. and Humlum, O. (1990). Paleoclimatic implications of an early Holocene glacier advance on Disko Island, West Greenland. *Boreas*, 19, 297-311.

Jost, W. (1940). Gletscherschwankungen auf der Insel Disco in Westgrönland. *Zeitschrift für Gletscherkunde, für Eiszeitforschung und Geschichte des Klimas*, 27, 20-28.

Knudsen, N.T., Yde, J.C. and Gasser, G. (2007). Suspended sediment transport in glacial meltwater during the initial quiescent phase after a major surge event at Kuannersuit Glacier, Greenland. *Danish Journal of Geography*, 107, 1-7.

Mackay, J.R. (1999). Cold-climate shattering (1974 to 1993) of 200 glacial erratics on the exposed bottom of a recently drained Arctic lake, western Arctic coast, Canada. *Permafrost and Periglacial Processes*, 10, 125-136.

Møller, H.S., Christiansen, C., Nielsen, N. and Rasch, M. (2001). Investigation of a modern glacimarine sedimentary environment in the fjord Kuannersuit Sulluat, Disko, West Greenland. *Danish Journal of Geography*, 101, 1-10.

Nielsen, N. (1982). Periglaciale former på kyster opbygget af løse sedimenter – nogle iagttagelser fra Disko, Vestgrønland. *Danish Journal of Geography*, 82, 67-73.

Pedersen, A.K., Larsen, L.M., Ulff-Møller, F., Pedersen, G.K. and Dueholm, K.S. (2001). *Geological map 1:100,000 69 V.2 N Pingu*. Copenhagen, Geological Survey of Denmark and Greenland.

Rasch, M., Nielsen, N., Christiansen, C., Balstrøm, T., Gilbert, R. and Desloges, J. (2003). Role of landscape parameters in riverine run-off, and sediment and organic matter yield on Disko Island, West Greenland. *Danish Journal of Geography*, 103, 1-11.

Roberts, D.H., Yde, J.C., Knudsen, N.T., Long, A.J. and Lloyd, J.M. (2009). Ice marginal dynamics and sediment delivery mechanisms during surge activity, Kuannersuit Glacier, Disko Island, West Greenland. *Quaternary Science Reviews*, 28, 209-222.

Steenstrup, K.J.V. (1883). Bidrag til kjendskab til braerne og brae-isen i Nordgrønland. *Meddelelser om Grønland*, 4, 69-112.

Steenstrup, K.J.V. (1901). Beretning om en undersøgelsesrejse til øen Disko i sommeren 1898. *Meddelelser om Grønland*, 24, 249-306.

Weidick, A. and Bennike, O. (2007). Quaternary glaciation history and glaciology of Jakobshavn Isbræ and the Disko Bugt region, West Greenland: a review. *Geological Survey of Denmark and Greenland Bulletin*, 14, 78 pp.

Whymper, E. (1872). Opdagelsesrejser i Grønland. *Dagbladet*, Copenhagen, 23th November 1872.

Yde, J.C. and Knudsen, N.T. (2005). Observations of debris-rich naled associated with a major glacier surge event, Disko Island, West Greenland. *Permafrost and Periglacial Processes*, 16, 319-325.

Yde, J.C. and Knudsen, N.T. (2007). 20th-century glacier fluctuations on Disko Island (Qeqertarsuaq), Greenland. *Annals of Glaciology*, 46, 209-214.

Yde, J.C and Knudsen, N.T. Surge-type glaciers on Disko Island, Greenland. In: Krugger, M.I. and Stern, H.P. (eds.). *New Permafrost and Glacier Research*. Nova Science Publishers, 283-297.

Yde, J.C., Knudsen, N.T. and Nielsen, O.B. (2005a). Glacier hydrochemistry, solute provenance, and chemical denudation at a surge-type glacier in Kuannersuit Kuussuat, Disko Island, West Greenland. *Journal of Hydrology*, 300, 172-187.

Yde, J.C. and Knudsen, N.T., Larsen, N.K., Kronborg, C., Nielsen, O.B. Heinemeier, J. and Olsen, J. (2005b). The presence of thrust-block naled after a major surge event: Kuannersuit Glacier, West Greenland. *Annals of Glaciology*, 42, 145-150.

In: New Permafrost and Glacier Research
Editors: Max I. Krugger and Harry P.Stern

ISBN: 978-1-60692-616-1

Chapter 6

MOUNTAIN PERMAFROST DEGRADATION IN THE NEPAL HIMALAYAS AND THE RUSSIA ALTAI MOUNTAINS

Kotaro Fukui* and Yoshiyuki Fujii

National Institute of Polar Research, 1-9-10 Kaga, Itabashi-ku, Tokyo 173-8515

ABSTRACT

Recent studies have described mountain permafrost degradation due to global warming in many mountain regions, such as European mountains, the Tibetan Plateau, the Tien Shan and mountainous areas of Mongolia. In this chapter we describe the recent mountain permafrost degradation in the Nepal Himalayas and the Russia Altai Mountains. The Nepal Himalayas is one of the largest mountainous areas of the world. In 1973, the permafrost lower limit was estimated to be 5200–5300 m above sea level (ASL) on southern-aspect slopes in the Khumbu Himal, the eastern part of the Nepal Himalayas. Using ground-temperature measurements, the mountain permafrost lower limit on slopes with the same aspect was estimated in 2004. The results indicate that the permafrost lower limit was 5400–5500 m ASL in 2004. The permafrost lower limit was estimated to be 5400 to 5500 m on slopes with a southern aspect in the Khumbu Himal in 1991 using seismic reflection soundings. Thus, it is possible that the permafrost lower limit has risen 100–300 m between 1973 and 1991, followed by a stable limit of 5400 to 5500 m over the last decade. The Russia Altai Mountains is located on the southern fringe of the Siberia Plain. The altitudinal range of sporadic/patchy permafrost zone and that of discontinuous/continuous permafrost zone are 1800–2000 m ASL and above 2000 m ASL, respectively. The mean annual air temperature at Russian meteorological stations in Russian Altai exhibited remarkable warming trends. We observed the phenomena relating to permafrost degradation, such as landslide influenced by antecedent permafrost degradation, and rapid degradation of pingos around the lower limit of discontinuous/continuous permafrost zone.

* E-mail addresses: fukui@nipr.ac.jp, pukuun@hotmail.com (K. Fukui)

PERMAFROST STUDIES IN NEPAL HIMALAYAS

The mountain permafrost lower limit has been studied in the Khumbu Himal, the Nepal Himalayas, since 1973 (Fujii and Higuchi, 1976; Jakob, 1992; Barsch and Jakob, 1993). Fujii and Higuchi (1976) measured ground temperatures and found that the permafrost lower limit on slopes with a northern aspect was about 4900–5000 m in 1973. Barsch and Jakob (1993) conducted seismic refraction soundings and estimated that the lower limit on slopes with a southern aspect was 5400–5500 m in 1991. Fukui et al. (2007a) examined the permafrost lower limit in 2004 based on ground-temperature measurements. We summaries the changes in the permafrost lower limit over the last three decades based on these permafrost studies.

KHMBU HIMAL

The Khumbu Himal is located in the eastern part of the Nepal Himalayas (Fig. 1). The permafrost studies focused on the upper Imja Valley (27°53–60'□, 86°48–50'E) located in the central part of the Khumbu Himal (Fig. 2). The Khumbu glacier extends from the south face of Mt. Everest (8850 m) to 4900 m ASL in this valley. The tree line limit forms at approximately 4000 m, above which alpine meadows extend up to glacier terminus (Fig. 3).

Automatic weather stations were established near Dingboche (4355 m) in 1987 and Lobuche (5050 m) in 1990. Mean annual air temperatures (MAATs) at Dingboche and Lobuche are –1.5 and –2.6°C, respectively (Grabs and Pokhrel, 1993; Tartari et al., 1998). Annual precipitation at Dingboche and Lobuche is 412 and 465 mm, respectively, and nearly 90% of the precipitation occurs during the summer monsoon season (June–September; Bollasina et al., 2002). Because little winter snow falls in this area (Fig. 3a), the permafrost distribution is unaffected by snow cover.

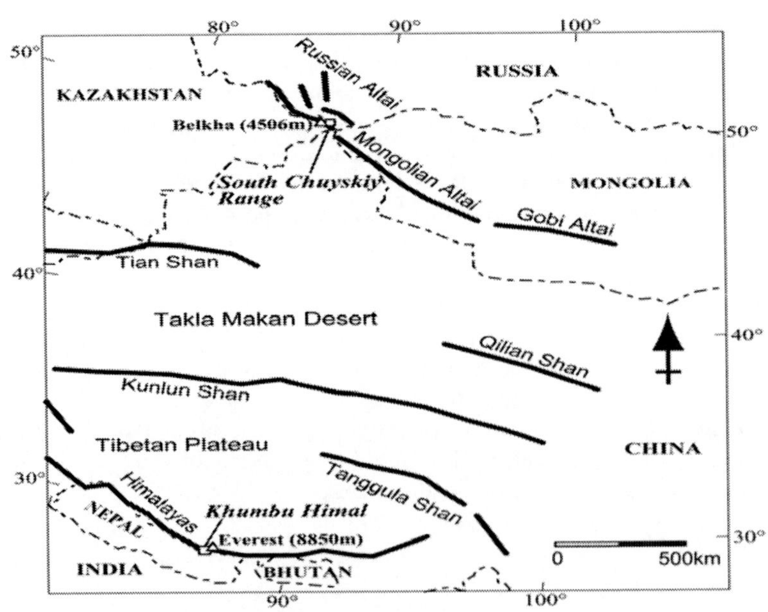

Figure. 1 Locations of the Nepal Himalayas and the Russian Altai Mountains.

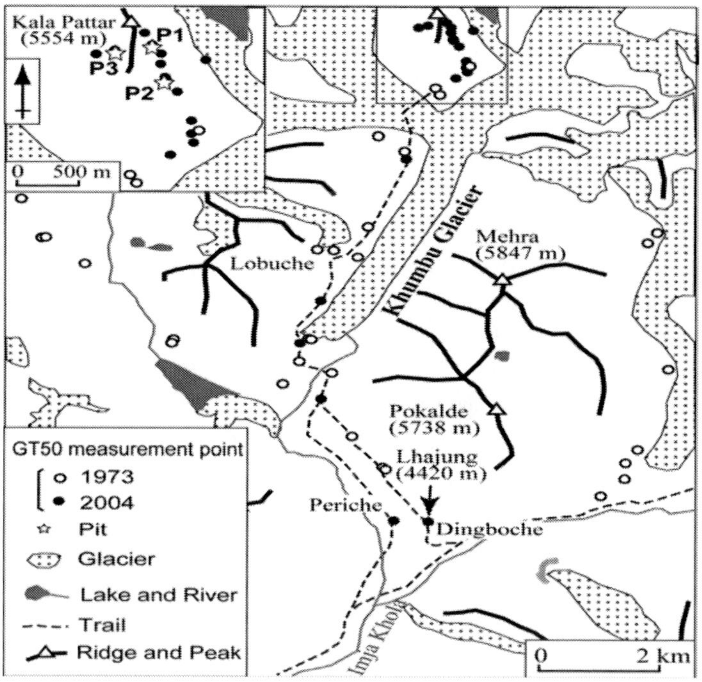

Figure. 2 The upper Imja Valley and ground temperature at 50-cm-deep (GT50) measurement points in 1973 and 2004 (modified Fukui et al., 2007a).

Figure. 3 The upper Imja Valley in late December 2001 (a) and the south facing slope of Kala Pattar in October 2004 (b).

Measurements of Ground Temperature Lapse Rates at 50-cm Depth and Pit Surveys

The diurnal variation in ground temperature is negligibly small at depths below 40 to 50 cm. The lapse rate of the ground temperature at 50-cm depth (GT50) during summer in the seasonal-frost zone is close to that of the air temperature, because the GT50 is principally controlled by ground heat flux for some days (Fig. 4). However, the GT50 lapse rate in the permafrost zone during summer is much larger than that of air temperature, because the GT50 is controlled by not only ground heat flux but cooling from permafrost. For example, Fujii and Higuchi (1972) and Fujii et al. (1999) estimated the permafrost lower limit based on the tendency for change in the GT50 lapse rate at Mt. Fuji, Japan, in August of 1973, 1998, and 1999 (Fig. 4).

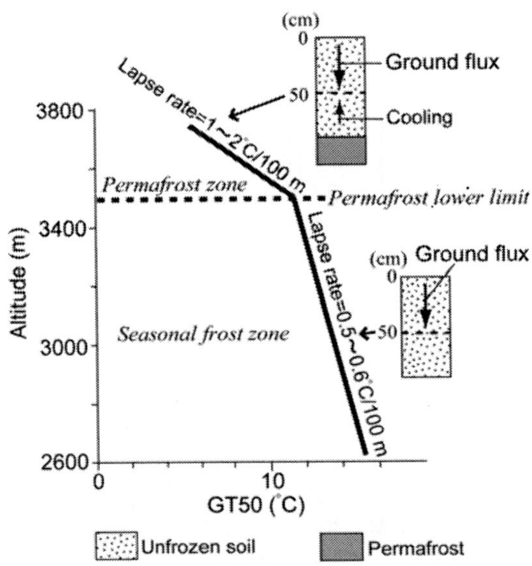

Figure. 4 Ground temperatures at 50-cm depths during August 1998 on a south-facing slope of Mt. Fuji (modified Fujii et al., 1999).

Fujii and Higuchi (1976) also measured the GT50 at approximately 100 movement points on slopes of all aspects between 4500 and 5300 m in the Khumbu Himal in summer 1973. Because research period of GT50 measurements reached several weeks, time series variations in GT50 were observed during the research period. The time series variations in GT50 were revised using fixed-point GT50 measurements at Lhajung (4420 m) at 3-h intervals; revised GT50 (T_{re}) values were calculated in °C using the following equation:

$$T_{re} = T_m - T_{ha},$$

where T_m is the GT50 at the movement points in °C and T_{ha} represents the time series temperature variations for GT50 at Lhajung in °C.

Fukui et al. (2007a) took GT50 measurements at 20 movement points from 8 to 21 October 2004 on a south-facing slope in the Khumbu Himal between 4200 and 5600 m ASL

(Fig. 2) with a digital thermometer (testo110; testo AG, Lenzkirch, Germany), which has a resolution of 0.1°C and accuracy of 0.2°C. Shallow pits were dug and the thermometer probe was inserted in the pit at 50-cm depth just after excavation. GT50 was also monitored at Lhajung at 15-min intervals from 8 to 22 October using a miniature temperature data logger (TR-52; T and D Corp., Nagano, Japan) and revised the time series variations in the GT50.

To determine the frozen layer in the permafrost, three pits (0.7–1.4 m deep) were dug around Kala Pattar (Fig. 3b; 5554 m) on 20 and 21 October 2004 (Fukui et al., 2007a). The ground temperature profiles and stratigraphy were observed in the pits.

THE LAPSE RATE CHANGE OF THE GT50 IN 1973 AND 2004

Figure 5a shows the revised GT50 for locations on southern-aspect slopes in 1973 versus the altitudes of the revised GT50. The data were derived from Fujii and Higuchi (1976). In 1973, while the lapse rate below 5200–5300 m was close to that of air temperature (0.5°C/100 m), the lapse rate above 5200–5300 m became much larger than that of air temperature (3.4°C/100 m). Thus, the permafrost lower limit of the south-facing slope was estimated as 5200–5300 m in 1973.

In 2004 (Fig. 5b), while the lapse rate below 5400–5500 m was close to that of air temperature (0.5°C/100 m), the lapse rate above 5400–5500 m became much larger than that of air temperature (2.5°C/100 m). Thus, Fukui et al. (2007a) estimated that the permafrost lower limit of the south-facing slope in 2004 was 5400–5500 m.

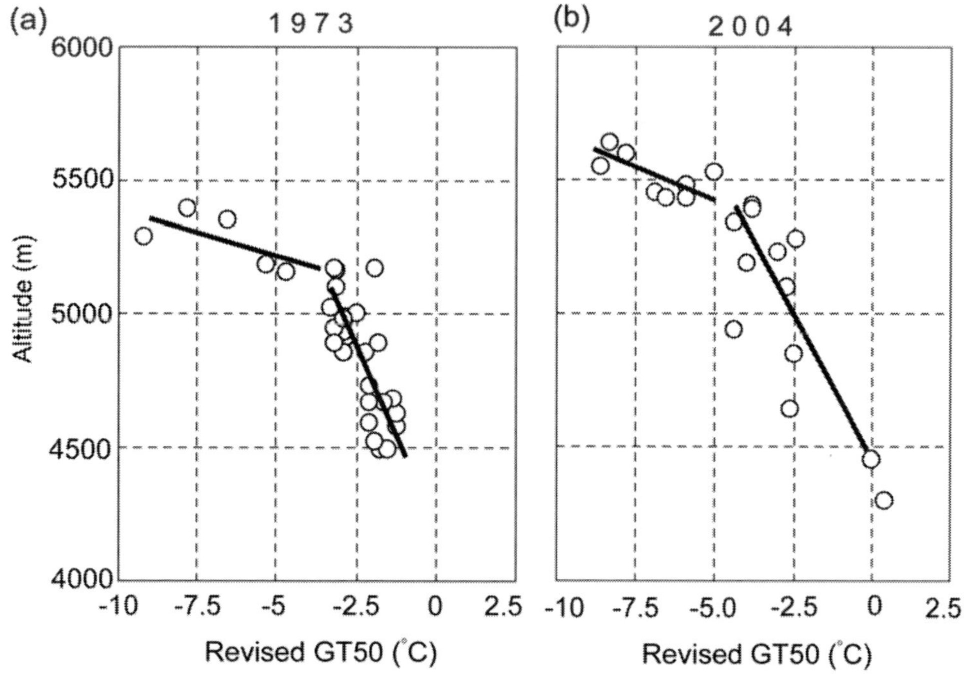

Figure. 5 Lapse rate of the GT50 for south-facing slopes, the Khumbu Himal, in 1973 (a) and 2004 (b) (modified Fukui et al., 2007a).

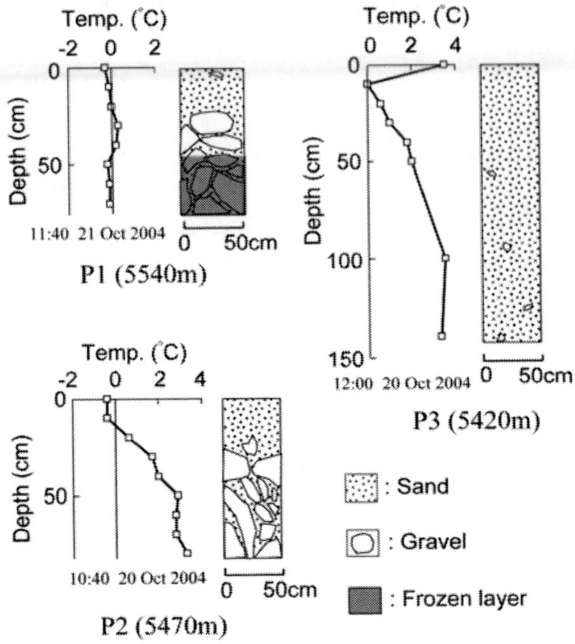

Figure. 6 Ground temperature profiles and stratigraphy of the pits around Kala Pattar in October 2004 (modified Fukui et al., 2007a).

Finding of the Frozen Layer

The surface layer of pit P1 had a 20-cm-thick sand-rich layer, underlain by a layer dominated by angular gravels (Fig. 6). The surface layer had been frozen in the morning, but melted in the afternoon. The frozen layer was found below a depth of 45 cm. The frozen layer was very dry and had few visible ice lenses. The ground temperature was slightly positive at 30- to 40-cm depths but slightly negative below 50 cm. The frozen layer was not found in the pits P2 and P3.

Permafrost Degradation Between 1973 and 2004

The lapse rate change in 2004 at GT50 indicated that the permafrost lower limit lies between 5400 and 5500 m on south-facing slopes of the Khumbu Himal. The frozen layer was found below 45 cm at P1 located at 5540 m. Because ground freezing had already started by 20–21 October, this frozen layer had likely survived from the preceding summer and could be considered permafrost. Judging from the above results, the permafrost lower limit on slopes with a southern aspect in the Khumbu Himal lay between approximately 5400 and 5500 m in 2004.

The permafrost lower limit on slopes with a southern aspect in the Khumbu Himal in 1973 was estimated as 5200 to 5300 m based on the ground temperature data collected by

Fujii and Higuchi (1976). These findings indicate that the permafrost lower limit rose 100–300 m between 1973 and 2004.

Barsch and Jakob (1993) and Jakob (1992) examined the permafrost lower limit using seismic reflection soundings and estimated the permafrost lower limit of 5400 to 5500 m on slopes with a southern aspect in the Khumbu Himal in 1991. Their estimation corresponds with our finding on the permafrost lower limit in 2004. Thus, it is possible that the permafrost lower limit has risen 100–300 m between 1973 and 1991, followed by a stable limit of 5400 to 5500 m over the last decade (Fig. 7).

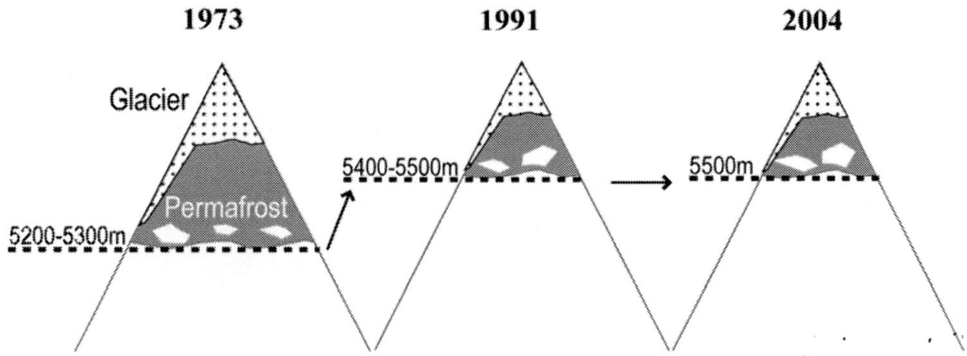

Figure. 7 Schematic diagrams of changes in the lower limit of permafrost in the Khmbu Himal.

Figure 8 shows the variations in average air temperatures during winter (January–February–March) and those during summer (June–July–August) at weather stations in Namche Bazar (3450 m) in 1971–1982 and Dingboche (4355 m) in 1987–2001. Namche Bazar is located in approximately 2.5 km south-west of Dingboche. The data were collected and published by the Department of Hydrology and Meteorology (DHM), His Majesty's Government (HMG) of Nepal and its Snow and Glacier Hydrology Unit (SGHU). Because these data were much fragmentary, we could not calculate annual means. In 1971–1982, variations in the air temperatures in summer were almost stable, however those in winter had clear warming trend. In 1987–2001, variations in the air temperatures in summer were stable and those in winter had large variety year to year and no warming trend. These should be indicate that the air temperature variations in the Khumbu Himal have a warming trend in 1971–1982 and have no warming trend in 1987–2001. These trends correspond with our finding on the changes in the permafrost lower limit over the last three decades.

It has been reported that the permafrost area of the Tibetan Plateau has experienced the most substantial climate warming (Jin et al., 2000). Wang et al. (2000) reported that the permafrost lower limit rose 40–80 m from the 1970s to 1990s in the Tibetan Plateau, according to an increase in MAAT of approximately 0.2 to 0.4°C. However, because the rise in the mountain permafrost lower limit in the Khumbu Himal is larger than that in the Tibetan Plateau, it is possible that climate warming in the Khumbu Himal has been more severe than that in the Tibetan Plateau.

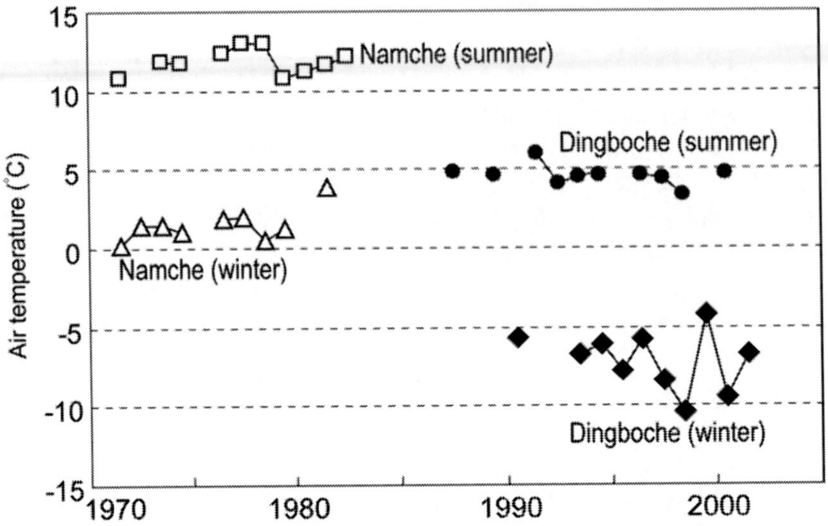

Figure. 8 Variations in average air temperatures during winter (January–February–March) and those during summer (June–July–August) at weather stations in Namche Bazar (3450 m) in 1971–1982 and Dingboche (4355 m) in 1987–2001 (modified Fukui et al., 2007a).

Permafrost Studies in Russia Altai

Rapid atmospheric warming has recently occurred in Siberia (e.g., Chapman and Walsh, 1993; Weller, 1998), and change in the lower limit of mountain permafrost in the Russian Altai Mountains on the southern fringe of the Siberia Plain may serve as an important indicator of past and future climate change in the region. Shats (1978) indicated that the lower limit of permafrost in the region was between 1400 and 2000 m ASL. Gorbunov and Gorbunova (1994), however, reported that sporadic permafrost occurred in burial mounds above 1000 m ASL. Recently Fukui et al. (2007b) determined that sporadic/patchy permafrost zone was 1800–2000 m and discontinuous/continuous permafrost zone was above 2000 m based on permafrost-indicator features in this area. We report possible climate warming impact on permafrost near the lower limit of discontinuous permafrost in the Russian Altai Mountains.

The South Chuyskiy Range

The Altai Mountains extend more than 2000 km through Central Asia (Fig. 1). Traditionally, these mountains have been subdivided into the Russian, Mongolian, and Gobi Altai. The Russian Altai Mountains are located in the border region of Russia, Mongolia, China, and Kazakhstan between approximately 48 and 52°N latitude and 83 and 91°E longitude. The highest peak is Mt. Belukha (4506 m ASL). Our study focused in and around the South Chuyskiy Range (49°40'–50°N, 87°30'–88°30'E), which is one of the highest ranges in the region.

The South Chuyskiy Range encompasses three large U-shaped valleys (Taldura, Akkol, and Karaoyuk) which were occupied by ice at various times in the Holocene (e.g., Okishev, 1982). Radiocarbon dates from the Akkol valley give ages for the moraines of the Sofiyskiy glacier ranging 5700–250 years BP (Okishev, 1982).

Three Russian meteorological stations are located in and around the study area (Fig. 9). The Kosh-Agach station (1757 m ASL) is at the centre of the Chuyskiy basin, the Aktru station (2121 m ASL) is in a mountainous district of the North Chuyskiy Range, and the Onguday meteorological station (832 m ASL) is located 150 km northwest of the study area. The MAAT at these stations exhibited significant warming trends for 1965-2000 (Fig. 10). Average MAATs at the Onguday and Kosh-Agach stations for 1966–1975, for example, were –0.4 and –5.3°C, respectively, while those for 1985–1994 were 0.6 and –4.2°C, respectively.

Annual precipitation for 1972–1995 at the Onguday station was approximately 360 mm and recorded winter precipitation (December to February) was only 22 mm. The minimal winter precipitation results in a characteristically shallow winter snow pack.

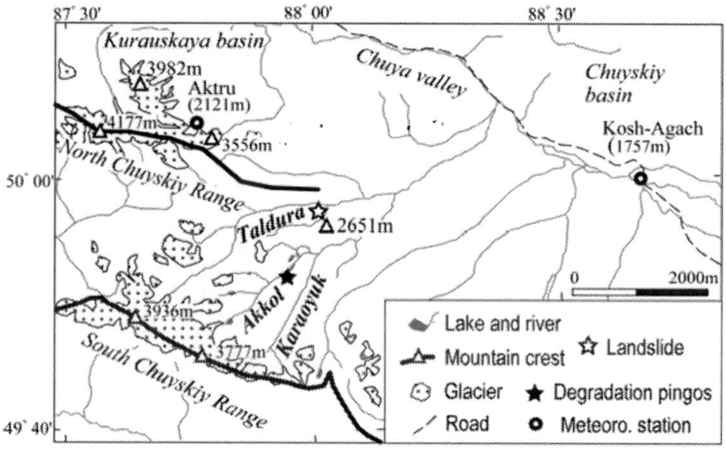

Figure. 9 North and South Chuyskiy Ranges, Russian Altai Moutains.

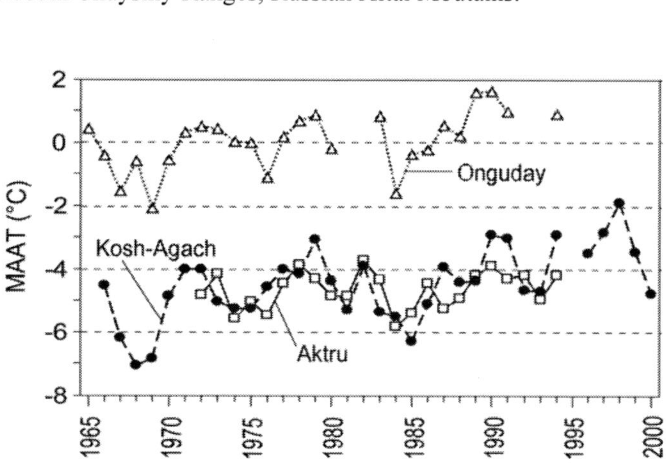

Figure. 10 MAATs at the Onguday (832 m ASL), Kosh-Agach (1757 m ASL) and Aktru (2121 m ASL) stations for 1965–2000 (modified Fukui et al., 2007b).

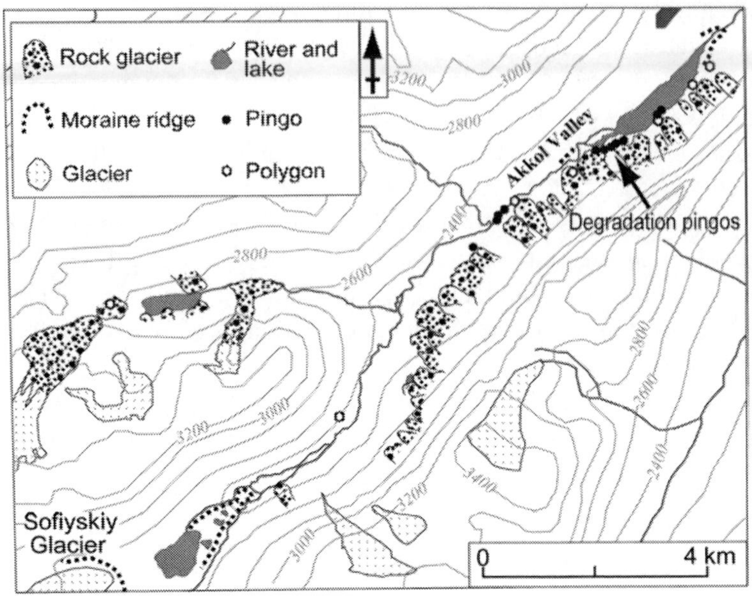

Figure. 11 Rock glaciers, pingos, and ice-wedge polygons in the Akkol valley (modified Fukui et al., 2007b). Contour interval = 200 m.

Permafrost-indicator features (rock glaciers, pingos, and ice-wedge polygons) occur at many places in and around the South Chuyskiy Range of the Russian Altai Mountains (Fig. 11). Based on the distribution of these features and the results of two years of air temperature monitoring, Fukui et al. (2007b) determined that the sporadic/patchy permafrost zones extend from 1800–2000 m ASL and the widespread discontinuous/continuous permafrost zones are above 2000 m ASL.

Glaciers in the Altai are of the summer-accumulation-type (Fujita et al., 2004; Pattyn et al., 2004; Smedt and Pattyn, 2004). The equilibrium line altitude (ELA) lies between 2500 and 2900 m ASL on north-facing slopes and from 2700 to 3100 m ASL on the south-facing slopes (Ostanin and Mikhailov, 2005). Steppes cover a large part of the study area. Larch forests extend up the mountains, with the timberline lying at approximately 2500 m ASL on north-facing slopes (Nakazawa et al., 2004).

Phenomena Related to Permafrost Degradation

A large earthquake (magnitude 7.3 according to the U.S. Geological Survey) struck the Russian Altai on 27 September 2003. This earthquake was the largest in the region since the earthquake of 20 December 1761, which is thought to have had a magnitude of approximately 7.7. A large landslide occurred on the north-facing slope (2000–2200 m ASL) of the Taldura valley after the 2003 earthquake (Fig. 12), forming a landslide scar along the boundary between the larch forest and the steppes. The wide, length, thickness and volume of the landslide mass were 800 m, 1000 m, 50 m and 35000000 m^3, respectively. The permafrost, approximately 30–40 m thick, was widely exposed on the landslide scar (Fig. 12b). We did not observe permafrost in the landslide mass, which covered the steppe, but we identified fossil ice-wedge polygons on the landslide mass (Fig. 12c). The fossil ice-wedge polygons imply permafrost occurrence not only in the larch forest slope but also in the landslide mass in the past.

Figure. 12 Landslide in the Taldura valley at 2000 to 2200 m ASL. The upper half of the landslide (a). Permafrost exposure on the landslide scar (b). Fossil ice-wedge polygon on the landslide mass (c).

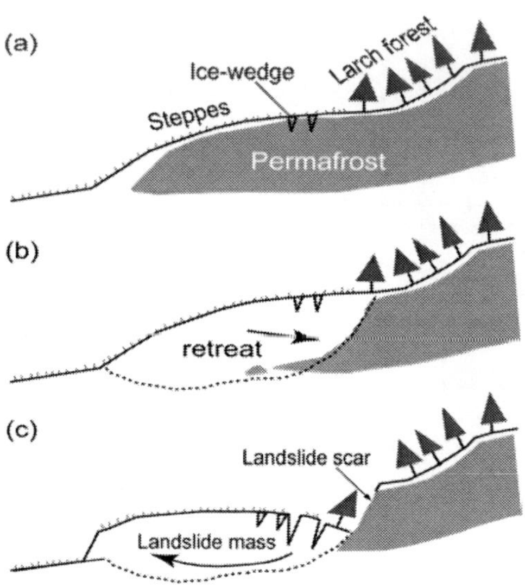

Figure. 13 Schematic diagrams of landslides in the Taldura valley. Before the onset of climate warming (a). Duration of climate warming (b). After the earthquake of 27 September 2003 (c).

Because the landslide was located near the lower limit of discontinuous permafrost, the following processes may have led to this landslide (Fig. 13). First, permafrost existed not only in the larch forest slope but also in the steppe slope before the onset of atmospheric climate warming. The ice-wedge polygons had developed on the steppe slope. Second, because the permafrost had thawed in the steppe slope due to recent atmospheric climate warming, the steppe slope became unstable. Third, the steppe slope slipped downward after the earthquake of 27 September 2003.

In the Akkol valley, a pingo ice core was exposed and had retreated; the debris above the ice core slumped to the foot of the ice scarp (Fig. 14), a phenomenon we refer to as "pingo thaw slump". We found this phenomenon around the thermokarst lake and observed two pingo thaw slumps in August 2004 and two more in August 2005.

Figure. 14 Pingo thaw slumps in the Akkol valley. August 2004 (a). August 2005 (b). Ice core and thawed materials of the pingo in August 2005 (c).

We consider four processes to be involved in pingo thaw slump: sliding of the active layer, thermal erosion on the exposed ice scarp face, slumping of thawed debris from above the ice scarp face, and the flowing of thawed materials into the thermokarst lake.

This phenomenon caused one pingo to nearly disappear in only 2 years. Such pingo degradation is much more rapid than the pingo decay observed in other permafrost regions (e.g., Mackay, 1973, 1990). The occurrences of these phenomena may imply a remarkable rising of the lower limit of permafrost in the study area.

CONCLUSION

The permafrost lower limit on slopes with a southern aspect in the Khumbu Himal was estimated as 5200 to 5300 m in 1973 and was determined 5500 m in 2004 based on the ground temperature data. In 1991 the permafrost lower limit on slopes with a southern aspect was estimated as 5400-5500 m using seismic reflection soundings. Thus, it is possible that the permafrost lower limit has risen 100–300 m between 1973 and 1991, followed by a stable limit of 5400 to 5500 m over the last decade.

The permafrost lower limit rose 40–80 m at the Tibetan Plateau as estimated by an increase in mean annual air temperature of approximately 0.2 to 0.4°C from the 1970s to the 1990s (Wang et al., 2000). The larger rise in the mountain permafrost lower limit in the Khumbu Himal than that in the Tibetan Plateau suggests that climate warming in the Khumbu Himal has been more severe than that in the Tibetan Plateau.

The sporadic/patchy permafrost zones extend from 1800–2000 m ASL and the widespread discontinuous/continuous permafrost zones are above 2000 m ASL in the Russian Altai Mountains. The MAATs at Russian meteorological stations in the Russian Altai Mountains exhibited significant warming trends for 1965-2000. We observed two notable phenomena related to permafrost degradation in the Russian Altai Mountains. First, slope instability caused by permafrost degradation induced a large landslide. Second, we noted remarkable pingo degradations. These phenomena suggest a rapidly rising lower limit of permafrost in the Russian Altai Mountains.

REFERENCES

Barsch, D. and Jakob, M. *Proceedings of the Sixth International Conference on Permafrost* 1993, 27–31.

Bollasina, M., Bertolani, L. and Tartari, G. *Bull. Glacier Res.* 2002, *19*, 1–11.

Chapman, W.L. and Walsh, J.E. *Bull. Am. Meteorol. Soc.* 1993, *74*, 33–47.

Fujii, Y. and Higuchi, K. *Seppyo* 1972, *34*, 9–22 (in Japanese).

Fujii, Y. and Higuchi, K. *Seppyo* 1976, *38*, 125–128.

Fujii, Y., Masuzawa, T., Hashimoto T., Onoda, M. and Ueno, K. *Reprints of the 1999 Conference Japanese Society of Snow and Ice* 1999, 36 (in Japanese).

Fujita, K, Takeuchi, N, Aizen, V. and Nikitin, S. *Bull. Glaciol. Res.* 2004, *21*, 57–64.

Fukui, K., Fujii, Y., Ageta, Y. and Asahi, K. *Glob. Planet. Change* 2007a, *55*, 251–256.

Fukui, K., Fujii, Y., Mikhailov, N., Ostanin, O. and Iwahana, G. *Permafr. Periglac. Process.* 2007b, *18*, 129–136.

Gorbunov, A. and Gorbunova, I. *Permafr. Periglac. Process.* 1994, *5*, 289–292.

Grabs, W.E. and Pokhrel, A.P. *IAHS Publ.* 1993, *218*, 3–16.

Jakob, M. *Permafr. Periglac. Process.* 1992, *3*, 253–256.

Jin, H., Shuxun, L., Guodong, C. and Shaoling, W. *Glob. Planet. Change* 2000, *26*, 387–404.

Mackay, J.R. *Can. J. Earth Sci.* 1973, *10*, 979–1004.

Mackay, J.R. *Can. J. Earth Sci.* 1990, *27*, 1115–1125.

Nakazawa, F., Fujita, K., Uetake, J., Kohno, M., Fujiki, T, Arkhipov, S.M., Kameda, T., Suzuki, K. and Fujii, Y. *J. Geophys. Res.* 2004, *109*, F04001, doi:10.1029/2004JF000125.

Okishev, P.A. *Dynamics of the Altai glaciation in the Late Pleistocene and Holocene*; Tomsk University Press: Tomsk, 1982.

Ostanin, O.V. and Mikhailov, N. *Ice and Climate News* 2005, *6*, 19–21.

Pattyn, F., Smedt, B.D., Brabander, S.D., Huele, W.V., Agatova, A., Mistrukov, A. and Decleir, H. *Ann. Glaciol.* 2004, *37*, 286–293.

Shats, M.M. *Permafrost Conditions in the Altai-Sainy Mountain System*; Nauka: Novosibirsk, 1978 (in Russian).

Smedt, B.D. and Pattyn, F. *Ann. Glaciol.* 2004, *37*, 143–149.

Tartari G., Verza, G. and Bertolami, L. *Mem. Ist. ital. Idrobiol.* 1998, *57*, 23–40.

Wang, S., Jin, H., Shuxun, L. and Zhao, L. *Permafr. Periglac. Process.* 2000, *11*, 43–53.

Weller, G. *Ann. Glaciol.* 1998, *27,* 543–552.

GLACIERS

In: New Permafrost and Glacier Research
Editors: Max I. Krugger and Harry P.Stern

ISBN: 978-1-60692-616-1

Chapter 7

THE FORMER GLACIATION OF HIGH- (TIBET) AND CENTRAL ASIA AND ITS GLOBAL CLIMATIC IMPACT - AN ICE AGE THEORY WITH A REMARK ON POTENTIAL WARMER CLIMATIC CYCLES IN THE FUTURE

Matthias Kuhle[*]
Institut der Universität Göttingen, Goldschmidtstr. 5, D-37077 Göttingen, Germany

SHORT ABSTRACT

Since 1973 one drillhole (5-14 warm times / 6-15 glacials) and 39 expeditions in the Himalayas, Tibet, the Karakorum, the Kuen Lun, the Tienshan, the Sayan Mountains, the Altai and other parts of Central Asia contributed to detailed knowlegde about extension (2.4 Mio km^2) and thickness (800–1000 m) of High Asian inland-ice. The data are best for the Last Glacial Period (Würm, Marine Isotop Stage (MIS) 4-2). The author thinks that during the last 2.75 Ma conditions that are comparable to the LGP ocurred several times in Central Asia. Geometric boundary conditions that resulted from the low latitude caused a substantial albedo–induced impact on the energy budget of the Earth during glacial times. The vast extension of the ice–sheets and the high elevation (~6000 m asl) contributed to this. A substantial albedo–induced cooling of the atmosphere is inferred.

EXTENDED ABSTRACT

Since 1973 one drillhole (5-13 warm times/ 6-15 glacials) and 39 expeditions in the Himalayas, Tibet, the Karakorum, the Kuen Lun, the Tienshan, the Sayan Mountains, the Altai and other parts of Central Asia contributed to detailed knowlegde about extension

[*] Address of the author: Prof. Dr. Matthias Kuhle - Geography and High Mountain Geomorphology; Geographisches Institut der Universität Göttingen, Goldschmidtstr. 5, D-37077 Göttingen/Germany; E-mail: mkuhle@gwdg.de

(2.4 Mio km^2) and thickness (800–1000 m) of High Asian inland–ice. This alternations of inland–ice and non–glaciated situations followed upper pliocene subtropical conditions. Currently the Tibet–Plateau has an altitude of approx. 4400-5600m. Measurements showed that at its high altitude the incoming radiation approximates the solar constant. The low latitude contributes to a high energy input. During glacials about 75% of the incoming radiation is reflected. Thus low latitude and high altitude contribute to a substantial albedo–induced cooling. During non–glacial times much of the high energy input from low latitudes and high altitudes remains in the system. Therefore in either environmental situation, Tibet inland–ice or warm non–glaciated the Tibet–Plateau contributes to stabilizing the climate. The formation of glacials from non–glaciated conditions is as well delayed as the formation of non–glacial conditions from glacials.

Upper Pliocene suptropical sediments show that conditions to support the formation of an inland–ice have been established after about 2.75 Ma. This coincides with the formation of large glaciations on the northern hemisphere. The altitude and latitude induced energy loss by reflection from a glaciated Tibet–Plateau is regarded as one important co–factor to this. A small number of outcrops show laterite deposits overlain by glacial sediments. It is thus inferred that, also in Central Asia, considerable differences between glacials and warm times existed. A series of drillsites that documents Quaternary and Pliocene environments is thus needed.

For the future two scenarios are possible:

Currently the Tibet Plateau uplifts by 10 mm/a. To support a reglaciation the Tibet–Plateau needs to be elevated above the snow line. An anthropogenic warming by 1°C causes a prolongation of the current warm phase by ca. 15.35 ka as an uplift by additional ca. 160 (140-200) m is needed to support a reglaciation.

Currently for Central Asia pliocene subtropical environments are interpreted as being linked to a Tibet–Plateau that is lower than at present. Thus in Central Asia the reestablishment of pliocene conditions requires a massive change of boundary conditions to counteract the uplift of the Tibet–Plateau. Therefore for Central Asia the conditions of warm interglacials are currently regarded as more probable for the future than upper pliocene conditions. New drillsites are needed to address this issue.

The uplift of Tibet to high altitudes is synchronous with the onset of large glaciations on the Northern Hemipsphere from ca. 2.75 Ma on. It is inferred that a causal link, namely the passing of a threshold value exists. Absolute datings with different methods assign MIS 4–2 to the observed glacigenic forms. Thus older glaciations are inferred to have comparable or reduced extensions. Using 13 climate stations radiation– and radiation–balance measurements have been carried out between 3800 and 6500 m asl in Tibet. They indicate that the incoming subtropical global radiation reaches its highest values on the High Plateau. The resulting radiation balance indicates that today Tibet is the most important heating surface on Earth.

In glacial times 70% of the incoming radiation was reflected by snow and firn of the glaciated 2.4 million km^2. Assuming that 100% of the non–reflected incoming radiation is transformed to heat, including latent heat, 32% of the global energy deficit during glacial times is caused by the Tibet plateau. If the radiation deficit caused also a temperature–deficit, e.g. in warmer times in Tibet the incoming radiation was transformed to heat, in glacial times the Tibet Plateau would have been the most important cooling surface of the Earth. Interglacial periods are explained by glacio–isostatic lowering of Tibet by 650 m. The extended foreland glaciations are interpreted to have moved below the snow–line. At least rapid deglaciations are supported by this boundary conditions. A subsequent ice–age is supported if isostatic uplift, caused by the molten foreland glaciations, raised the Tibet–Plateau above the snow–line. The comparably small energy differences caused by the variation of orbital parameters are regarded as modifying component of climate changes. If a glaciated, and isostatically uplifted Tibet is subject to a warming of 4 C° a loss in glaciated surface would be observed but no deglaciation. The same impact on an isostatically depressed Tibet Plateau can contribute to a deglaciation. Thus the Tibet Plateau has a stabilizing effect for the global climate. Both warm intervals are stabilized once they are established (time–lag of isostasy) and glacials are retained after their establishment as well. Currently the Tibet Plateau uplifts by 10 mm/a. An anthropogenic warming by 1°C causes a prolongation of the current warm phase by ca. 15.35 ka. The reason is that the Tibet–Plateau has to be lifted isostatically to an altitude that exceeds those from the preceding glacial by ca. 160 (140-200) m. Based on the Vostok–Curve we should currently observe temperatures of 4.5 C° below preindustrial values if the conditions from the preceding deglaciation/glaciation are applied. The author interprets that the uplift since the LGP was insufficient to iniate a glaciation. Potential changes of geological boundary conditions are not included. Thus the stabilizing contribution of the Tibet–Plateau for the climate system, both for cold and warm time–intervals is one of the main results of this work.

Key Words: Quaternary and Neogene Climate, Tibetan Inland Ice, Stabilizing Climate

1. Introduction

Both during the Neogene and the Pleistocene cold times with cold and warm conditions occurred. During the Pleistocene the interglacials had by present–day nomenclature been greenhouse–climates with temperatures 2.5–3.6°C above pre–industrial values.

Especially after 2.75 Ma from these pronounced glacial times started.

While the last glacial maximum (LGM) and the last glacial period (LGP, lasting from about ca. 60.000 a to 18.000 a) have been studied extensively, the warm times before and after the LGP and in betweeen glacials are an area of a wealth of upcoming possibilities. This applies particularly as the future is generally expected to be warm and not cold.

If orbital parameters did not change during the earth history such as by meteoritic impacts or fly–bys their impact on the global climate as expressed by the ice–volume varied through time. This applies both to the time before and after 2.75 Ma and before and after 1 Ma. A change of other important boundary conditions, such as in the ocean circulation, important parts of orography (Andes, Tibet Plateau) is to be tested.

Another related question that is of importance for the future is the persistence of the Holocene climatic conditions.

Once the preceding pleistocene greenhouse climate (inter–glacials) had been established glacials conditions had been approached again. If the conditions of the second but last glacial would have ocurred from 11.600 a on, then today values of –4.5 C should be observed. If the conditions of the third but last glacial are applied (Vostok ice–core, Petit et al. 1999) then today values of -8.5 C should be observed.

As compared to earlier times the Holocene is remarkably stable at least one boundary condition for a reglaciation was not met again. If orbital parameters did not change 11.600 years ago those parameters of the geosystem need to be tested that are able to have a respective impact on climate.

Based on this a prediction for the duration of the current warm times is made. The author regards this prediction as one possible co–answer that sheds light on important aspects of the geosystem. There might have been times were other parameters had been more important.

There might also have been times (see the details of the Vostok ice–core) where deglaciations ended abruptly. Thus all important aspects of climate change including the deep ocean circulation, submarine landslides, orograhic changes including isostasy at various parts of the earth need to be considered.

One important factor of climate change is the global radiation balance.

To assess this based on hard data, both during glacials and interglacials, not only the extent of ice covers needs to be known but also their impact on the radiation balance. High latitude ice-sheets, possibly also in low altitudes, have for geometrical reason a much smaller impact on the radiation balance than low latitude ice-sheets in high altitudes.

This means that works that study Neogene and Quaternary climate change need to know the extent of subtropical ice–sheets and, for reasons of isostasy, also their thickness, that for example atmospheric general circulation models need to meet as milestones.

Knowledge on all Neogene and Quaternary ice-sheets that might have existed in low latitudes is difficult if not impossible to obtain. Given the regular pattern of the Vostok ice–core, at least for the last 400 000 years, ice extensions and thicknesses that have similarities to that from the LGM (LGP) can be expected.

Thus, for subtropical latitudes from High Asia (Himalaya and Tibet to Sayan Mts., Siberia) field evidence on extensions, thicknesses and timing of ice–sheets have been determined. The field evidence differs remarkably from the reconstructions of CLIMAP (Cline 1981). As also older field data are in agreement with the data of the author, the field evidence is presented as reference data–base.

After that the radiation–balance is discussed.

Following that the isostatic impact of that ice–sheet is outlined.

Independent of the interpretation of the isostatic impact the hard data of the extension of the ice–sheets that have to be met by atmospheric general circulation models are of value beyond this contribution.

This work discusses the following aspects:

2. Empirical Data
2.1. Overview of existing knowledge
2.2 Evidence of a large Inland–ice on the Tibet Plateau (2.2.1.-14.)
2.3. Ice-thickness, lowest postion of ice–margins around the Tibetan Plateau and the depression of the equilibrium line during the Last Glacial Period (LGP) (2.3.1. – 2.3.18.)
3. Synopsis of the extent of the Inland–Ice in Tibet since the earliest LGP
4. The Tibet Plateau – A stabilizing factor for the climate system
4.1. The Radiation Balance
4.2. Isostatic Impact of the Tibetan Inland Ice
4.3. Impact of orbital Parameters in low Latitudes
4.4. Impact of ELA Depressions on global Climate
4.5. Duration of ice buildup from warm Conditions
4.6. The relief–-induced Decay of Glaciers at the End of a Glacial
4.7. Climatically steered Tectonics
4.7.1. Neogene and Pleistocene uplift as in the Holocene
4.7.2. Pronounced uplift since the Holocene
4.8. Tropical Weathering in Tibet and Pre–LGP Interglacials
5. Start and End of warm Climates
5.1. The Panama Seaway

2. EMPIRICAL DATA

This chapter presents the empirical field data. Ready to use data (maps) can be found in Figs.1 and 3. Readers wishing to modify details are referred to the following detailed sections.

Other readers can proceed directly to the maps and from that to the discussion in sections 3ff.

An important point is, that extending the data from CLIMAP the Tibet Plateau was during the LGP and possibly during older neogene glacials covered by a large inland–ice that extended 2.4 Mio km^2. Paragraph two presents these. These data are part of boundary conditions atmospheric and environmental simulation models, such as CSM from NCAR can use. If all components of the earth system are modeled, these data are part of the milestones the models have to meet.

Photo 1 and 2: Striations on phyllites (stick 145 cm) and some far-travelled erratic boulders (⇧) (granite) in 3760 m asl indicating the LGP-glaciation in the very arid Chapursan Valley (Fig.1, No.62). Analogue-photos M. Kuhle, 24/8/2006

2.1 Overview of Existing Knowledge

A synopsis of older results and views on the Pleistocene glacier cover in Tibet has been provided by von Wissmann's compilation (1959). The glacier cover of Tibet is echoed in recent Chinese literature by Shi and Wang (1979), and has also been reproduced by the LGM reconstrcution of CLIMAP (Cline 1981). These authors speak of a 10% to maximum 20% ice cover of the mountains and plateaus of Tibet. Contrasting to the CLIMAP reconstruction Loczy (1893), v. Handel-Mazzetti (1927), Dainelli and Marinelli (1928), Norin (1932), de Terra (1932) and others (such as Kuhle 1987b, 1988e pp416/417) describe ancient ice margin sites scattered throughout the high regions of Asia. Other early researchers, such as Tafel (1914), Prinz (1927), Trinkler (1932) and Zabirov (1955) (cf. Kuhle 1988e, pp.416/417), making more or less direct use of the data they obtained by observation, reconstructed larger glacier areas. Despite their observations above authors inferred only a few hundred meters of depression of the equilibrium line altitiude (ELA).

The older literature implies a much larger glacial ice–cover during the LGM and possibly also during older Pleistocene and Neogene times than inferred by CLIMAP. According to the author's calculations, the above work represents ELA–depressions of more than 1000 m and indicate, locally, significantly more glacier cover than the von Wissmann´s (1959) scheme had acknowledged. However, above–mentioned authors, interpreting their own findings did not draw the necessary conclusions (ice–extension, ELA–depressions).

The author has been fortunate in being able to carry out 39 expeditions and research visits since 1973, some of which extended to seven months, with the purpose of reconstructing the extent of glaciers in Asia during glacial periods. Two of these were to the arid East Zagros Mountains, one to the Sayan Mountains in South Siberia, the others to Tibet and its flanking mountain systems (Fig.1). The location and large number of areas under investigation permit reconstructions of the glacier areas to be made for entire Tibet. Reconstructions are supported by data from some earlier authors. They modify reconstructions with negligible ice cover as published by CLIMAP (Cline, 1981) and related authors (see Ehlers and Gibbard 2003, p.1562).

Apart from the Tien Shan, the glaciation of Tibet during the last ice age is given as approximately 2.4 million km² (Fig.3a), and is, based on field–data, estimated to include a central thickness of about 2.7 km (Fig.3b). There was, therefore, inland ice with a central dome of about 7000 m asl in Tibet, the details of which are to be demonstrated below. Breaking up on the edges, ice discharged through the surrounding mountains as steep outlet glaciers (Kuhle 2004) (cf. among others Hughes 1998).

2.2. Evidence of a Large Inland–Ice on the Tibet Plateau

Twenty areas covered by till and erratics, or by erratics alone, ground moraines and central areas of roches moutonnées, give evidence of a former glacier cover (Figs.1 and 3).

In addition a 150 m deep cored drillsite at 36°48'N/99°04'E documents 6-15 glacier advances and retreats (see below).

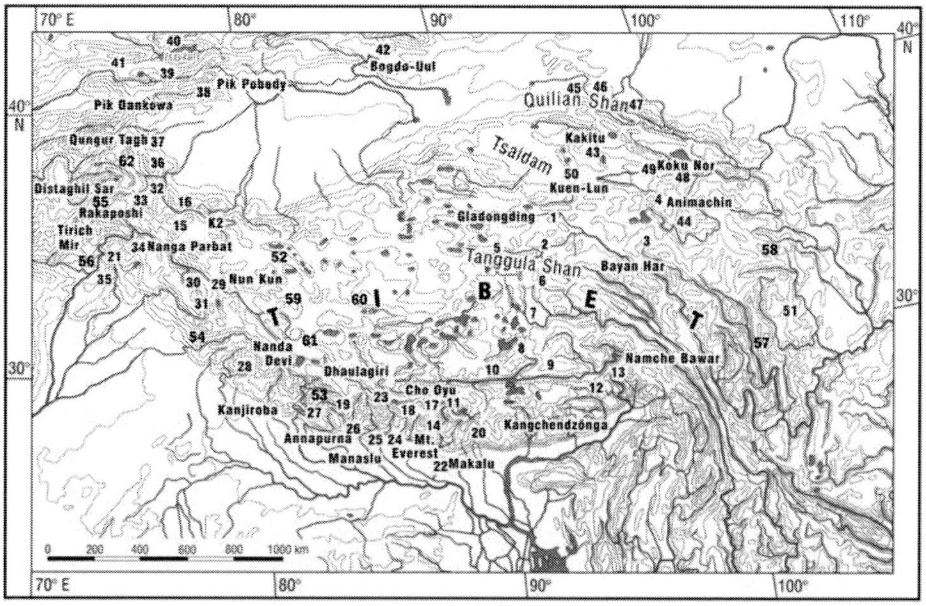

Figure 1: Index map of sites referred to in the text. The localities, quoted in the text as evidence of inland–ice, have numbers.

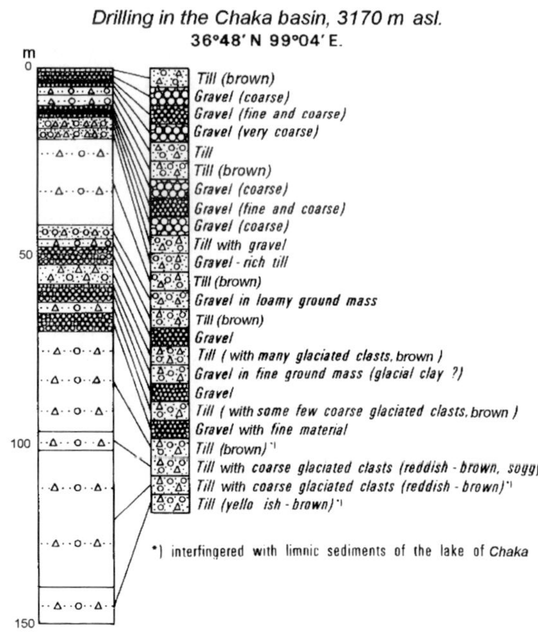

Figure 2: 150 m deep drilling (11 cm diameter core) 0,5 km N of the settlement Chaka (Fig.1, No.49) originating from a position on a ground moraine (till); the surface elevation is 3170 m asl. Till layers with different coloured matrix (brown, reddish-brown to terra fusca coloured, yellowish-brown) are interbedded at least 5 times with gravel and interposed limnic sediments. The gravels consist of red and white mica (biotite-rich) granites and syenite-like rocks; they are partly rounded, rounded at the edges and seldom angular; but also facetted gravels, derived from a moraine, have been found. Each change of fabric represents a period of occupation by a glacier tongue. In the meantime the site has been covered by glaciofluvial gravels which accompanied the glacial advances and retreats and also by lacustrine deposits which developed in lakes dammed up by the terminal moraines.

2.2.1. 150 m Deep Cored Drillsite

A 150 m deep core was drilled in the foreland of the Kukunor Shan (Qinghai Nanshan; Chaka Basin, Fig.1, No.49; 3170 m asl, 36°48'N/99°04'E; Kuhle 1998 Fig.15, No.VIII). It exhibits alternating deposits of stratified advance scree, ground moraine and limnic sediments deposited in terminal basins. The interpretation of the core renders likely the passage of 6 to 15 Pleistocene glaciations in the foreland and, in consequence, in the interior of Tibet (Kuhle 1987b, p.261 Fig.6).

2.2.2. Northern Edge of the Central Tibetan Plateau

On the northern edge of the Central Tibetan Plateau (Fig.1, No.1) large quantities of erratic granite blocks have been deposited on ridges up to heights of 5300 m asl (Kuhle 1993, Photo 2). They lie on outcropping slatey rocks up to 700 m above the thalweg. These materials, such as basal or lodgement till, have to be regarded as glacier deposits. Thus an ice thickness of 700 m is needed to explain the deposition of the erratic blocks. At the edge of the plateau, individual outlet glaciers branched off and may have flowed down the adjacent, steeply descending valleys. This datum, from the Kuenlun Shankou (35°39'N/94°05'E), is supplemented by information further to the west, where (35°38'N/93°52'E) decameter–thick masses of granite blocks, having been transported from the granite barriers in Central Tibet (i.e. from the south/west), overlie outcropping crystalline schists. The granite blocks have been superimposed upon the currently unglaciated marginal chains of hills or mountains in Tibet in such considerable quantities, that they form own erratic hills (Kuhle 1987c, p.188 Fig.7). This can only be explained as result of an extensive glaciation with inland ice.

A northern inland ice margin not only explains these erratic hills, but validates them as typical for moraines which have been deposited on the plateau face. By considering the steep valley gradient, from the former glacier feeding areas to the wastage areas, the ice flow velocity must have increased sharply and thus resulted in a reduction of the glacier's cross-section. This variable ice–flow velocity in the glacier, in turn, was the reason for the breaking–up of the ice margin into separate outlet glacier tongues. The space between these bodies of ice allowed the above–mentioned moraines to accumulate.

For reasons of topography or ice–flow dynamics mentioned above, the moraines extended far above the glacial snow line. During the Late Glacial melt–down of the inland ice (i.e. the upward advance of the snow line) this accumulation received additional, and probably more substantial, supplies of moraine.

2.2.3. South of the Kuenlun Shankou

To the south of the Kuenlun Shankou 10 to 30 km wide plateau–areas are covered with boulder clay containing large, sometimes rounded, erratic blocks. Further south, this boulder clay facies thins and makes way for a slightly undulating landform with more fine-grained ground moraine (Fig.1, No.2) that is decimeter– to several meters thick. 1-8% matrix–supported gravels, with sharp to rounded edges, are embedded in the groundmass. Similar ground moraines the author observed in the extensive inland ice areas of Canada. Neither the slightly undulating morphology nor the deposits of far–travelled, sometimes centimeter long clasts and pebble–sized stones, scattered in loamy matrix, can be interpreted as fluviatile. In accordance with the North American pattern, they are to be understood as a ground moraine plain.

2.2.4. Bayan Har and Animachin Massif

Further east (Fig. 1, No.3) wide-spread granite erratics do occur from 5 km N of the Bayan Har Pass (34°10'N/97°42'E) to beyond the northern edge of the Yen Yougo Basin. They lie right in the from W to E running valley of Yeh Matan, S of Madoi (Mato) (34°41'N/98°03'E), and on the NW edge of Chaling Hu (Ngorin Lake 35°04'N/97°42'E; see also Kuhle 1982c, pp.74-76 Tab.1; 1987b, pp.302–311 Fig.9; Kuhle 1997; 2003 Fig.15).

These are granite erratics with large crystals of sanidine, which have been transported from the 5160 m high threshold of Bayan Har in Central Tibet over a distance of at least 70-140 km. At present this threshold is completely non–glaciated. Bedrock granite can be observed in places. The erratic blocks, up to the size of a room, are facetted and belong to typical ground moraine with loamy groundmass. They have been deposited above of reddish and brown sandstones and crystalline schists. The groundmass, containing polymict blocks, shows fist– to head-sized striated quartzite boulders, locally broken out. The ground moraine covers polish depressions of meter- and polish thresholds of decimeter thickness.

This is proven by glacially streamlined hills. Evidence can also be given by the plateaus of the environment of the Haschi Scha settlement (Fig.1, No.4; see Kuhle 2003 Fig.15, No.13).

The plains, at about 4150 m asl, consist of calcitic limestones. On these plains classical ground moraines (35°01'–12'N/98°49'–59'E), with striated quartzite blocks and erratic granite blocks as indicator–boulders are observed. They prove, that the western bedrock granites of the Animachin Massif (Figs.1, 3) had connection to the ice sheet. During 1981 the glaciation–history of this massif was investigated (details: Kuhle 1982c, 1987b). The granite blocks have been transported from the S, from the area near the Ischikai Station (Kuhle 2003 Fig.15, Nos.13–14), over a distance of ca. 80 km. Thus here at Haschi Scha settlement was a confluence of two tributary ice–streams: One which initially came from the Mt. Payen Khola (with Bayan Har Massif) from the S, has been diverted by Mt. Burhan Budai (E-Kuen Lun; Fig.1, No.3, Fig.3a) to the E. The other came immediately from the S, i.e. from W of Mt. Animachin (Fig.1, No.44; Fig.3a). The N to NE adjacent outlet glaciers and their lowest ice margin positions have been reconstructed, too (Kuhle 1997; 2003 Fig.15) (see 2.3.15).

2.2.5. Central Tibet, Eastern Foreland of the Geladaindong Massif

In Central Tibet, in the eastern foreland of the Geladaindong Massif, north–western Tanggula Shan (33°30'N/91°17'E; Fig.1, No.5; Fig.3a) there are large erratic blocks of granite overlying polished and abraded round mountain ridges of outcropping metamorphosed sedimentary rocks. However, they have not been transported far, since they originated from the central peaks of the Tanggula Shan Massif (see Kuhle 1991a Fig.43, Nos.1 and 2; 2003 Fig.1, No.3). The Tanggula Shan Massif rises up to 6640 m, and is only a few kilometers to decakilometers away. These mountain ridges, though, with erratic blocks at 5800 m asl, reach 500–600 m above the Tibetan Plateau. They are evidence of a plateau ice thickness of more than 500–600 m. Since the ELA now lies at 5600–5800 m asl, the minimum glacier level must have been above 5800 m asl, at or somewhat above the present level of the snow line. This also implies that the ice level was markedly above the then much depressed ELA. Thus Central Tibet was part of the feeding area of this inland ice.

This implies that the snow feeding surplus must have caused the thickness of the central Tibetan ice–dome surface to exceed 1000 m by far. The 5800 m high mountain ridges and even mountains which reached more than 6100 m asl (Kuhle 1991a Photos 1-4 and 6; 2003

Fig.1, No.3) have been abraded and polished round by a glaciation that covered them completely. It is thus for the LGP and pleistocene/neogene glacial intervals inferred that there was an inland ice thickness of 1000 m or more (Fig.3b), considering the basal altitude of the Tibet Plateau as 5000–5300 m asl.

2.2.6. Tanggula Shankou

Immediately south above the Tanggula Shankou (Tanggula Pass 32°50'N/91°50'E, Fig.1, No.6; see Kuhle 1991a Fig. 43, No.4; 2003 Fig.1, Nos.2 and 4) there are moraine ledges with erratic granite blocks on bedrock metamorphites which run in N–S direction at 5500-5600 m asl. They are evidence that a continuous ice–cover stretched across the watershed of this central Tibetan pass. The ice surface was about 400–500 m above the origin of the valley on both sides of the pass. It belongs to the LGP (Table, I-IV). During the LGM (0 in Table) the ELA was too low to permit the formation of such lateral moraine ledges, i.e. the ice was much thicker during the LGP (see section 2.2.4). On both sides of the Tanggula Pass ground moraines containing coarse blocks are extending over tens of kilometers (Fig.1, from No.2 up to 7; see Kuhle 1991a Photos 9–13; 2003).

2.2.7. Pangnag and the Northern Nyainqentanglha

Between the settlement Pangnag and the northern Nyainqentanglha Mountains, (30°30'–32°N/91°–92 E; Fig.1, No.7, see also Kuhle 1991a Fig.43, Nos.6-16; 2003 Fig.1, Nos.4-8) there are extensive polished and abraded areas, evidence of which is found within topographical lows and on roches moutonnées. The roches moutonnées sit on the high plateau level between 4400 and 4600 m asl, are from meters to several hundred metrers high, with scattered patches of ground moraine (Kuhle 1991a, Photos 14–18, 21; 2003 Fig.1, Nos.5–8). The alignment of the roches moutonnées tends to be north to south, then veering from north/west to south/east with the flatter scour–sides facing north to north/west. Thus, they provide evidence that the central inland ice followed the general gradient of the high plateau. In many places abrasion and polishing marks as well as exaration rills on the surface of these roches moutonnées are proof of ice, scouring the outcrops of the sedimentary rocks, so that characteristic outcrop strip-polishings are wide-spread (Kuhle 1991a, Photos 16, 22, 23; 2003 Fig.1 Nos.6-8). In the same area of Central Tibet, ground moraine cover, with large facetted and striated blocks of granite and other material, is common (Kuhle 1991a, Photos 14, 15, 20; 2003 Fig.1 Nos.5–8). Thus during the LGP a complete glacier cover results also for this area.

Table.

glacier stadium		gravel field (Sander)	approximated age (YBP)	ELA-depression (m)
-I	= Riß (pre-last High Glacial maximum)	No. 6	150000 - 120000	c. 1400
0	= Würm (last High Glacial maximum)	No. 5	60000 - 18000	c. 1300
I-IV	= Late Glacial	No. 4 – No. 1	17000 - 13000 or 10000	c. 1100 - 700
I	= Ghasa-stadium	No. 4	17000 - 15000	c. 1100
II	= Taglung-stadium	No. 3	15000 - 14250	c. 1000
III	= Dhampu-stadium	No. 2	14250 - 13500	c. 800 - 900
IV	= Sirkung stadium	No. 1	13500 - 13000 (older than 12870)	c. 700
V-'VII	= Neo-Glacial	No. -0 – No. -2	5500 - 1700 (older than 1610)	c. 300 - 80
V	= Nauri-stadium	No. -0	5500 - 4000 (4165)	c. 150 - 300
VI	= older Dhaulagiri-stadium	No. -1	4000 - 2000 (2050)	c. 100 - 200
'VII	= middle Dhaulagiri-stadium	No. -2	2000 - 1700 (older than 1610)	c. 80 - 150
VII-XI	= historical glacier stages	No. -3 – No. -6	1700 - 0 (= 1950)	c. 80 - 20
VII	= younger Dhaulagiri-stadium	No. -3	1700 - 400 (440 resp. older than 355)	c. 60 - 80
VIII	= stadium VIII	No. -4	400 - 300 (320)	c. 50
IX	= stadium IX	No. -5	300 - 180 (older than 155)	c. 40
X	= stadium X	No. -6	180 - 30 (before 1950)	c. 30 - 40
XI	= stadium XI	No. -7	30 - 0 (=1950)	c. 20
XII	= stadium XII = recent resp. present glacier stages	No. -8	+ 0 - + 30 (1950-1980)	c. 10 - 20

Draft: M. Kuhle

Glacier stadia of the mountains in High Asia, i.e. in and surrounding Tibet (Himalaya, Karakoram, E-Zagros and Hindukush, E-Pamir, Tien Shan with Kirgisen Shan and Bogdo Uul, Quilian Shan, Kuenlun with Animachin, Nganclong Kangri, Tanggula Shan, Bayan Har, Gandise Shan, Nyainquen Tanglha, Namche Bawar, Minya Gonka) from the pre-Last High Glacial (pre-LGM) to the present-day glacier margins and the pertinent sanders (glaciofluvial gravel fields and gravel field terraces) with their approximate age (after Kuhle 1974-2005). The author infers comparable data for older pleistocene/neogene glaciations as well. Older times with a lower elevation of the Tibet Plateau are inferred to have also reduced glaciations.

2.2.8. East and West of Nyainqentanglha

A very wide valley, which acted as a channel for the south/west discharge of the inland ice, is situated east of Nyainqentanglha Mountains (Fig.1, No.8, see Kuhle 1991a Fig.43, Nos.15–20; 2003 Fig.1, Nos.5–8).

This 15 km wide space is accordingly marked by glacigenically abraded valley flanks (Kuhle 1991a, Photos 22, 23, 24, 27, 29; 2003 Fig.1, Nos.6–8). The flanks (preceding photos 24 and 29) of the orographic cuspate areas have been scoured out up to 6000 m asl. The flank abrasion and polishing is interrupted by tributary valleys and cirques that are presently glaciated. This presents a similar picture to mountain and valley forms in Scandinavia which were also formed by inland glaciation. Since the valley floor lies at altitudes varying between 4200 and 4600 m asl, the truncated spurs are evidence of an inland ice stream thickness of 1400–1800 m. Erratic blocks of granite (30°18'N/90°36'E; Fig.1, No.8) have been deposited up to 5270 m asl (Kuhle 1991a, Photo 30, Fig.43, No.20; 2003 Fig.1, No.8), whilst the valley floor has ground moraine deposits with erratic granite blocks (30°25'N/90°56'E and 30°02'N/90°26'E) extending to several meters (Kuhle 1991a, Photos 27, 28 and 39; 2003 Fig.1, No.8). The valley–filling LGP glaciation received tributary ice flows from the Nam Co (lake) (Tengri Nor) in the north within 5200–5500 m high transfluence passes. Evidence of the tributary ice occurrs as classic U–shaped valley profiles (Kuhle 1991a, Photo 25, Fig.43 Nos.17; 2003 Fig.1, No.8). One example is a valley leading from the Nyainqentanglha Mountains south/eastern longitudinal valley in S/E direction down to the Tsangpo River (i.e. S/E is the direction of Lhasa) which has been preserved as a glacigenic trough–valley. Up to the junction with the main valley, its tributaries are also trough–shaped valleys, thereby providing evidence of former glacier thicknesses of more than 800 m. On the valley floor of the central confluence area there are classic roches moutonnées with flat scour–sides and steeper lee–sides at 4120 m asl, which are preserved in the granite (30°02'N/90°37'E) (Kuhle 1991a, Photos 36 and 37; 2003 Fig.1 Nos.8-9).

In the area of the Lhasa Valley (Fig.1, No.9) and extending as far west to the east–trending furrow of the Tsangpo Valley, flank abrasions and polishings, polished ledges and polishing lines are preserved in the adjoining tributary valleys. They are evidence of ice thicknesses exceeding 700-1000 m (Fig.3b; Kuhle 1991a, Photos 72, 73; 2003 Fig.1, No.9).

2.2.9. Southern Tibet, North of the Tsangpo Valley

Another example is provided from an observation in southern Tibet. At 29°41'N/90°12'E, north of the Tsangpo Valley, there is a 5300 m high pass known as the Tschü Tschü La (Chalamba La) (Kuhle 1991a Fig.43, No.27, Photos 42, 43; 2003 Fig.1, below No.8). There the Trans–Himalaya valley network opens out onto the central plateau of Tibet to the E (Fig.1 No.10).

Up to at least 200 m above the depression that forms the pass, lying on dark rhyolite bedrock with chlorite but without potassium feldspar, there are light–coloured tectonically marked (i.e. with broken and on shear faces faulted crystals) granite erratics with potassium feldspar components, but lacking chlorite (Kuhle 1988c Figs.4 and 5). Thus even at the pass the ice–thickness amounted to 200 m. The U-shaped valleys on both sides of the Tschü Tschü La pass reach a depth of about 1000 m below the pass. Ground moraine observed in the valleys indicate that they may have been filled with glacier ice well above the pass, up to at least 1200 m above the valley bottom. This suggests the existence of a network of ice streams. The nearest bedrock granite is known to occur 20-50 km further north/east. The direction of transport and the glacier direction, given by the valley system, point to numerous quasi–parallel outlet glaciers of substantial thickness. This is also suggested by roches moutonnées, which form hill– and mountain chains of several hundred metres height. Their rounded ridges rise up to 5600 m asl. Rough crest lines and peaks which begin above them, allow to estimate the thickness of the ice to 1500 m. The outlet glaciers left the Central Tibetan inland by way of an interposed section of the ice stream network interrupted by mountains of the Trans-Himalaya (Fig.3a, I2). At this longitude (about 89°E) they just missed the Tsangpo isobath at 3800 m (Kuhle 1985, 1987d). The final extremity of this particular outlet glacier ended at 4020 m asl within a side valley. It ends 6-10 km from the Tsangpo River. It was similar in the parallel Orio Machu valley (29°27'N/89°38'E; Fig.1, No.10; see Kuhle 1988c Fig.2, No.8, Fig.6).

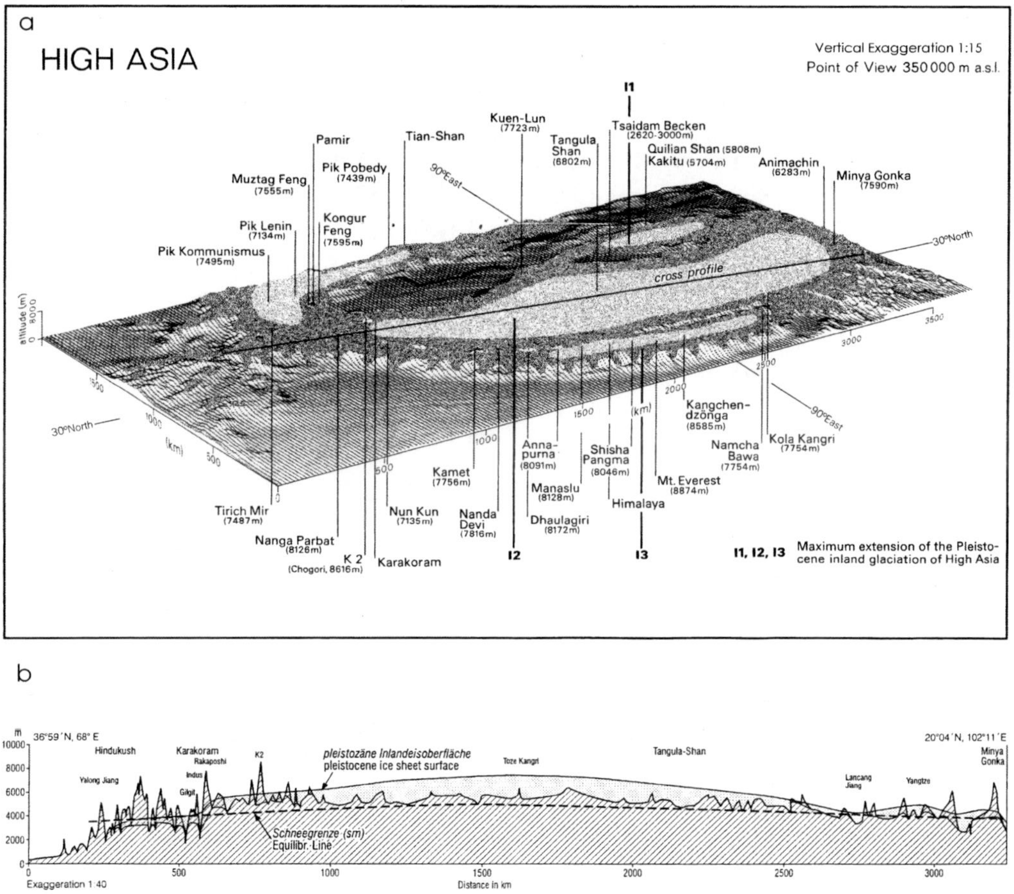

Figure 3: (a) The reconstructed 2.4 million km² ice sheet and ice stream network at the edges covering the Tibetan plateau (data from Kuhle, 1980, 1982a, 1982c, 1985, 1987a,b, 1988c,d, 1990d, 1991a,b, 1993, 1994, 1995b, 996a,b, 1997, 1998, 1998a, 1999, 2001, 2003, 2004, 2005, 2005a). The centers are marked I 1, I 2, I 3. Only peaks higher than 6000 m rise above the ice surface. See also Fig.1 (b) Cross section through the central ice sheet from Hindu Kush in the west to Minya Gonka in the east. Note that the ice–sheet is considerable north of the High Himalayas and the Mt. Everest. It is not a mountain glaciation but a large inland–ice in high altitudes and low latitudes. See also Pliocene temperatures and wind circulation shown in this volume.

Evidence of the ice stream extension may be found in the over 120 m high lateral moraines which coalesce into terminal moraine walls. From these a sandar or outwash cone emerges (Fig.1, No.10). The moraine deposits with outwash cone are located in that side valley that joins the main valley near Lhasa (29°43'N/91°04'E; Kuhle1988c Fig. 3). They are situated at 4250 and 3950 m asl. This implies an ELA depression of about 1075-1200 m in southern Tibet (Fig.1, No.9).

2.2.10. Lulu Valley

More to the south of Tibet, at 28°50'N/87°20'E is the Lulu Valley. It cuts deeply into hydrothermally decomposed basalt (50% pyroxenes, pseudomorphically replaced by dolomite and chlorite).

Between 4400 and 4950 m the valley floor is filled with diamictites, which extend 170 m up the valley flanks in some places. In the very fine groundmass there are isolated clasts of very coarse micaceous granite (Kuhle 1988c Figs.18-23). Transported over long distances from the north, dozens of meters thick, spreading over tens of kilometres, these diamictites are regarded as ground moraine (Kuhle 1988c Fig.2 Nos.11–16; 1991a, Photos 76, 77).

A convergence with a debrisflow is ruled out for geomorphological, sedimentological, petrographic and topographic reasons (details in Kuhle 1988c, 1991a). In addition there are no topographic conditions on the plateau for former ice-dammed lakes, and sudden outburst of them which may have resulted in debrisflows. In case of debrisflows the erratic granite blocks for instance, deposited on the orographic right valley flank 170 m above the thalweg would require (a) a 170 m high filling of the valley with debrisflow and then (b) its complete erosion with the exception of the erratic blocks. Finally the diamictite cover passes continuously into the extensive ground moraines of the plateau, which lie above 4950 m asl and are rich in blocks.

2.2.11. The Himalaya North Slopes (Cho Oyu–Gurlamandata, Fig.1, Nos.11, 17 up to 61) to Central- (Transhimalaya–Gangdise Shan–Nganclong Kangri, Fig.1, No.60) and West Tibet (Lingzi Tang-Aksai Chin) (Fig.1, No.52): More Evidence of a Tibetan Inland Ice During the LGP

In 1996 detailed geomorphological work, including several cross–sections through representative reliefs of Tibet from the Central Himalaya to the Kuenlun have been carried out (Kuhle 1999). These included the mapping of ice–margin positions. ELA depressions between decameters and ca. 100–250 m have been inferred. The inter– and extrapolated lowest ice margin positions allowed the reconstruction of pertinent depressions of the snow line which, owing to the altitude of the Tibetan plateau reached a maximum of 400–700 m. Accordingly the early Late Glacial (Stadia I to II; Table) and High Glacial glacier traces (Riss or pre–LGP and LGM (Stadium I and/or 0) occurred over a horizontal distance of 1620 km across the plateau with an average height of 4700 m asl without the characteristics of ice margin positions.

From this profile, running from the Cho Oyu (Central Himalaya) in the SE via Gertse (Kaitse; Central Tibet; Fig.1, No.60) up to the Lingzi Thang- and Aksai Chin and from there into the Kuenlun and also from a parallel section of the Gurla Mandhata (Fig.1, No.61) in the SW up to the at present very arid Nako Tso (Fig.1, No.59), located centrally in the W, 20 sediment samples have been analysed, which provide evidence for a ground moraine genesis and therefore confirm the macroscopic field observations (Kuhle 1999 Fig.2). Only the relatively very small basin of Ali might have been – like the Indus valley chamber of Leh – free of ice even during the (LGM). Forms of glacial horns as well as roches moutonnées, flank polishings, backward abrasion of mountain spurs at intermediate valley ridges, glacial streamlined hills, but also high–lying erratics prove the wide–spread ice cover (Kuhle 1999 Fig.2). Important thicknesses of ice have been recognized by means of transfluences. Especially by and in the Nako Tso (lake) the limnic undercutting of roches moutonnées shows the Post Glacial infilling of the lake.

Summing up, the glacio-geological and geomorphological observations concerning the Tibetan investigation area 1996 (Fig.1, from No.10 up to 52; Kuhle 1999 Figs.2 and 3), provide evidence of a glacier cover forming an inland ice, which marginally passed into outlet glaciers (Fig.3a, I3 between Shisha Pangma and Mt. Everest). The presentation of the cross section of the Tibetan margin north of Aksai Chin (Fig.1, No.52; Fig.3b left of Toze Kangri) explains the exponential increase of the glacier feeding areas as a result of the uplift of the plateau and the mountains on top of it above the ELA, i.e. as a result of the climatic drop of the ELA below the level of Tibet, or rather the relative movement of snow line and height of the plateau to each other. At a drop of the ELA from 5600 to 5000 m asl (i.e. a depression of 600 m) the feeding area already nearly triples and at a drop to 4400 m (i.e. a depression of 1200 m), the glacier feeding area even reaches the margin of the plateau. This figure is regardless of the self–increase, which happens during the glacier development, and, thus, further enlargement of the feeding areas. In addition to this, we have to consider the expansion of the glacier ablation area, lying below the snow line. This enlarges the glacier surfaces by a further third. This means that already at an ELA depression of 600 m nearly the total test area of 57.581 km² – in spite of its marginal plateau position, reaching far down – is covered with ice. With this in mind, one could perhaps get an idea how the partial uplift of Tibet above the snow line has led to a gradual expansion of the ice. Thus large–scale glaciogeomorphological findings become understandable (Kuhle 1982c; 1999).

Observations that lead to above conclusion are now described region by region. This enables future researchers to add new details, for example thicknesses of inland ice and outlet glaciers without redoing work. Readers who need only the results of this contribution can refer to the maps and tables and continue with section three, interpretation.

2.2.12. Tsangpo valley

In Central Tibet the most south eastern area under investigation is the Tsangpo Valley. The area from the valley to the 7751 m high Namche Bawar Massif (Namcha Bawa) in the Tsangpo knee was studied for traces of past glaciations (Fig.1, No.12, 13) (Kuhle 1991a Fig.43, Nos.36-45). In the area of the junction with the Nyang Qu (valley), near the Pula settlement, there is a ground moraine and a lateral moraine ledge (Kuhle 1991a, Photo 56; 29°27'N/94°02'E) on the orographic left flank of the Tsangpo Valley. The ledge reaches up to 250 m above the valley floor at 2950 m asl (Kuhle 1991a Fig.43, Nos.39, 40). The ledge of ground and lateral moraine can be followed down–valley for another 10 km, where it joins the lateral moraine of a tributary valley (Kuhle 1991a, Photo 57, 29°29'N/94°46'E). Reaching the Tsangpo Valley at 2950 m asl, this tributary valley moraine is evidence of an ELA depression of 900 m, using a mean height of the valley catchment area of 4750 m asl. During the LGP the Tsangpo Valley may had a far more substantial ice–filling than a mere 250 m from the valley floor (see 2.2.8–10 and Kuhle 1999). It also might have been glaciated during parts of the Late Glacial.

The following field observations support this: On the S side, that is the orographic right-hand side of the valley, 15 km up the Tsangpo Valley, there is an 80 m high exposure (Kuhle 1991a Fig.43, No.38), containing glaciolimnic sands, covered by 8 m thick varved clay (Kuhle 1991a, Photos 53-55, near the Ganga Bridge; 29°18'N/94°21'E). These are the sediments of an ice–dammed lake, which existed during the LGP. This lake existed up to the Latest Ice Age (Late Glacial) when the Nyang Qu glacier advanced again into the main valley, damming–up the Tsangpo Valley. Its dating is based on the large coniferous trunks found at the base of the sediments. The lowest sample at ca. 3060 m asl has a C14-age of 9820+/-350 Yrs.BP. (sample No.26.9.89/1; laboratory No.17654; cf. Kuhle 1998 Tab.2). This minimum age indicates that during this period the Nyang Qu Glacier flowed towards a boreal coniferous forest. Such a forest continues to grow there. This proves that a minor climatic change led to the glaciation of the Central Tibet Plateau (see below 2.3.18).

The historic glacier fluctuations in the Namche Bawar area (Fig.1, No.13) provide evidence of, possibly, extensive glaciation. In 1989 the western Namche Bawar Glacier terminated at 3900 m asl (Kuhle 1991a, Photo 68). In the 1950s its tongue advanced to the main valley at 2900 m asl, damming the Tsangpo River as a result of it. End moraines (Kuhle 1991a, Photos 70, 71) are found beyond the river. This advance extended 1000 m down the valley. Evidence of the event exists in the catastrophic discharge of the impounded lake, which destroyed a settlement further down the valley, killing more than 100 people. Further evidence of this event is the 30 years old birch found growing on the end moraine as well as the still existing dead ice complex in the moraine valley (Kuhle 1991a Fig.43, Nos.42-45). An earlier advance, down to less than 3000 m asl, created 200–450 m high lateral moraine terraces on the floor of the main valley. These terraces are (near to the surfaces) C14 dated at 4490+/-95 YBP (Kuhle 1998 Tab.2). In the lowest section of the Tsangpo bend, on the northern edge of the Namche Bawar Massif (Kuhle 1991a Fig.43, No.44) remnants of ground moraine (Kuhle 1991a, Photos 61, 62) have been preserved on the valley floor (2800 m asl, Kuhle 1991a, Photos 58, 59) beside roches moutonnées formed of bedrock gneiss. Possible LGP polished and abraded bands in trough valleys (Kuhle 1991a, Photos 64-66,69) extend up to 4800 m asl, and are evidence of glacier thicknesses of about 2000 m (Fig.1, No.13).

2.2.13 Mt. Everest, Lhotse and Makalu Area

The southern edge of Tibet, the area north of Mt. Everest, Lhotse and Makalu (Fig.1, Nos.14, 24, 22) shows many traces of glaciation by an ice stream net. This net (Fig.3a, I3) with large, continuous ice capped areas, represents the southern continuation of the inland ice area of Central Tibet (I2). It runs south of the Tsangpo Valley (see above section 2.2.8.9.11). The relevant observations concern the area south of the Panga La (5200 m, 28°30'N/87°06'E) up to the crest of the Himalayas (Kuhle 1988c Fig.30). There are (Fig.1, No.17) trough valleys (Kuhle 1988c Fig.2, No.20, Fig.29), which join the Dzakar Chu Valley (Kuhle 1988c Figs.53–59), which is a continuation of the Rongbuk Valley. On the valley flanks glacial polished bands are preserved on outcropping strata up to very great heights, i.e. 600–800 m above the thalweg (Kuhle 1991a, Photos 78, 79). In many places morainic materials with erratic quartzite blocks have been preserved (Kuhle 1988c Fig.28, 4400 m asl, 28°27'N/87°09'E). On the bottom of the Dzakar Chu Valley, decameter thick ground moraine deposits have been preserved (3950 m asl, 28°22'N/87°10'E, Kuhle 1991a Fig.43, No.55). Further down–valley, in the Kada (or Kharta) Valley chamber, lateral moraines, containing polymict erratic blocks, are preserved on outcropping phyllites on the orographic left- hand, that is the E valley slope at 4760 m asl, up to 1130 m above the valley floor, which reaches 3630 m asl (28°10'30"N/87°22'E, 4760 m, Kuhle 1991a Fig.43, No.56; Photos 84-86). The author interprets them as having formed during the Late Glacial times because during the Main Ice Age of LGP the glacier surface may have run above the ELA (above 4760 m asl). This could be the reason for the absence of LGP lateral moraines. This speculative interpretation is based on the following observations: in this valley cross profile, the glacigenically rounded mountain–forms, abrasions and polished bands extend up to 5680 m asl (Kuhle 1991a, Photos 82, 83 ,84 ,86 ,87). Evidence of a 2000 m thick main valley glacier is also suggested by the high scour marks of adjoining tributary valleys such as the Kada (or Kharta) Valley (Fig.1, No.14), for example (Kuhle 1991a Fig.43, Nos.57, 58 Photos 88–91). The Kharta Valley ice stream joined the Karma Valley ice stream of Mt Everest, Lhotse, Chomolönzo and Makalu via the 4890 m high Tsao La (Pass). Evidence of ice transfluence from southern Tibet in the north to the Himalayan south slope is provided by abraded and polished bands between 4900 and 5300 m asl and by an extended landscape of roches moutonnées in the vicinity of the pass (Fig.3a north of Mt. Everest; 28°00'30"N/87°16'E, Kuhle 1991a Fig.43, Nos.57, 58 Photos 92–94). With a characteristically steep surface gradient (dropping from 5200 to 3500 m asl) the Karma Valley and its tributary valleys had maximum glacier thicknesses of up to 1500 m (Kuhle 1991a Fig. 43, Nos.59–61, Photos 95–101, 104). The common source area of the Kharta and Karma Valley glaciers was also shared by the north and east of Mt Everest, Changtse and Gyachung Kang (Kuhle 1991a, Photos 102–105; Kuhle 1988c Figs.58, 59 and 64–74). Here the ice reached levels of 6300–6800 m asl, and transfluences took place from the upper Rongbuk Valley (upper Dzakar Chu) via the 6010 m high Lho La (Pass) (Kuhle 1988c Figs.66, 69, 74) into the Khumbu Valley of the Himalayan south side (see below section 2.3.2), as well as via the 6548 m high Rapui La (Pass) (Fig.3a near Mt. Everest; Kuhle 1988c Fig.2, Nos.40, 41, Figs.67, 68) and the 6084 m high Karpo La (Pass) (Kuhle 1991a, Photo 105). Like all the valleys leading out of South Tibet and down through the Himalayas the LGP glacier discharged into the 2000 m thick Arun outlet glacier (see above 2.2.12 and below 2.3.4). Evidence of ice confluence (Fig.1, No.14 and 20; 28°18'N/87°22'E) is found in kilometer–wide valley junctions (Kuhle 1991a Fig.43, No.55, Photos 81, 82 very left, 83, 86 left).

2.2.14. Shaksgam Valley, Latzu Massif, Menlungtse Group, Northern Shisha Pangma Foreland

Erratics found in the Shaksgam Valley on the western edge of Tibet (36°06'N/76°28'E, Fig.1, No.15) provide evidence of glaciation. Today this is the most arid part of High Asia with less than 40 mm of precipitation annually at 4000 m asl.

The area studied is a cross–section of the Muztagh Valley in the region of its confluence with the Karakorum north slope, to the north of the 8616 m high K2. The valley has been shaped into a glacial trough up to an altitude of 1200 m above the valley floor. At altitudes between 4400 and 4700 m gneiss and granite, as well as dolomite erratics (90% Do, 5% Ca, micritic and sparitic, Kuhle 1994 Figs.117, 118) have been found on roches moutonnées that consist of 90% pure calcite (Kuhle 1994 Figs.37, 38, 51, 73). The roches moutonnées are located 600 m above the valley floors on a pass that leads from the Shaksgam Valley to the Muztagh Valley. More than 1.5 m long, they occur both as single specimens and in the context of bands of lateral moraine material. Gneiss and granite erratics appear on the inner side of the Shaksgam Valley. Thus the blocks required transport along the valley at a high elevation in order to be deposited here. On the outward side of the valley to the west the pass is bounded by a 4730 m high glacial horn polished at its top (Kuhle 1994 Figs.52, 73). Up–valley and to the east of the saddle, the polished micritic calcite rises to more than 500 m above the floor of the pass. Its uppermost section is roughened. Substantial thicknesses of ice are deduced from softly polished roches moutonnées on the pass (Kuhle 1994 Fig.38). Ground polishing occurs at temperatures near the melting point as a result of increased ice pressure. Now, as well as during the glacial period, the mean annual temperature at the glacier surface was about –10 to –12°C at the equilibrium line. At the time of the LGP (LGM) the glacier surface in this area was in the accumulation zone and was located approximately 1000 m above the equilibrium line. This is an indication that the roches moutonnées within the pass are more likely to be part of the Late Glacial (e.g. younger than the LGM) since the ELA at this time was substantially higher.

Corresponding grooves can be found on the N side, that is the orographic right-hand side of the Shaksgam Trough (Kuhle 1994 Figs.69, 71) and on the 5466 m high "Shaksgam Horn" (Kuhle 1994 Fig.84) some 25 km up-valley and up to 1400 m above the gravel floor of the large West Tibetan longitudinal valley. This is the area in which a distributary stream of the Shaksgam Glacier system buried the 4863 m high Aghil Pass (36°11' N/ 76°36' E) under a 500 m thick cover (Kuhle 1994 Figs.86, 87, 116). It communicated with the Yarkand ice stream network (see below section 2.3.12.). A minimum elevation of the ice–surface is estimated for this point at ca. 5300 m asl. Glacial polishings in the massive limestones on the W, that is the orographic left-hand side of the Aghil Valley, and in the granite on the E, that is on the right, provide evidence of Aghil Valley glaciation (Kuhle 1994 Fig.90). Below 3700 m, well preserved striae in the quartzite bedrock of the Aghil-Surukwat Valley are encrusted with ferro-manganese (Kuhle 1994 Figs.40, 41, 93).

Striae have also been observed on sandstone outcrops at about 3600 m, and on roches moutonnées. Polished bands on metamorphosed schist outcrops were found at 3400–3600 m near Illik (36°23'N/76°42'E, Kuhle 1994 Fig.128). This is the trough-shaped confluence area of the upper Yarkand Valley. Many other forms of glacial polishing (Kuhle 1988e; 1988d; 1994a) are also observed in this area (Fig.1, No.16). This data provide evidence of a West Tibetan Karakorum – Aghil-Kuenlun ice stream network at the time of the LGP (LGM). The glaciers of the central longitudinal valleys (Shaksgam and Yarkand Valleys, Fig.3 near K2) were large outlet glaciers which flowed down from the western margin of the Tibetan Plateau ice dome.

In addition to the observations of erratics and polished bands mentioned above, roches moutonnée fields in Central Tibet are evidence of an extensive ice cover (Kuhle 2003). In North– Central Tibet (Kuhle 1982c; 1987b), south of the Kuenlun Pass (35°33'N/93°57'E, Fig.1, from No.1 up to 2), there are roches moutonnées consisting of metamorphic sandstone and crystalline bedrock schist at 4800–5350 m asl. In the South–Tibetan Latzu Massif (28°55'N/87°20'E; Fig.1, No.11; see also Kuhle 1988c Fig.2 No.16, Fig.24) roches moutonnées, consisting of basalt, occur at 5000–5500 m asl. 100-120 km further to the west, i.e. north of the Menlungtse Group (28°32'N/86°09'-25'E; Fig.1, No.17), roches moutonnées, consisting of metamorphic sediments, occur between 4400 and 5100 m asl (Kuhle 1988c Fig.2, No.30). Another 50 km to the west there are roches moutonnées (metamorphites with quartzite) in the northern Shisha Pangma foreland (28°37'N/85°49'E; Fig.1, No.18; Fig.3a: Shisha Pangma). The predominantly trough–shaped valleys of northern and southern Tibet, although pebble fillings frequently lend them the appearance of box profiles, are of an appropriate character to suggest the existence of an in places greater than 2000 m thick inland ice cover (Fig.3b).

2.3. Ice-Thickness, Lowest Position of Ice–Margins around the Tibetan Plateau and the Depression of the Equilibrium Line during the Last Glacial Period (LGP)

2.3.1. Dhaulagiri- and Annapurna Himalaya

Determined from the position of 46 glaciers, the present ELA on the southern edge of Tibet, south and north of the Dhaulagiri– and Annapurna Himalaya (Fig.1, No.26), lies at about 5550 m asl. It was calculated following the established methods of von Höfer (1879), Louis (1955) and Lichtenecker (1938), outlined also in Kuhle 1982a, 1983, 1988a. In particular, the ELA north of the main crest of the Himalaya is at 5620 m and south of it at 5490 m. Moraines of the LGM can be found on the south slope down to 1100 m in the Mayangdi (Khola) Valley (28°23'N/83°23'E) and the Thak (Khola) Valley (1060 m asl, 28°24'N/83°36'E) (Fig.1, from No.19 to 26; Fig.3a: between Dhaulagiri and Annapurna, see Kuhle 1980, 1982a, 1983; and below section 2.3.4). On the basis of five valley glaciers the equilibrium line during the LGM was calculated as being at 4060 m asl.

The ELA–depression is calculated as follows:

$$\text{ELA–depression (m)} = \text{tp-ti}/2 \text{ (m asl)} \quad (1)$$

$$\text{Si} = \text{Sp-ELA–depression (m asl)} \quad (2)$$

where

tp = recent terminus of the glacier tongue (m asl)
ti = past terminus of the glacier tongue (m asl)
Sp = recent equilibrium line (m asl)
Si = past equilibrium line (m asl)
S = equilibrium line = ELA (m asl)
ACF= average altitude of the crest fringe of the glacier (m asl)
ELA, S i.e. Sp or Si = ACF – tp (m asl)/2 + tp (m asl).

By means of the glacial position of ice margins, evident in 31 terminal moraines, it was possible to establish the Paleo equilibrium line north of the main crest as being at about 3980 m asl. The large number of glacier margins is explained by the very deeply incised transverse valley of the Thak Khola, together with a catchment area of the South Tibetan Mountains that exceeds 6000 m only slightly. The transverse valley cuts through at almost 2000 m asl. North of the main crest it has only reached 2700 m altitude, that is very far below the recent ELA (Sp = recent equilibrium line). It follows that today in this area (Fig.1, No.19) hanging glaciers and those from longitudinal valleys that formerly have been outlet glaciers of the South Tibetan inland–ice and ice–stream network (Fig.3a, I3) do not coalesce into one single tongue (Kuhle 1982a Fig.184). An integral equilibrium line for this section of the South

Tibetan Plateau edge during the LGM was at 4020 m. This is based on the depression of the equilibrium line by 1530 m (see Table). This calculation includes the altitude of the ice margin of the Jhong Khola Glacier (Muktinath Basin) during the Late Glacial Stadium at 3250 m (near Kingar: 28°49'N/ 83°50'E; see map in Kuhle 1982a Figs.184, 38). It is situated in a very dry valley north of the main Himalayan crest. It is evidence of an equilibrium line at about 4450 m, the result of a local equilibrium line depression of 1210 m (Kuhle 1980, 1982a). This moraine has also been observed by Iwata et al. (1982, p.87) and Yamanaka (1982/83). The C14 dated minimum age of the corresponding glacier basin is being 8760 +/- 210 YBP. This Late Glacial age and the equilibrium line depression brings it in line with other moraines, such as the terminal moraine at Ghasa ("Ghasa Stadium I" after Kuhle 1980, 1982a Figs.91, 92) in the Thak Khola transverse valley (Fig.1, No.19; Table).

Ice marginal positions and the extent of outlet glaciers flowing down from South Tibet are based on riverside moraine terraces that rise up to 570 m above the valley bottoms. They reach down to 1060 m asl on well–marked terminal moraines with erratics.

In addition there are trough profiles with glacier striae descending as far as 1700 m asl to the evergreen forest (with quercus semecarpifolia) and grooves as evidence of glacier thicknesses of 1600 m in the central Mayangdi Khola (Valley) area (28°08' N/83°23' E, Kuhle 1980, 1982a, 1983). It must be emphasized, that the depression of the ELA to 3980 m asl north of the main crest is more than 1000 m below the average level of the valley floors of the South Tibetan Mountains (Tibetan part of the Himalaya, Kuhle 1982a Fig.184). Even with a Last Glacial line altitude of 4450 m asl (see Table) there was still a difference of 600 m to the valley floors. This implies, necessarily, glaciation of the entire Tibetan Himalaya which overwhelmed the relief. The detailed hints to glacial striae, through valleys, moraines and references to comprehensive expedition–reports with photos underline this. The ice extended further north (Fig.3a, I3 north of Dhaulagiri and Annapurna). There it contributes to an inland ice.

In 2005 sheets of lodgement till containing erratics have been investigated in the upper Thak Khola transverse valley in Mustang (29°25'-28°53'N, 83°45'-84°15'E) (Fig.1, above No.19). They indicate flow of a 700-1000 m-thick outlet glacier over the 4661 m-high Kore La across the water divide from Tibet towards south. This outlet glacier received local influx from the two flanking mountain groups of the Sangda- and Damodar Himal (Tibetan Himalaya). It reached the glaciations of Dhaulagiri and Annapurna, so that a joint lowest ice margin position at less than 1100 (1060) m asl has been developed between the settlements Ranipauwa and Beni (cf. above). During the LGP this Mustang-Thak Khola outlet glacier coming down from the inland ice margin in S Tibet flowed on a very thick pedestal-groundmoräne. It filled the bottom of the Thak Khola between Dhaulagiri and Annapurna up to 3000 m asl, i.e. it was 400-600 m thick.

2.3.2. The Outlet Glaciers between Shisha Pangma-, Cho Oyu-, Lhotse- and Mt. Everest Massifs and the Former Bo Chu (Valley)- and Dhudh Khosi (Valley) Glacier Network and some Remarks on Interglacial Weathering (Figs.1 and 3)

Approximately 340 km further west on the northern side of Mt. Everest conditions are similar. In this area (28°–29°50'N/85°24'–91°13'E; Fig.1, No.14) investigations established a recent macro-climatic equilibrium line at almost 5900 m asl. This is the average of 15 values which are based on field work (Kuhle 1988c). The Chinese Map for snow lines and glacier equilibrium lines (Xie Zichu, 1984, 1:2000 000) indicates values between 5500 and 5900 m asl for this area. Evidence indicates that the maximum ELA depression in this area amounts to 1180 m (Kuhle 1988c). This is based on eight ice margin positions (see 2.3.3 below). The climatic equilibrium line, accordingly, ran during the LGP at 4700 m asl (Kuhle 1984/85, 1985, 1987a, 1988c).

This ELA is based on ice–margin positions of the Late Glacial, e.g. after reatreat from the LGM positions. The reason is that documents of the LGM positions have not been observed. The LGM ELA was in this area most likely lower. It is likely, that ice margin positions from the LGM can only be found near the terminal points of outlet glaciers which formerly flowed through the High Himalaya. The Bo Chu (or Sun Kosi) Glacier of the LGP (28°50'N/ 86°09'E) may serve as an example (Fig.1, No.18) for such a terminal point of an outlet glacier. Located on the Tibetan Plateau, this glacier was supplied from the north, east and south-west flanks of the Shisha Pangma massive, as well as from the Menlungtse massif. The lowest LGP ice–margin was below ca. 900 m asl. (Kuhle 1999). Concerning research - carried out from the lowest glacial ice margin position valley-upwards - resulted in the following observations: The Sun Kosi valley chamber (e.g. valley chamber of Lamosango, 27°40'–48'N/85°45' –55'E) into which the Bo Chu outlet glacier extended, is situated at a height of the thalweg of ca. 700 and 900 m asl. There are 3 to 4 m long, rounded erratic gneiss boulders (augen–gneiss). These occur as bedrock on the Himalaya main ridge and also on the Shisha Pangma massif (Kuhle 1988c, p. 483 Fig.43). They lie on bedrock schists and metamorphic siltstones. Some of them are in the riverbed. Potholes in this augengneiss–boulders can be observed (Fig.1, No.24). For their development the boulders might have been held by the hanging glacier ice and then flushed out through the subglacial meltwater, which flowed under high pressure what resulted in such kind of cavitation corrasion. On the valley floor the in this low altitudes usually observed red weathering is lacking.

The mainly 0,5 - 2 m deep weathering (soil type: laterit, ferrosol) was developed under monsoontropic warm humid climatic conditions with summerprecipitation. Thus, during various warm interglacials with temperatures 2.5–3°C above pre–industrial values as observed in the Vostok ice–core (Petit et al. 1999) the study area was characterized by evergreen vegetation and 2500-6000 mm precipitation per year.

It is evidence of an ELA at 3900 m asl during the LGP. This is nearly the same level as in the Dhaulagiri– and Annapurna areas. Sometimes these large boulders have been dislocated by glacier water (high energy flows) or mudflows. Here they are observed in the moraine–like matrix formation. Downslope of this valley chamber the lateritic red weathering from the monsoon tropic periods auf the Pleistocene, probably the interglacial periods, is observed on the valley bottom and on the debris slopes. (Kuhle 1999).

In the further E, in the Khumbu Himalaya, an ice stream network and valley glacier system has been reconstructed (Fig.1, S of No.14) for the LGP (Würmian, Last Ice Age, MIS 4-2, 60-18 KaBP, Table Stage 0) with glaciogeomorphological and sedimentological methods (field- and laboratory data; Kuhle 1987a; 2005 Figs.3, 11, 19). It was a part of the glacier system of the Himalaya and has communicated across transfluence passes with the neighbouring ice stream networks toward the W and E (Kuhle 2005 Fig.2 No.1, Fig.4) The ice stream network has also received inflow from the N, from a Tibetan ice stream network, by the Kyetrak-Nangpa-Bote Koshi Drangka in the W (Kuhle 1999 Figs.2 and 3), by the W-Rongbuk Glacier valley into the Ngozumpa Drangka, by the Central Rongbuk Glacier valley into the Khumbu Drangka (see below; Kuhle 1988c Fig.2) and by the Barun-Arun Glacier from the antecedent Arun Nadi transverse-valley in the E of this investigation area (see 2.3.4.;

Kuhle 1991a Fig.43). The ice thickness of the valley glacier sections, the surface of which was situated above the ELA, amounted to 1000-1450 m (Kuhle 2005 Figs.3 and 4). The most extended parent valley glaciers have measured approx. 70 km in length (Dudh Koshi Glacier). The Dudh Koshi parent glacier tongue was from the confluence of the Imja- and Nangpo Tsangpo Valley (Bote Koshi-) Glacier up to its terminal 38 km-long.

Heuberger (1986, p. 30ff) postulated the lowest past ice margin of the Dudh Koshi glacier at 1580 m asl below the Khari-Khola inflow.

The author's results show that the glacier terminal was situated near the Inkhu Khola (Valley) - junction in ca. 900 m asl (27°28'30"N/86°43'20"E; Kuhle 2005 Fig.4 and 11). This lowest ice margin position is an extrapolation from glaciogemorphological indicators, verifying in detail a decrease in ice thickness of the Dudh Koshi Glacier from ca. 1300 m, via 1200 m and 1000 m to ca. 850 m (Kuhle 2005 Fig.52-55) in the junction area of the Deku Khola (Valley) (Kuhle 1986, 1987a, 1988d, p. 587, 199; 2001, p. 389-391; 2005 Fig.4 Pro.31, Fig.11). At the Deku Khola (Valley) the altitude of the valley ground amounts to only ca. 1500 m asl. The ice thickness of 850 m has existed 12-15 km up-valley of the extrapolated lowest ice margin positions. Down to the Khari Khola (Valley) inflow, the Dudh Koshi Nadi (Valley) shows sections of a trough-like form (Kuhle 2005 Fig.52, 53, 54), i.e. of a trough-shaped glacial gorge. This is typical of steep Himalayan cross-valleys, created by glaciers with a strong, decakilometre-long subglacial meltwater erosion, which have flowed down far below the ELA (Kuhle 2005 Fig.11 on the left below No.73 up to Deku K.).

In addition near the Jubing settlement, situated in the confluence of Deku Khola (Valley) and Dudh Koshi Nad (Valley), the author has observed smooth denudation forms on the rock flanks, neither corresponding to linear erosion nor to crumblings. The interpretation is to regard them as glacigenic abrasion forms. Alternatively, as above–mentioned lateritic weathering, they formed during warm interglacials. Additional indicators to support this interpretation are the bedrock properties (phyllite).

The Dudh Koshi parent glacier tongue has received an inflow by two orographic right and three left, mostly short and steep tributary glaciers. A longer tributary glacier was the 18 km-extended Lumding Khola (Valley) Glacier, which as the lowest tributary stream has joined the parent glacier from the still glaciated right tributary valley of the same name (Kuhle 2005, Fig.11 on the right of No.16; Photo 131 below No.16). During the Late Glacial Dhampu Stage (III, Table), the tongue-end of the parent glacier has passed the Kyashar Khola(Valley) -junction by ca. 1.5 km and reached ca. 2750 m asl (Kuhle 2005 Fig.4 above Pro.28, cf. Fig.11: III on the right below No.17). During the Taglung Stage (II, Table) the end of the parent glacier was situated ca. 2 km down-valley from the Thado Koshi Khola (Valley)-junction about 2500 m asl (Kuhle 2005 Fig.4 in Pro.29; cf. Fig.11: II above Deku K.) and during the oldest Late Glacial Stage, the Ghasa Stage (I, Table), the tongue end has approx. reached the Lumding Khola (Valley) -junction at 1800 m asl (Kuhle 2005 Fig.4 in Pro.30; cf. Fig.11: I above Deku K.)

The tongue end of the Dudh Koshi Glacier has flowed down to ca. 900 m asl. At heights of the catchment areas of 8205 m (Cho Oyu massif), 8501 m (Lhotse massif), i.e. 8848 (or 8872) m (Mt. Everest, Sagarmatha, Chogolungma) this is a vertical distance of the Ice Age glaciation of 7300-8000 m. The steep faces towering up to 2000 m above the névé areas of the 6000-7000 m-high surfaces of the ice stream network were located 2000-5000 m above the ELA. Accordingly, their temperatures were so low, that their rock surfaces were free of flank ice and ice balconies. From the maximum past glacier extension up to the current glacier margins, 13 (altogether 14) glacier stages have been differentiated and in part 14C-dated (Kuhle 2005 Tabs. 2-4). They were four glacier stages of the late glacial period, three of the neoglacial period and six of the historical period. By means of 130 medium-sized valley glaciers the corresponding ELA-depressions have been calculated in comparison with the current courses of the orographic snow-line. The number of the glacier stages since the maximum glaciation approx. agrees with that e.g. in the Alps and the Rocky Mountains since the last glacial period. Accordingly, it is interpreted as an indication of the LGP (Würmian age) of the lowest ice margin positions. The current climatic, i.e. average glacier snow-line in the research area runs about 5500 m asl. The ELA–depression of the LGP calculated by four methods has run about 3870 m asl, so that an ELA-depression of ca. 1630 m has been determined. This corresponds to a lowering of the annual temperature by ca. 8, i.e. 10°C according to the specific humid conditions at that time.

Just like the Mayangdi, Thak Khola and Bo Chu Glaciers, the Dudh Koshi glacier was an outlet of a large inland ice glacier or ice stream network which descended from the Tibetan Plateau (Kuhle 1999). The Dudh Koshi Glacier was formed by the confluence of the Nangpa La (pass) (28°06'N/86°35'E; 5716 m asl) west of Cho Oyu massif, and the rerouted Central Rongbuk glacier, which had come down from the Lho La (pass) (28°00'N/86°53'E; 6010 m asl) west of Mt. Everest (see 2.2.11.). Glacial abrasion and undercutting below the west shoulder of Mt. Everest (Kuhle 1988c Figs.38, 66, 69) at an altitude of 6500 m provides evidence of overflows from the north side to the Himalaya south side. In the northern forefield of the present northward draining Rongbuk Glacier the gradient of the lateral moraines tilts, so that at an ice level of 900 m above the valley floor the gradient begins to face south. This 180° reorientation of the Rongbuk Glacier explains the absence of older endmoraines more than 8 km down–valley from the present glacier tongue. There, 8 km away from the recent terminal of the Rongbuk Glacier, the lowest endmoraine of the valley is situated near the Rongbuk Monastery. It can be classified as belonging to Neoglacial-Stages V–'VII (Table). During the LGM and the LGP the glacier drained to the south (Fig.3a, north of I3). The overflows onto the steep southern ramp of the Himalaya may also explain the slight thickness of the LGM and Late Ice Age Rongbuk Glacier. Its lateral moraines run about 600 m above the recent glacier surface (Kuhle 1988c Fig.58). Further thickening of the ice became impossible due to overflow onto the steep southern side, only 5 km away. In the southern part of the highland, with a mean altitude of 4800–5000 m and valley floors of 4200 m asl at the lowest, the reconstructed equilibrium line depression of at least 4720–4300 m asl led to a relief–filling glaciation.

Ice levels were determined by topographic proximity to the steep southern edge of Tibet. The result was an approximately 900–1200 m thick ice stream network.

Glacier filling of the Tibetan mountain landscape can be compared to the Rhone Valley during the LGP. The Rhone Valley, having been filled with 1800 m of ice, finds its valley floor to be 1500 m below the LGP equilibrium line (Fig.3a, north of I3). The ELA position was here 300 m below the ice–surface. - Here on the southern edge of Tibet the ELA ran even ca. 1300-1800 m below the reconstructed ice-surface.

2.3.3. The Lowest Ice Margin Positions of Tibetan Outlet Glaciers in the Kangchendzönga Massif (South/East Himalaya; Fig.1, No.20) and the ELA Depression of the LGP

The 8585m high Kangchendzönga Massif (Fig.3a) lies ca. 160 km E of Mt Everest, Lhotse and Makalu Massifs on the south–edge of Tibet. Tied to the feeding areas above 6000 m asl the present glaciers in the Kangchendzönga area no longer exeed 10–20 km (Kangchendzönga and Yalung Glaciers, Kuhle 1990 Fig.9). During the LGP the Ghunsa and the Simbua Valleys discharged the decakilometers long ice streams of the massif. They were connected with Tibetan Ice by way of the 6114 m high Jongsang La (pass). Evidence of their dimensions is provided by trough valley forms with scouring extending up to more than 500 m (Kuhle 1990 Figs.1-4). Both valleys, the Ghunsa– and the Simbua Valley, join the Tamur Valley leading down from South Tibet. Here, at a confluence of 1500 m asl near the Hellok settlement, the oldest Late Glacial terminal moraine (Kuhle 1990 Fig.5) of the Ghasa–Stadium (I, Table) is located. In the lower 4 km of the Simbua Khola, the Late Glacial lateral moraines of the Ghasa–Stadium I have been found, extending at the Hellok settlement into the Tamur main valley (Kuhle 1990, p. 418-421). During the field work in 1999, the author (Kuhle 2001, p. 391) has encountered a ground moraine pedestal in an increasing thickness in the upper to middle Simbua Khola, which has been multifariously modified and fluvially dissected since the deglaciation. Down–valley this pedestal peters out into the level mapped as lateral moraines (Kuhle 1990 Fig.9 (I) NE of Hellok). Accordingly, it is also a dissected ground moraine pedestal. In the orographic left flank of the Simbua Khola at 3400–3430 m asl, decameter–thick ground moraine deposits with rounded components and polymict boulders have been observed as far as the Lamite Bhanjyang pass (27°30'59"N/87°53'45"E; Kuhle 2001, Photo 196). Thus the Simbua Khola glacier has reached the Tamur parent glacier and, until the Late Glacial, was one of its tributary streams. It was at least 600 m-thick in its middle course. It thus overflowed the Lamite Bhanjyang (pass), situated just 400 m above the valley bottom, down into the SE–adjacent Kabeli Khola at a thickness of at least 200 m. Above observations show that the Simbua Khola glacier functioned as an overspill from a the Himalaya ice stream network into a Himalaya fore-chain–valley without a self–glaciation worth mentioning. It must have remained rather constant from the LGP (Stadium 0) as far as into the Late Glacial (Ghasa stadium I or even II).

The lowest glacial ice terminus (endmoraine, LGP, Stadium 0) at 890 m asl lies 610 m below that of the Ghasa–stadium (I). The horizontal distance is 19 km. No endmoraines occur in this section but traces of accumulation near the settlement Marijam (near Hellok, Kuhle 1990 Figs.5 and 9), only 350 m above the lower limit, suggest an intermediate terminal position between the settlements of Tapethok and Chirwa at 1240 m altitude. These are assumed to be part of the Pre–Ghasa ice stagnation (Kuhle 1982a, p.153 Fig.184 1/2).

The lowest past glacial terminal (glacier tongue position) is witnessed by much higher ice marginal forms lying above the settlements Tapethok and Kkejinim (Kuhle 1990 Fig.9). Clear evidence for this higher marginal forms are the slope abrasions and polishings on the orographic left slope and the "glacial mills" (potholes) 500 m up-valley from the settlement Chirwa and above the settlement Mitlung (Kuhle 1990 Fig.6). 1.5 km up-valley from the endmoraine at 890 m asl on the orographic left are striae and roches moutonnées up to 250 m above the valley bottom (Kuhle 1990 Fig.7).

The lowest past glacial terminal (LGP) occurs at the hanging bridge at Thuma (890 m asl; Fig.3a Kanchendzönga) and, in a well developed glacial tongue basin, includes moraines 40-

120 m high and classical earth pyramids (Kuhle 1990 Fig.8). There are abraded and polished features 1.2 km down-valley but there is no clear evidence of moraines. In view of the considerable altitude of the confluence region (6900 m) the ELA calculated from the ice terminal at Thuma of 3900 m asl represents an ELA-depression of 1660 m. The Tamur network of ice streams functioned as the outlet system for South-Tibetan ice, north of the main range of the Himalayas (see 2.3.1. and 2.3.2.) with glaciers, as in the Tamur Valley, reaching a thickness of more than 1000 m (cf. Kuhle 1982a, pp.57-63).

The present–day climatic ELA in the Kangchendzönga Group lies at 5560 (to 5540) m (N: 5720 m, S: 5650 m, W and S down to 5250 m asl), about 10–70 m higher than in the Central Himalayas (Dhaulagiri–Himal; Kuhle 1980, p. 246; 1982, p. 168). The difference in exposure between the southern and northern sides (i.e. between windward and leeward sides of the Himalayas) and southern Tibet results in the ELA difference of 70 m (i.e. 5650 m versus 5720 m asl). This windward/leeward effect can be recognized in the whole arc of the Central Himalayas. The depression of the equilibrium line of 1660 m to 3900 m asl, is the largest calculated depression found for the whole mountain system of the Tibetan Plateau and corresponds with glacial tongues at 890 m asl. It is interesting to compare this with the lowest LGP ice terminus below 870m asl, 1750 km to the west in the Indus Valley (Fig.1, No.21; Kuhle 1988d, p.588; 1988f, p. 606; 1996a, p. 153; 1997, p. 123 and 239). It is five to ten times drier in the Karakorum and Nanga Parbat Massif at the altitude of the Ice Age-ELA (3500-4000 m asl) than in the eastern Himalayas. In the Karakorum and Nanga Parbat Massif the monsoon is absent. Therefore the similar extent of glaciers is an indicator of the lack of summer–monsoon in the Himalayas, too, during the LGP and possibly also during older glacials.

If a lapse rate of 0.6°C/100 m is used, the observed ELA depression allows the reduction of summer temperature to be calculated as at least 9.6°C. The very probable reduction in summer precipitation clearly suggests an even greater cooling than this.

2.3.4. Extension of the Barun-Arun Glacier System (Khumbakarna Himal) and of Further Outlet Glaciers From South Tibet During the LGP

This section discusses the LGP (Stadium 0, Table) extent of further outlet glaciers (Fig.1, Nos.24, 22) which flowed down from South Tibet through the Central Himalayas (Fig.3a between Nanda Devi and Kangchendzönga), following the transverse valleys, as well as lowest ice margin positions of exemplary valleys of the Himalaya S-slope in the Kumbakarna Himal, Langtang-, Ghanes-, Rolwaling-, Manaslu-, Annapurna-, Dhaulagiri- and Kanjiroba Himalaya (Dolpo). These are the lowest LGP terminal positions of the Barun-Arun Nadi (Valley), Bhote Kosi (Valley), Tamba Kosi (Valley), Buri Gandaki (Valley), Marsyandi Khola (Valley), Madi Khola (Valley), Seti Khola (Valley), Modi Khola (Valley), Thak Khola (Valley), Mayandi Khola (Valley) and Barbung-Bheri Khola (Valley) (details in Kuhle 1980, 1982a, 1983, 1998, 1998a, 2001, 2004).

During the LGP a dendritic valley glacier system has joined in the Arun parent glacier (Fig.1, No.22). It has been fed by the S-Tibetan ice stream network (Kuhle 1991a Fig.43 Nos.56, 59) as well as by the Karma-, Barun- and Irkhuwa glaciers and, accordingly, has also been nourished by the High Himalaya (Fig.3a I3 between Everest and Kangchendzönga; details Kuhle 2005 Fig.4). This composition of the parent glacier resulted from the arrangement of the Arun Nadi as an antecedent Himalayan transverse valley leading down from the Tibetan Plateau. The Arun oulet glacier descending from there, i.e. from the valley chamber of the settlement Kada, was ca. 110 km long and flowed down to ca. 450 m asl up to the inflow of the Sankhuwa Khola (27°27'N/87°08'45"E; Kuhle 2005 Fig.11). The tributary streams and glaciers of the Arun parent glacier were the Barun- and Irkhuwa glacier (Kuhle 2005 Fig.4). They flowed down from the Khumbarkana Himalaya SE- to SSE-slope, from the Makalu- and Chamlang massif, up to the junction with the Arun parent glacier over a length of 61, i.e. 40 km, as far as 1100 and 700 m asl (Kuhle 1998a; 2005 Figs.4, 11). The whole length of the Barun glacier, including the fully 33 km-long tongue of the Arun parent glacier from the confluence of the two ice streams up to the lowest joint ice margin position at 450 m asl (see above), amounted to 94 km.

During the High Glacial (LGP maximum) the ice thicknesses of the Barun- and Irkhuwa glacier amounted to at least 1300 (Kuhle 2005 Figs.10, 12), i.e. 1400-1600 m according to the underlying thickness of the ground moraine and 1100 m (Kuhle 2005 Fig.14, cf. Fig.4). At low-lying valley floors of merely ca. 1100 and 720 m asl, the lower Arun parent glacier has even just reached 1100 (Kuhle 2005 Fig.15, cf. Fig.4) and 700-830 m (Kuhle 2005 Fig.16, Photo 47). In the feeding areas of the source branches of the Barun glacier, the Barun- i.e. Upper- and Lower Barun substream, the altitudes of the valley glacier levels lay at 6200-6450 m (Kuhle 2005 Figs.7-10), so that ice transfluences have taken place across 6070-6275 m-high passes into the N-adjacent Kangchung Nadi (Valley) or Karma Chu, into the W-adjacent Imja Khola (Valley) and into the E-adjacent Arun Nadi (Valley) (see 2.3.2.; Kuhle 2005 Figs. 3, 11). Due to additional transfluences from or into the Kharta valley via the Karma Chu (Valley) - and also via the upper Arun Nadi (Valley)- a further connection to the S-Tibetan ice stream network (Kuhle 1991a, p. 216) has existed in the N. There was also a most important, a fully 500 m-transfluence to the SW-adjacent Irkhuwa glacier across the Iswa La (Pass) (Kuhle 2005 Figs.3, 4, 11 No.47). Here, the joint Barun-Irkhuwa glacier level situated above this pass has dropped from the Barun glacier surface at ca. 5900 m asl down to the Irkhuwa glacier level at ca. 5300 m.

The current height of the snow-line in the catchment area of the Barun- and Irkhuwa glacier, being the source areas of the LGP Arun glacier in the Himalaya SSE-slope, amounts to 5450 m asl. This is a 200 m lower value than has been calculated for the S-side of the Kangchendzönga Himal situated 100 km further to the E (Kuhle 1990, p. 420). The calculation is based on the current orographic ELA of the S-exposed hanging glacier of the Chamlan group in the left Barun valley flank at 5300 m asl and of the Irkhuwa glacier in a SE-exposition at 5600 m asl.

The glacier tongue of the S-glacier of the Chamlang-group discussed here, ends at 4590 m asl. From this follows a difference in height of 4140 m to the lowest ice margin position of the Ice Age Barun-Arun parent glacier at 450 m asl (see above). Accordingly, a snow-line depression of 2070 m has existed (calculation of the ELA-depr.: 4590-450=4140; 4140:2=2070). As to the Irkhuwa glacier terminus at 4100 m asl the difference in height to the end of the parent glacier at 450 m asl is 3650 m. Thus, the ELA-depression was 1825 m (calculation of the ELA- depr.: 4100-450=3650; 3650:2=1825). For the S- to SE-exposition an ELA-depression of 1950 m can be calculated (2070+1825=3895; 3895:2=1947.5).

This is to say that the LGP snow-line of the Arun glacier system, i.e. in the relevant area of the Himalaya S- to SE-exposition, has run at ca. 3500 m asl (5450-1947.5=3502.5) - an ELA-value confirmed by the reconstructed LGP cirque glaciers and the corresponding altitude of the cirque level between 3300 and 3600 m asl (Kuhle 2005 Photo 26, Fig.11 on the right and half-right above No.46). The Kasuwa glacier, which in an extreme windward position is exposed to precipitation, testifies to a local Ice Age ELA at only just 3025 m asl.

The Arun parent glacier is at the same time an outlet glacier from the margin of the S-Tibetan ice stream network (Fig.3a, I3). Thus, the Himalaya lee-side with its snow-line about 4200-4300 m asl, has also been considered with regard to the joint lowest ice margin position at 450 m asl. An averaging of the orographic snow-lines in windward- and leeward positions yields an ELA at ca. 3640 m asl (4250-3025=1225; 1225:2=612.5; 3025+612.5=3637.5). A favourable factor, which despite this snow-line - which against the 3500 m-ELA (see above) calculated for the Himalaya SSE-slope was 140 m-higher - enabled the Arun outlet- and parent glacier to extend down to 450 m asl, is the ca. 1300-2000 m-thickness of the LGP ice stream network on the S-margin of Tibet. This feed-back self-heightening of the cold-based ice stream network due to the flat gradient of discharge, must have been the cause for a secondary heightening and extension of the feeding areas.

As for the lowest ice margin position at 450 m asl (see above) the LGP snow-line (ELA) of the Arun glacier established about 3500 m asl (see above) corresponds to a medium height of the glacier feeding area of 6550 m (3500-450=3050; 3500+3050=6550). This applies approximately to the catchment area built up from S-Tibet and the High Himalaya which here rises up to 8481 m asl (Makalu Massif). Comparable heights of catchment areas made up of N- and S-slope have also been calculated for the Dhaulagiri- and Annapurna Himalaya (see 2.3.1.; Kuhle 1982a, p.150-152) and the Mt. Everest- and Shisha Pangma Massifs (see 2.3.2.; Kuhle 1986, p.443-452, Tab.3; 1988c, p.468-470; 2005).

In the Kangchendzönga massif 100 km further to the E, the reconstructed High Glacial ELA ran at about 3900 m asl, i.e. ca. 400 m-higher than in this investigation area (see 2.3.3.) According to the current state of knowledge the reconstructed snow-line in the Kangchendzönga Massif amounted to ca. 1660 m (Kuhle 1990, p. 420), i.e. 290 m less than the ELA-depression of 1950 m calculated for this research area.

The LGP ice–margin positions of the Langtang-Himal SW- and Ganesh Himal SE-slope and in the S-slope of the Menlungtse-group, Rolwaling-Himal and of the Manaslu Himalaya (Fig.1, No.25) are described as follows: Langtang- Ganesh-Himal: On the W-flank of the Bhote Kosi (Valley) as far down as its talweg (Trisuli river), decametres–thick ground moraines have been preserved (Kuhle 2001 Photo 192). The lowest occurrences of ground moraine have been met at ca. 900–1000 m asl near the Donga settlement. It proves that the terminus of the LGP Langtang glacier tongue has reached the junction of the Mailung Khola (Valley) (28°05'N/85°13'20"E). This valley glacier tongue was a joint outlet glacier tongue of the connected Langtang- and Ganesh Himal ice stream network (Kuhle 2001).

Rolwaling-Himal: The LGP–glacier had in the Tamba Kosi, in the cross-profile near the location of the Jagat settlement, a minimum thickness of 800 m (Kuhle 2001, Photo 193). Down-valley at 900 m asl, a kame near the Malepu settlement provides evidence of an ice thickness of still 200 m (Kuhle 2001, Photo 194). Thus the glacier tongue end might have been situated 4 km down–valley at 860 m asl (27°38'N/86°06'E; at the start of the valley narrow SW below the Marbu settlement; Kuhle 2001).

Manaslu-Himal: The Buri Gandaki outlet glacier also discharged the S–Tibetan ice stream network (Fig.3a, I3 near Manaslu). It passes the Himalaya E of Manaslu (Fig.3a). Its glacio– geomorphological clear traces are completely mapped from the Chinese border in S–Tibet (N of the Himalaya) up to the southern Himalaya foothills. "Glacier mills" or potholes occur on the glacially polished valley flanks at 200–300 m above the thalweg, i.e. without being bound to a run–off rill of the slope. The same applies to flank abrasion and roches moutonnées. The glacier flowed down to at least 680 m asl at 28°08'N/84°51'E (Fig.1, No.24; Kuhle 1998a).

The westward adjacent Marsyandi outlet glacier flowed down from S-Tibet (Fig.1 between Nos.19 and 23) through the Himalaya Range between the massifs of Manaslu– and Annapurna (Kuhle 1980, 1982a, 1983). Cross–sections of trough valleys, glacially shaped horns, flank abrasion and polishing provide evidence of the glacier's discharge during the LGP. Overall this is proven by a very extended landscape of ground–, lateral– and endmoraines with large to very large erratic augen–gneiss and granite blocks, reaching up to the settlement of Dumre at the Himalaya foothills (28°07'20"N/84°26'E, Kuhle 1998a). The lower course of this young outlet glacier is very well set off against the deep tropical red weathering (ferraltic or laterite weathering) of the adjacent low lying areas. The glacier tongue terminal reached down to 460 m asl. (Fig.1, No.25). That is the lowest LGP glacier margin position the author was able to observe on the edge of High Asia.

Now the lowest LGM ice margin positions of the Annapurna Himalaya (Fig.3a) are discussed (details: Kuhle 1982a).

In the Madi Khola (Valley), located to the W of Marsyandi Khola (Valley), another lowest glacier end was reconstructed. More than 2 km long lateral moraine on the E flank and large erratic gneiss blocks on phyllite bedrocks on the W flank are found down to ca. 630 m asl (28°12'20"N/84°05'20"E, Fig.1 No.26; Kuhle 1998, p.87; 2001, Photo 191) in the Madi Khola (Valley). The glaciers that formed these flowed down from the Annapurna IV, II and Lamjung Himal,

The LGP Seti Khola glacier has been mapped as being uncertain (details in Kuhle 1982a). Based on two up to 3 m-long erratic gneiss boulders on a rock head of schist bedrock W of the Ghachok settlement at 1500–1540 m asl (Kuhle 2001, pp. 383/384, Photo 190), 350-390 m above the current talweg, its LGP terminal position can be extrapolated at ca. 1000 m asl. At an ice thickness of ca. 300–400 m and a glacier width of 2.5–3 km at Ghachok, the LGP Seti Khola glacier has most probably reached the junction of the Seti– and Yamdi Khola situated 6 km further down–valley (28°16'20"N/83°57'30"E). The erratic boulders derive from the Himalaya main crest (Annapurna III–IV). According to the arrangement of their positions they can neither be explained by mudflows (or related phenomena) nor by a Late Glacial glacial extent (Kuhle 2001, 2004).

The tongue of the Modi Khola glacier received an influx from the Chomrong Khola- and Kyumnu Khola glacier during the LGP (Stage 0) and flowed down as far as ca. 800 m asl up to the Dobila locality at the junction of the Jare Khola (Valley) (28°13'50"N/83°43'E; Kuhle 2001, Photo 188). The trough valley cross-profile of the lower Modi Khola (Valley) reaches up to there, as does an orographic right kame-terrace. Down-valley the glacier mouth gravel floor terraces of Chuwa (Kusma) set in (Kuhle 1982a Bd. I, p. 71/72, Bd. II, Abb. 7 and p. 105-108; Kuhle 198, .p 193-196). Among others, the following observations testify to this past glacier extension: ground moraines up to at least 2400 m asl at the spur between the Chomrong- and Modi Khola (Valley) and on the orographic right side at the exit of the junction area of the Kyumnu Khola up to 2000 m asl (near settlement Udi); prehistoric subglacial potholes on the orographic left between settlement Landrung and Tolka at a height of 300-500 m above the current talweg and an orographic right ground moraine cover at the inflow of the Bhurungdi Khola at ca. 1450 m asl. Accordingly, the Modi Khola glacier near

Birethanti (see Kuhle 1982a, Bd. II Abb.8,d) still had a thickness of approximately 400 m (Kuhle 2001, Photo 189; 2004).

The South-Tibetan Ice Age glacier areas of Dolpo and Kanjiroba (Fig.1, No.53) were located WNW of the Thak Khola and Mayangdi Khola outlet glaciers (mentioned above), which flowed down to at least 1100 m asl (Fig.1, from No.19 up to 26; Kuhle 1982a Fig.8h and e). During the LGP the Barbung–Bheri Khola outlet glacier discharged these areas. Preserved in exposed positions, fresh ground moraines and large erratic granite blocks at the settlement of Tripurakot (29°02'N/82°46'E) suggest a minimum thickness of this glacier of still 400 m at 1900 m asl (Fig.1, No.27; Kuhle 1998a). The lowest glacier end was located down–valley in the Bheri Gorge. It could not be visited.

2.3.5. The Ice Age Glaciation of the Garhwal (Chamoli) Himalaya

In this marginal area of southern Tibet investigations focussed on the valley system of the Nanda Devi–Kamet Group (Fig.1, No.28), which converges in the Alaknanda Nala (Valley). The lowest ice margin position in this main valley is at Pipalkoti (30°34'N/79°25'E; Kuhle 1997, p. 127) at 1000–1100m asl. The lowest glacier tongue of the Nanda Devi–Kamet ice stream net, which received supplies from inland ice outlet glaciers from southern Tibet via the ca. 5000 m high passes of the Tun Jun La (Marphi La) and the Shashal La, ended near this village in the Alaknanda Valley (Fig.3a: I3 between Kamet and Nanda Devi). Branching out into hundreds of kilometres of valley glacier, this ice stream net had preserved all these lowest end– and lateral moraine exposures. The mean altitude of its catchment area is at about 6000 m asl. The Ice Age ELA on this precipitation–rich SSW edge of Tibet is calculated to be at ca. 3500–3600 m asl (implying an ELA depression of ca. 1400 m). In the Gohna Nala (Valley), leading down from the 6300 m high Nanda Ghunti (western Trisul Massif), LGP (Table, Stadium 0 or I) moraines of an ice marginal position were found at 1800m asl. These moraines confirm a climatic snow line at about 3600 m asl

The calculation of the ELA is explained above. Using above principle: Mean altitude of the catchment area: 5400 m minus altitude of the observed ice–margin (1800 m) is 3600, divided by 2 is 1800 plus 1800 (altitude of the observed ice-margin): The result is 3600 m asl.

This implies that tectonic uplift that might have taken place in the last 18000 years or 2.75 Ma is not accounted for. If the tectonic uplift of 10-25 mm/a that is observed today, applied also to the LGP, the value would be ca. 200-500 m lower. The same applies in a modified form to the end of the preceding glacial.

In order to confirm this value, the Nandakini Nala (Valley) on the Trisul south-western slope was investigated. The present Trisul South-Glacier extends down to 3500 m asl. The mean altitude of its catchment area is 6500 m asl, indicating an ELA at about 5000 m asl:(6500-3500)/2 = 1500+3500 = 5000. The lowest former ice margin position in the valley, evidenced by polished rock surfaces, lateral and ground moraines with erratic blocks, was at 1200–1400 m asl (near the settlement Khunana: 30°16’45’’N, 79°24’20’’E). It represents a snow line depression in the Nandakini Valley of ca. 1050 m - 1150 m to 3850 m asl (Kuhle 1998, 1998a, 2004).

During the LGP, in the western parallel valley of the Alaknanda Nala (Valley), the Mandakini Nala (Valley), in the S-slope of the 6940 m-high Kedarnath massif, the dendritic Mandakini valley glacier network reached down to 1100 m asl up to the Okhimath settlement (30°31'N/79°06'E). This could be verified by geomorphologic field investigations and sedimentological analyses in 2004 (Kuhle 2004, Map 1:1 Mio; 2005 Fig.2).

In the main valley, the Bhagirathi Nala, situated still further west, the LGP Bhagirathi main glacier tongue of the ice stream network west of the 7138 m-high Chaukhamba massif, as the continuation of the currently 31 km-long Gangotri glacier, has reached down to ca. 1050 m asl. As indicated by moraine material and abruptly starting large gravel terraces, this glacier tongue end has reached the Slalam Gad (tributary valley) (30°44'N/78°24'E; Fig.1 No.28) 3 km down-valley of the city of Uttarkashi (Kuhle 2004, Map 1:1 Mio; 2005 Fig.2).

2.3.6. Ladakh Range and Zanskar Himalaya (Upper Indus Valley), Lahaul valley and SE Pir Panjal Range (Fig. No.54)

There are two parallel valleys on the south slope of the Ladakh Range (Fig.1, No.29), the Phyang Valley and the Leh (Puchu Chu) Valley. In both of them well preserved end moraines extend down to 3400 m asl. They are regarded by relative dating as being remnants of the Late Glacial (Table, Stadium I). These ice margin positions are evidence of a substantial snow line depression of ca. 1200 m. Assuming the mean altitude of the catchment area to be 5300 m asl, the altitude of the former snow line is calculated to be 4350 m asl. Ground moraine and roches moutonnées, outside and below these tongue basins, are evidence of LGP glaciation in the upper Indus Valley (altitude at more than 3000 m asl). Scouring bands prove, that in the junction area of the Taglang La North-Valley, which runs down from the 5100 m high Taglang La (Pass) on the 6401 m high Ruberung Massif, the valley floor of the Indus Valley had been covered by glacier ice approximately 1000 m thick (see below 2.3.9; 2.3.10).

On the flanks of the Taglang La North–Valley lateral moraines with erratic blocks on outcropping crystalline slates (33°40'N/77°43'E) extend to above 4000 m asl, thus confirming corresponding ice thicknesses for the Indus tributary valleys.

Considering the glacially abraded flanks and truncated spurs above the moraines, this ice thickness might have been still 400–600 m more, i.e. the polished slopes were lying above the pertinent snow line. Ground moraines, up to decameter thicknesses, occur in the upper Indus Valley between the Nimu (mouth of the Tara Phu (Valley)) and Khalsi settlements (34°08'–20'N/77°25'– 76°50'E). There are up to 900 m high abraded and polished mountain forms (roches moutonnées). Up and down valley from the Khalsi settlement large erratic blocks of granite on outcropping metamorphites are widespread. In many places they appear to have been washed out from the ground moraine matrix, possibly, by the former (Holocene) Indus River. These observations are evidence that the upper Indus Valley between Upshi and Khalsi had been filled by a valley glacier that was the main branch of the Ladakh Zansgar ice stream net (see below in addition 2.3.9 and 2.3.10).

Research in the western Zanskar Himalaya, in the Nun Kun Group (Fig.3a; 33°59'N/76°01'E), yielded provided data about a very large ice stream during the LGP. The entire Nun Kun Group, the highest peak of which reaches 7315 m, was filled with glaciers up to a level of at least 4200 m asl, as shown by granite erratics on slates. The floor of the Karcha or Suru Valley has roche moutonnée forms at Parkachik that lie between 3200–3400 m asl (Fig.1, No.30; details Kuhle 1998, p. 89, Fig.10). In the Kargil Basin, 70 km away from the Nun Kun Massif, ground and lateral moraines, as high as 3150–3200 m asl, provide evidence of LGP glacier thicknesses of at least 600 m. A NNW facing corrie, with end moraines at 3700 m asl, is evidence of a snow line at 3900–4100 m asl.

Between the Ruberung Massif and the Takh settlement (North–Lahoul), 100 km further south (Fig.1 between No.29 and 31), there is an elevated area with peaks of 6000–6632 m and valley floors at about 4400–4700 m asl. According to the character of its landscape, this area is part of Western Tibet (at 33°N/78°E). This Western Tibetan area has a network of trough valleys, separated from one another by flat transfluence passes. These valleys tend to be without valley heads (i.e. forming a widely–branching network of merging trough valleys) similar to those found in Southern Scandinavia (Jotunheimen). This is characteristic of a high valley landscape scoured by inland ice. There the ice–filled valley–network is comparable. The mountains themselves pierced well above the inland–ice.

Further south, the mountain groups known as Lahaul (Lahoul) or Pir Panjal, up to the Manali settlement in the Kullu Valley or Solang Nala had also been filled during the LGP (Fig.1, No.31). The 3980 m-high Rothang (Jot) pass (32°21'45"N/77°14'50"E) was a transfluence pass over which the Lahaul glacier (Chandra glacier) has flowed into the Kullu valley adjacent to its S (Kuhle 2001, pp.381/382, Photo 185). This is evidenced by glacigenic abrasion forms like trough valley profiles with steep flanks, polished bands, transfluence passes and roches moutonnées and their ground moraine covers. Due to this transfluence the level of the Lahaul glacier has run at a height just about 4300–4400 m. Its ice thickness has amounted here to ca. 1100–1200 m (Kuhle 1998, 2001).

The two parallel valleys which from the 4971 m-high SE Pir Panjal massif NE of Dharamsala peter out into the Himalaya foreland to the SSW, have been glaciated down to the foreland. Their lowest ice margin positions, recognizable by well–preserved lateral– and end moraines, can be considered as belonging to the LGM. They are situated below the exits of the Tori valley at 1250 m (32°12'30"N/76°22'30"E) and the Triund valley (32°12'50"N/76°21'10"E) at 1200 m asl (Kuhle 2001, pp. 383/384, Photos 186, 187).

2.3.7. Extent of Glaciers and Snow Line Depression in the North-Western Karakorum (Fig.1, No.55; Photo 1 and 2)

The Shimshal Pass (4600 m asl; 36°26'N/75°41'E) is a broad saddle in the Ghujerab Mountains north of the Karakorum main crest. The pass leads from the north slope (Shaksgam Valley; see also 2.3.10) to the south slope via the Shimshal and Hunza Valleys, and down to the Indus Valley (also 2.3.9 and 2.3.10; details Kuhle 1988d). It was a transfluence pass, with a minimum ice thickness flow of 250–400 m (Fig.1, between No.55 and 32). Evidence of this is found as elongated moraines SSE of the pass, containing blocks of granite, limestone and metamorphites which extend up to 4850 m asl. In some places the moraines interfinger with glacio–fluvial drift, implying that at the time of deposition the ELA must have been above 4850 m asl. The ELA is now running at about 5200–5300 m asl, indicating that transfluence over the Shimshal Pass is part of a snow line depression of, at most, 500 m (350–450 m). Only based on the ELA depression, it is tentatively dated as Late Glacial (Table, Stadium IV). In the LGP, however, the snow line was 1100–1300 m lower (see 2.3.8, 2.3.9, 2.3.10) than at present. Thus during the LGM (Table, Stadium 0) transfluence ice thickness of the Karakorum north– and south slopes must have been several hundred metres more than 250–400 m (Fig.3b Karakoram). Ground moraine is observed in the Shimshal Pass. Twenty–seven kilometres down–valley towards the west, on the N side near the junction with the Shimshal Valley, there are erratic blocks of granite on a limestone mountain spur at 4300–4450 m asl (Chatmerk Pass, 36°28'N/75°26'E), about 1300–1400 m above the floor of the Shimshal Valley. The parent rock of the erratics outcrops approximately 20 km up–valley in the area of the Shimshal Pass. Evidence of ice thickness in the Shimshal Valley is observed as abraded valley flanks and truncated spurs on the W side of the Shimshal Valley. However, this relatively large ice thickness may be regarded as Late

Glacial, since the erratics observed must have been deposited below the snow line. The Late Glacial ELA occurred at more than 4450 m asl and, most likely 4450m asl, not more than 750–850 m below the present ELA. Preserved 1400–1600 above the valley bottom, a ledge of lateral moraine remains on the E side in the Shimshal Valley, down-valley from the camp Ziarat, at 4100-4300 m asl (36°32'N/75°04'E). Thus the maximum ELA depression is 700–900 m (present ELA level at this location: 5000 m asl) to allow for moraine deposition below the snow line. This places the ca. 1400–1600 m thick Shimshal Glacier, as deduced from above–mentioned moraines, tentatively into Late Glacial times (Table, Stadium II-III). Quartzite roches moutonnées on the dividing ridge between the Lupghar and Momhil Valleys (4300m asl; 36°28'N/75°02'30"E) infer an equivalent or even more substantial LGP thickness in the valleys. The longitudinal axis of the roches moutonnées, transverse to the mountain ridge, is evidence of the transfluence of the glacier over the ridge from east to west with an ice thickness of several hundred meters. The occurrence of glacially abraded valley flanks, truncated spurs and ice scour limits, indicates an LGP (LGM) surface at 4500–4800 m asl. The considerable ice thickness in the lower Shimshal Valley area matches interpretations made in the Hunza Valley. Schneider (1959, p.209) observed erratics on the Shanoz Ridge at a height of 4000–4150m asl (36°35'N/74°41'E). The erratics occur south of the Batura Glacier (1000 m above the present–day surface), and 1500 m above the floor of the Hunza Valley.

West of the junction of the Shimshal Valley with the Hunza Valley (36°28'30"N/74°00'50"E) (Fig.1, between No.55 and 33; Kuhle 1988d) large polymict erratic blocks – such as gneiss, quartzite, porphyries and granite – were found on a round-polished limestone (calcite and marl) mountain near its peak at 3530 m asl (details Kuhle 1998, p. 92, Fig.12). Adjacent to the erratics are relicts of fine–grained moraines that provide evidence of a young, probably Late Glacial (Table) age ice stream. These moraines and erratics are evidence of a Hunza Glacier of more than 1000m thickness and a glacier surface at 3500m asl. As erratics indicate minimum glacier thickness in the area below the snow line, it is assumed that in the area of the former Hunza Glacier net with a glacier surface at around 4250–4500 m asl, the LGP glacier thickness was at least 1750–2000 m (see below 2.3.8–2.3.10). The Hunza ice stream joined the Gilgit Glacier and contributed to the supply of the Indus Glacier enabling it to reach its lowest ice margin position down valley of the Sazin settlement at below 870 m asl (Fig.1, No.21; Kuhle 1988d, p.588; 1988f, p. 606; 1996a, p.153; 1997, p.123 and 239). In discordance with the here presented results Haserodt (1989) and Derbyshire et al. (1984) have recorded the Late Glacial ice margins of the Hunza Glacier (see below 2.3.9). Thus according to Haserodt and Derbyshire et al. the Hunza glacier was much shorter than the here summarized evidences prove. As no doubts about the observations of both the author and Haserodt (1989) and Derbyshire et al. (1984) exists, the following interpretation intergares both observations: their ice margin positions belong to the Late Glacial.

Indications of a thick ice stream in the upper Hunza Valley were found in the intramontane basin of Sost, in the Chapursan Valley (Photo 1 and 2) and further north/east as far as the Kunjerab Pass (see below 2.3.10 and 2.3.13 and Kuhle 1988d).

2.3.8. Reconstruction of the LGM Glaciation in the Nanga Parbat Massif (35°05'-40'N/74°20'-75°E, Fig.1, No.34; Fig.3a)

Details about this massif can be found in Kuhle 1988d; 1996a; 1997, Fig.28; 2001. Ground moraines and polished flanks are preserved on both sides of the upper Rupal Gah

(Valley) above the Toshain Glacier at 4000–4670 m asl (35°10'N/74°27'-30'E). A mountain spur, leading down to the south from the Mazeno crest, has been eroded back to 4680 (or 4700) m by the Ice Age Rupal Glacier. It got the shape of a truncated spur (35°10'40"N/74°32'20"E). Late Glacial lateral moraines of Stadia III–IV (Table), which are preserved in this valley, provide evidence of an ELA depression of 700–800 m, i.e. to ca. 4400–4500 m asl. On the flanks of the Rupal Valley, in the section between Shaigiri– and Bazhin Glacier (Kuhle 1996a Fig.1; 1997 Fig.28), former glacial flank polish margins and moraine remnants with large blocks (35°11'N/74°34'35"E; 35°12'N/74°35'30"-74°37'E; 35°12'40"N/74°37'10"E) have been observed. They prove Rupal Glacier levels between 4460 m and 4680 m (Kuhle 1997 Fig.28, No.5–8). Ground moraines and traces of glacier scouring, reaching still further up, are proof of Rupal Glacier levels between 4550 and 4650 m asl between the junction of Bazhin– and Chhungphar Glacier (ibid. Fig. 28, No.9–11). Evidences also have been mapped on the SE side of the Rupal valley flank (ibid. Fig.28, No.19–21). The down–valley joining Chhungphar Gah shows very well preserved glacially polished flanks up to 4600 m asl, so for instance at 35°16'39"N/74°44'40"E, below the Sharsingi Peak (ibid. Fig.28, No.26). This provides evidence of an ice thickness of more than 1400 m (Kuhle 1996a Fig.2; 1997). Further down to 3700 m asl, there are preserved meter– to decameter thick ground moraines. This is also true of the confluence with the Rupal Gah (Valley) (ibid. Fig.28, Nos.27, 28). An E side lateral moraine remnant (ibid. Fig.28, No.29) is located far below of the earlier polish margins of the LGM (see above). Due to its altitude relative to the ELA it probably can be classified as belonging to the Late Glacial Taglung Stadium II (Table).

The described observations in the Rupal Gah (Valley), provide evidence of the Ice Age glaciation of the Nanga Parbat SE face (details see ibid. Fig.28). The observations show, that the Rupal Glacier joined the Astor Valley with increasing thickness. The ice flow directions of the tributary glaciers of the Astor ice stream net, which fringed the Nanga Parbat to the E, are also recorded (Kuhle 1988d, 1997, 2001). We now omit a large number of observations and turn to the junction of the Astor Valley with the largest main valley, the Indus Valley (ibid. Fig.28 Nos.57-64). In the Astor Valley at the junction with the Indus Valley polished flanks are preserved on the NNE side at 2900 m asl (35°33'30"N/74°43'E). Below that altitude there are in places concrete–like remnants of ground moraine (ibid Fig.28 No.58). From here, down to the Indus Valley, these ground moraines cover the NNE lower slopes extensively (35°34'40"N/74°41'30"E). Glacial rock roundings are preserved up to 2800 m asl (Kuhle 1988d; 1991b; 2001). Following the SW side of the Astor Valley from the Indus Valley upward, a decameter thick ground moraine (35°34'N/74°40'E) is deposited above the Ramghat Pul (bridge) at ca. 1250–1950 m asl. Further above, at the spur peak Hattu Pir, moraine sediments occur at 3000 m asl. One km upward the Astor Valley glacially polished outcrops of the stratum occur up to almost the top of Hattu Pir. Above the thalweg, the LGP Astor Glacier has attached thick ground moraines to the abraded rock surface at 1000–1200 m asl (35°32'50"N/ 74°40'30"E). Further up this SW side valley flank, continue decameter to more than 100 m thick ground moraines, which are interrupted by glacially polished rock (35°32'30"–32'N/74°40'30"-42'30"E). The ground moraines can be easily diagnosed by funnel erosion in the area of the settlements of Doian and Mang Doian. The highest, i.e. LGM Astor glacier level can be reconstructed with the help of the upper edges of these ground moraines: In in upward direction at ca. 3100–3300 m asl they become lateral moraine remnants at Hattu Pir in the area, where the Astor Glacier joins the Indus Glacier. The classification as LGM is based on the moraines (Kuhle 1996a Fig.3), which, though exposed to erosion, because they stick to the very steep valley flanks, are to a great extent well preserved. The author received absolute 14C-datings only from the historic lateral moraines of the Chhungphar Gah Glacier (Table, StadiaVII–XI). Here, the thickness of the Astor Glacier was 1500–1600 m. This allows to draw the conclusion, that the Indus main glacier was of similar thickness (see Kuhle 1996a). W of the Astor Glacier, the Lichar–, Buldar– and Rakhiot (Tato Gah) Glaciers joined the Indus Glacier. As a representative example, we concentrate on the reconstruction of the Rakhiot Glacier. There are two lateral moraine ledges on the left-hand side of the upper Rakhiot (Tato) Gah, the higher one lies at 4370 m asl, i.e. 370 m above the recent Ganalo Glacier surface (35°19'43"N/74°34'17"E; ibid. Fig.28 No.69). On the orographic right-hand side valley flank (ibid. Fig. 28 No.70) is a corresponding lateral moraine or kame-terrace complex at 4380 m asl (35°19'30"N/74°38'E) ca. 500 m above the present glacier surface. These moraines have been formed at an ELA depression of ca. 400-500 m (actual ELA 4900 m to ca. 4500-4400 m asl) and are to classify as being of the latest stadium of the Late Glacial (IV) (Table) (Kuhle 1996a; 1997 1988d;). Down-valley glacial polishes and abrasions on both valley flanks reach higher than the other moraine remnants, i.e. up to max. 4350 m asl (orographic left-hand side, cf. ibid. Fig.28 No.69, 78, 81; orographic right-hand side, No.72, 73, 76, 77). They provide evidence of a Rakhiot Glacier thickness of ca. 1000-1400 m during the Main Ice Age (Table, Stadium 0). In this context the observation of a ground moraine cover on the 750 x 450 m valley shoulder at the 3822 m high point E of Bezar Gali (ibid. Fig.28 No.81; 35°25'23"N/74°33'48"E) is of importance. The fine ground mass contains isolated large granite blocks, which are round at the edges or round. This ground moraine overlies forms of roches moutonnées and is cut off from any supply with slope debris (Kuhle 1996a; 1997). It is evidence of a Rakhiot Glacier thickness (LGM) in this valley cross profile of at least 1400 m. The ground moraine position is only 8,4 km away from the merely 1150 m asl high Indus Valley thalweg (ibid. Fig.28 No.82). From the valley head below the Nanga Parbat-N-face up to near the junction with the Indus Glacier, the level of the 26 km long Rakhiot (Tato Gah) Glacier has decreased from 4350 m to 3822 m, i.e. merely ca. 500 m. Therefore, an Indus glacier level at ca. 3000 m can be supposed. This is confirmed by the Astor glacier level in its junction area with the Indus Glacier at Hattu Pir (ibid. Fig.28 No.61, 87; see 2.3.7; 2.3.9; 2.3.10) at about 3000-3100 m asl (Kuhle 1996a). Numerous direct observations provide evidence of the thickness of the Ice Age (LGP, LGM) Indus main glacier of ca. 1800 m (cf. ibid. Fig.28 No.82-105). Here are given some examples: on the orographic left-hand side the trough valley flank of the Indus Valley has been polished and abraded by the Indus Glacier up to 3000 m asl (ibid. Fig.28 Nos.84-90, 92). Overlying ground moraines are preserved in numerous, partly decametre thick remnants (ibid. Fig.28 No.83, between Nos. 85 and 86, Nos.82, 101, 102; Kuhle 1996a Fig.4). Ground moraines of metre to more than 100 m thickness, partly up to an extent of 4 km, have been mapped at many localities on the orographic right-hand side flank (ibid. Fig.28, Nos.94-98, 100); there are also glacial flank abrasions (e.g. ibid. Fig.28 between Nos.93 and 94) and roches moutonnées, jutting out of the valley slope (ibid. Fig.28 above Nos.96 and 98) (see Kuhle 1996a). One of these ground moraines is deposited at point 1758 m asl, 620 m above the Indus thalweg (35°34'19"N/74°36'55"E; ibid. Fig.28, No.94); another one up to point 2401 m asl (35°31'10"N/74°35'58"E; ibid. Fig.28 No.95) at the T.P. Gor Gali Peak NE-slope. On the S-slope of this mountain, the ground moraine reaches up to ca. 2650 m asl (ibid. Fig.28 No.97). The most significant Main Ice Age (LGM) Indus Glacier thickness and -width is indicated by the ca. 1 km long lateral moraine ridge, which culminates at 2850 m asl above the Dirkil settlement (35°32'30"N/74°33'11"E; ibid. Fig.28 No.105). The moraine fringes the "Dead Valley", that is a classic glacier lateral valley. It has developed between the Indus Glacier and the orographic right-hand side Indus Valley flank at 1700-1750 m above the present thalweg. Since the deglaciation it is fossilized, i.e. without water-flow, and therefore bears its name rightly. Haserodt (1989, p. 208) also identified these moraines, but, in contrast to the author (Kuhle 1996a; 1997), understood them as local moraines of a small S-exposed hanging glacier from the Luthi Gal Chamuri flank, down from Terimal. In this case, however, there would be necessary an ELA depression of 1800-1900 m, i.e. 600 m more than the author suggests for the LGM (Kuhle 1988f; 1988d). - Further bank deposits of the Indus Glacier occur at the settlements Gor and Dirkil; so for instance moraines, kames and bank outwash (sandar) (ibid. Fig.28 Nos.100, 103, 104). The highest round-polished mountain ridges reach 3037 m asl at Gor Gali (ibid. Fig.28, No.106). - These observations point to a complete ice stream net glaciation of the Nanga Parbat Massif (cf. ibid. Fig.28) with an ice surface between 3000 and 4850 m, ice thicknesses of 800-1400 m and an ELA depression of at least 1200 m according to a snow line altitude of ca. 3600 m asl during the Last High Glacial Maximum (Table, Stadium 0). On Nanga Parbat the Indus main glacier reached ice thicknesses of even 1800-1900 m (Kuhle 1996a; 1997) (see. 2.3.9 and 2.3.10).

2.3.9. The Southern Slope of the Karakoram (Fig.1, between No.33 and 34)

The present, almost 60 km long glaciers terminate in the catchment area of the Indus Valley at altitudes below 2500–2700 m (Hunza-Karakorum). These are the glaciers that descended furthest in High Asia. According to Ward (1926) the Namche Bawar north glacier (Fig.1, No.13, humid eastern Himalaya) makes a similar steep descent to the Tsangpo Gorge. On Nanga Parbat the glaciers reached altitudes between 2900 and 3600 m asl (Kuhle 1997, Fig.28). During the LGP, however, all the tributary glaciers from the Karakorum south slope and the Nanga Parbat Group united in a 120000 to 180000 km² (ibid. Fig.28; 3 above) ice stream network, the Indus ice stream network (Kuhle 2001 Figs.2; 2/1; 2/2; see 2.3.7; 2.3.8; 2.3.10). At an altitude of 980 m or lower the largest outlet glacier reached the lowest ice marginal position at Sazin at the mouth of the rivers Daret and Tangir, a little E ward the bend of the Indus (35°34'N/73°28'E, Fig.1, No.35 and Kuhle 1988d). At (35°30'N/73°25'E) the author observed ground–moraines down to 850 m asl (Kuhle 1997).

There are two normally graded lateral moraines, more than 100 m thick (Kuhle 1991b, Photo 3), with a final bend exhibiting the terminus of the moraine (Kuhle 1988d; 1988f). These moraines occur at a distance of about 10 km from one another, each flanking the Indus Valley over several kilometres. The moraines consist of typical glacier diamictites of polymict composition (Kuhle 1991b, Photo 3). Down–valley, the valley cross section becomes a V–shaped valley gorge as a result of former subglacial meltwater erosion. Up–valley from the position of the ice margin a concave, worn–down U–profile is observed that eventually expands to a trough profile with well–preserved abrasions and polishes on the flanks (Kuhle 1991b, Photo 3). Sixty km up–valley, at about 1100 m asl, there is a basin–shaped opening - the "Chilas Chamber". Here, remnants of lateral moraines on the southern bank provide evidence of the LGP ice thickness. NW of the Chilas settlement, glacial abrasion and polishings on the NW side provide evidence of a minimum glacier thickness of 450–550 m. Between Chilas and the lowest moraines, the valley glacier edge during the Late Glacial (Table) can be traced by stable lines of light coloured glacio–limnic bank sediments.

In some places the limnites were dammed back into the tributary valleys. The next 60 km, along the north/western slope of Nanga Parbat (Fig.1, No.21), are characterized by a cover of ground moraine that coats roches moutonnées fields, and by several hundred meter deep deposits of ground and lateral moraines (see above 2.3.8; Kuhle 1997, Fig.28, Nos.83–105, Kuhle 1988d).

This wealth of moraines may be explained by the immense supply of scree from the Nanga Parbat Glaciers. Apart from abrasion and polished marks on the flanks, which in many places reach upward hundreds of metres, the ground moraines near the settlement Bunji in the Indus Valley are an important feature of the valley section from the confluence of the LGP Astor- and Indus Glacier up to the mouth of the LGP Gilgit Glacier where it flowed into the Indus Glacier (1228 m asl) (see 2.3.7 and 2.3.12) (Kuhle 1988d; 1991b; 2001). Thrust into the Indus Valley flank of Bunji, this 400 m thick moraine deposits on the E bank correspond to those at the mouth of the Astor Valley (Kuhle 1996a, 1997). The Gilgit- Hunza part of the glacier stream was reconstructed (Kuhle 1988f, 1988d, 1991b; 1993c) up to the 58 km long Batura Glacier (Fig.1, No.55) by observing diamictites, glacial abrasions, striations and polishings. Below the junction of the Hunza with the Gilgit Glacier, the tributary glaciers of Batkor and Bagrot joined. Their Late Glacial advances are provisionally classified by the author as belonging to the Ghasa Stadium (I) (Table) (Kuhle 1982a, 1986, 1987c). They appear to have reached the Gilgit Valley outside of the terminal basin of Gilgit at an altitude of 1300–1380 m (Fig.3b), because of the steepness of the valley and the high altitude of the catchment area, reaching 7788 m asl (Rakaposhi). Using glacial abrasions and polishings, along with striations extending as far down as 1900 m asl at Rakaposhi (36°15'N/74°24'E, Kuhle 1991b, Photo 2), the Hunza Glacier can be reconstructed up to 1600–1800 above the floor of the valley (Fig.1, No.55).

It is a demonstration of Pleistocene alpine glaciation, with a central montane thickness of 2000 m.

The approaching Shimshal Glacier (see above 2.3.7) provided a link with the north side of the Karakorum (Fig.1, Nos.55, 33, 32, 15, 16) and presented a connection with the Shaksgam Glacier system (see 2.2.13 and 2.3.11; Kuhle 1994 Fig.138). The Pleistocene predecessors of the Hispar–Biafo Glacier produced the southern transverse connection with the Muztagh–Karakorum Ice (Fig.3b; Kuhle 1988d, 1998, 2001 Figs.2-2/2). The Pleistocene network caused a great deal of friction. Therefore the ice streams tended to be dome-shaped. On the eastern edge of the Karakorum, approximately at the Pangong Tso (33°45'N/79°E) and further north on the Depsang Plateau (35°25'N/78°20'E), this Karakorum ice stream network gradually merged with the compact inland ice cover of Central Tibet (Fig.3; Kuhle 1999).

By contrast with the recent glaciers of the Indus Valley, which terminate at about 3400 m asl, the ELA depression for the lowest LGP ice margin, at 980 to below 870 m asl (Kuhle 1988d, 1991b, 1996a, 1997 p. 123), is calculated to be 1200-1250 m. Porter (1970) also found the depression of the equilibrium line in Swat Kohistan, 100 km farther west, to have been about 1200 m. Compared with a contemporary ELA at 4600–5000 m asl, this implies an LGP equilibrium line at 3400–3800 m asl in this area (Fig.1, No.56). The increase in the area under glaciation, along with the depression of the ELA by only 500 m is shown in Kuhle (1998 Fig. 22), by 1200 m in Fig.3.

On the extremely arid western edge of Tibet (Fig.1, Nos.15, 16, 32-34, 52, 59, 62), where today the mean annual precipitation at meteorological stations (Misijiaer, Gilgit, Chilas;

1961-1970) in the valley floor is only 142.5 mm per annum, there is evidence of a relief–filling glaciation which completely masked the topography (Kuhle 1988d, 2001, 2004).

The ELA ran 1200–1600 m below the average altitude of western Tibet (Kuhle 1999).

2.3.10. Summary of the Glacier Extensions and Ice Thicknesses in the Central Karakorum, on the Deosai Plateau and of the Past Hunza Glacier During the LGP as Parts of a Pleistocene Karakoram Ice Stream Network

Geomorphological and Quaternary–geological field- and laboratory data are used to interpret the LGP (ca. 60–18 ka) glaciation of the Central-, South- and Northwest Karakorum. The investigated Indus drainage areas include the Hunza Valley with its tributary sytems of Batura-, Shimshal-, Hispar-, Hassanabad-, Bar- and Jaglot Valley, the Braldu-, Basna-, Shigar Valley as well as the Deosai Plateau between the Skardu Basin and the Astor Valley. They show that between ca. 60 and 20 ka the Karakorum was covered by a continuous, approximately 125,000 km²–large glacier network (see 2.3.9). These glaciers converged into a common glacier, the Indus Glacier, the tongue of which descended to 850–800 m asl (down from 35°32'N/73°18'E).

In its central part the Indus glacier network reached an altitude of at least 6000 m asl. Its maximum ice thickness was about 2400–2900 m. In contrast to the author's investigation results summarized in this paper, several scientists suggest a non–continuous LGP Hunza Glacier, i.e. a glacier that had just come as far as the Saret settlement, that had not reached the Gilgit Valley and – even less – had not joined the Indus Valley.

Furthermore, there is an intermediate position based on the assumption of a Hunza Glacier flowing down from the Batura Glacier through the Gilgit Valley up to above the junction of the Astor Valley. The author's investigations, however, which include the large tributary valleys, provide evidence of (1) a 2000–m-thick Hunza Glacier reaching down from the Khunjerab–pass as far as into an Indus trunk glacier. The joint glaciers flowed up to the Sazin settlement and then 20 km further down the lower Indus Valley (down from 35°32'N/73°18'E); (2) a Hunza Glacier which was part of a continuous ice stream network. In the N, over the Khunjerab–pass, it was connected to the ice of the E–Pamir as well as via the Shimshal–, Hispar– and Barpu Glacier–systems to the Shaksgam– and Shigar Glacier-sytems. Owing to this, it has also formed a common ice surface with the middle Indus Glacier across passes of the Destigil Sar- and Haramosh Chains.

2.3.11. Karakoram North Slope and K2

During the 1986 expedition to the north slope of the Karakoram and to K2 (Fig.1, Nos.32, 16, 15) the author investigated the even more arid north/western Tibet as far as the northern ramp of the Aghil and Kuen Lun Ridges and down to the desert–like Tarim Basin (Fig.3a; Kuhle 1994 Fig.138, Nos.1–45; see 2.2.13). Thus the long–term mean annual precipitation of today at meteorological stations is 67.3 mm per annum. At 2500 m asl it is 4–5°C colder than on the Karakoram south side. At an altitude of 3090 m the station of Tashikuergan (Taxkorgan; Fig.1, No.32) measured an average temperature of 3.1°C during the period 1957-1980.

The lowest moraines of the LGP are at 2000–1900 m asl (Kuhle 1994). Tens of kilometres long, these chains of medial moraines extend the transverse valleys of the Kuen Lun (37°20'N/77°05–35'E) south of the settlement Yeh Cheng far out into the mountain foreland (Fig.1, No.16). The thickest moraines are 400–700 m high. They consist of very large, polymictic, facetted blocks with rounded edges, embedded in a loamy ground mass. They also include, in many places, stratified glacio–fluvial drift and rhythmic limnites that have been thrust into the moraine (Kuhle 1988e; 1988b). Thirty to forty percent of the large blocks are adjusted to the direction of movement. The extreme thickness and the flexures within these diamictites exclude an interpretation of convergent debrisflow deposits, as does their granite, phyllite and limestone composition. Their morphology is one of elongated ramps, which bend to form frontal moraines around terminal basins (Kuhle 1994 Fig.138, Nos.43, 44, 45). Glacial exaration rills and striations, at the foot of the inner slopes of the moraines, have been preserved. These moraines were deposited by the large outlet glaciers in the course of repeated Pleistocene glaciations. During interglacial periods renewed development of scree took place in the mountains and highland. About 200 m down-valley of the recent moraines there are more extensive wide–ranging terminal glacier basins. These basins were not glaciated during the LGP, but during the previous glaciation (Riß?: see Table).

The glacier cover of the LGP of the area is shown in Fig.3. The depression of the ELA during the LGP amounted to 1300 m. Thus the ELA ran in the area under investigation (Fig.1, Nos.15, 16) at about 3900 m asl (3700–4100m). The ice retreat to the approximate position of the present glacier margins had largely been completed before 12870+/-180 YBP (Table). Evidence of this was supplied by 14C dating of a peat deposit in the soil of the Muztagh Valley at 3940 m asl in the forefield of the 43 km long Skamri Glacier (Kuhle 1994).

2.3.12. Reconstruction of the Lowest Preserved Ice Margin Position and the Maximum Glaciation During the LGP in the Chitral, Hindukush (Fig.1, Nos. 55, 56) and Eastern Zagros Mountains

Haserodt (1989) has suggested that the Ice Age Chitral glacier terminated ca. 8 km up–valley of the Buni settlement, so that the Mastuj– or Chitral main valley should have been free of ice. Kamp (1999), in contrast, assumes that the former Chitral glacier has reached at most 100 km further down-valley, as far as the Drohse settlement at 1300 m asl. According to the author's reconstruction (Kuhle 2001, 2004) the LGP-Chitral glacier has even flowed down 15 km further, i.e. as far as the valley chamber of Mirkhani at 1050–1100 m asl (35°28'40"N/71°46'30"E). This ice stream was not only thicker than 500 m, as has been supposed by Kamp (1999), but three–times thicker. Whilst Kamp (1999) does not assume a transfluence of the Tirich Mir glacier over the Zani pass, the author's findings (Kuhle 2001) of granite erratics above the depression of the pass argue in favour of this transfluence, thus providing evidence of a larger ice thickness.Accordingly, here too, an over 1000 m–thick ice stream network existed during the LGP, which was connected to the Karakorum ice stream network across the Schandur pass. From there is was connected down through the Gilgit Valley (Kuhle 2001 Fig.2; see above 2.3.7; 2.3.10).

Further west in the semiarid Zagros Mountains evidence has been provided of two mountain glaciations in the 4000 m-high, currently non–glaciated Kuh-i-Jupar: an older period around 120 ka, and a younger one during the LGP, MIS 4–2: 60-18 ka (Table; see Kuhle 1974; 1976 Fig.162–164; 1987a; 2004a). During the glaciation around 120 ka the glaciers reached a maximum of 17 km in length; during the LGP they were 10–12 km long. They flowed down into the mountain foreland as far as 2160 asl, i.e. 1900 m. The thicknesses of the valley glaciers reached 550, i.e. 350 m.

During the glaciation around 120 ka a 23 km extended, continuous piedmont glacier lobe has been developed parallel to the mountain foot. There the ELA ran at 2960 m asl, that is 1590 m below the current theoretical snow line. During the LGP the glacier termini mainly remained isolated in the mountain foreland. The ELA ran at 3060 m asl, that is 1490 m below the current theoretical snow line.

The observed laterite horizons are not interpreted as having endured overriding by a glacier. Thus it is inferred that the laterite horizons were formed during warm interglacials which, according to the Vostok ice–core, reached temperatures of 2.5–3°C above pre–industrial values. Drillsites in the North Atlantic (ODP Site 642B, W of Norway) document ice–rafted debris from about 5.8 Ma on. There today no ice–rafting is observed. Accordingly, further north an ice–cover can be expected. Thus also these terrestrial observations point to very pronounced latitudinal temperature gradients during warm interglacials.

So, a temperature depression of ca. 11–16°C, i.e. 10–15°C is proven for the SE–Iranian highland, ca. 120, i.e. 60-18 Ka ago.

2.3.13. Maximum Glaciation Between the Kunjerab Pass (Ghujerab; Karakorum) and the East-Pamir on the Western Edge of the Tarim Basin (Fig.1, Nos. 33, 32, 36; Fig.3a; Kuhle 1997 Fig.14)

It remains unresolved whether this pass (36°50'N/75°25'E) had been an LGP transfluence pass running, from south to north, from the Karakorum to the East-Pamir, as no erratics have been observed. A north to south discharge is less likely, since the mean altitude of the relief on the Karakorum side is greater than that on the Pamir side. It is, however, probable that the Kunjerab Pass was the corridor between the Karakorum ice stream, which drained to the Indus, and the East-Pamir and Kuenlun ice stream, which drained to the Tarim Basin (Fig.3a). It is certain that the Kunjerab Pass contained a glacier of at least several hundred meters in thickness, which, in whatever direction it drained, brought about the connection of the two ice streams to a single large mass west of the edge of the Tibetan Plateau. Abraded valley flanks and truncated spurs occur on both sides of the pass. The 120 km long Taxkorgan Valley, which leads from the Kunjerab Pass (4730 m asl) down to Taxkorgan (Tashikuergan, 3090 m asl; 37°49'N/75°13'E; Fig.3, No.32), is a glacigenic U–shape trough. Its floor is covered with the usual sequence of historical, neoglacial and late Late Glacial moraines (Table). It is divided into tongue basins. The relative dating of this sequence is possible by the forms in which the moraines are preserved. On the flanks of the middle section of the valley there are erratic blocks of granite several hundred metres above the ca. 4000–3800 m high valley floor (37°N/75°30'E). LGP glacial deposits, however, are necessarily absent in this area due to glacial erosion. Abraded valley flanks are clearly preserved and extend up the valley beyond a corrie terrace and past the level of a hanging valley. Thus an LGP ice thickness of 600–1000 m can be inferred. An ice margin position with hummocky end moraines and outwash fans lies at 3500–3300 m asl (37°25'N/75°22'E), 25 km below the mouth of the Mingtieh–Kai Ho, itself a glacigenic U–shaped valley approaching from the west. The Taxkorgan Glacier and the eastern edge of a large plateau ice–complex, which had been superimposed upon the adjacent plateau toward the west (37°25'N/75°12'E) terminated at this position (Fig.1, No.32). Further down the 45 km long Taxkorgan Valley appears to have been ice–free. In this case the ice margin position at 3300 m asl would have been part of the LGP. Fresh forms of end moraines are observed at the western edge (3600–3400 m asl) of the Taxkorgan Valley floor.

However, if the Taxkorgan Valley was filled with ice during the LGM (Table: 0), the ice margin positions mentioned above belong to the older Late LGP (Ghasa Stadium I: Table; after Kuhle 1982a–2004). Roches moutonnées (37°53' N/75°11' E) on the valley floor between Taxkorgan (Tashikuergan) and the Muztagh Ata (Mustag Feng) Massif, 35 km further north, support this interpretation (Kuhle 1997 Fig.14, No.34; Kuhle 1995b). During the LGP the Muztagh Ata and the Qungur Tagh (Kongur Feng) Massif, as well as the adjoining Pamir High Plateau (Fig.1, No.36) were completely covered by an ice cap between 3400 and 5000 m asl (Kuhle 1995b). Evidence of this exists as a basal till (ground moraine) cover (Kuhle 1997 Fig.14, Nos.13, 18, 20, 26, 29) with a gently–rolling surface in which later roches moutonnées (Kuhle 1997 Fig.14 between Nos.20, 22, 29) have pierced (Fig.3). A Late Glacial (Table) end moraine landscape consisting of numerous separate glacier tongues flowing down from the 7620 m high Muztagh Ata Massif and the 7830 m high Qungur Tagh Massif dissects the earlier moraine (Kuhle 1997 Fig.14, Nos.23, 21, 22, 24, 30, 31, 32). Leading down from the East–Pamir Plateau (Sari Koi Ling, Karakol Lake, Muji Basin) and past the Qungur Tagh Massif, the Gez Valley contained an outlet glacier with a thickness of at least 550 m. Evidence of this is provided by ground– and lateral moraines that begin at the plateau edge near the valley head (upper edge 3600 m; 38°45' N/75°02'E). The feeder streams of the local Qungur glaciers, i.e. the Kaiayayilak, which currently extend to 2820 m asl (Qungur–North-Glacier) increased the thickness of this outlet glacier in the middle section of the valley (2400–2200 m asl) to more than 800 m. The lowest ground moraine material observed is at 2180 m asl (Kuhle 1997 Fig.14, No.13). Schroeder–Lanz (1986 Figs.4, 5) also observed lateral– and ground moraines in this valley chamber. In view of the substantial ice thickness during the LGP (Table: 0), however, the glacier terminus during this period would have been at least 20 km further down-valley (Kuhle 1995b). Though the valley floor contains only drift deposits, moraine–like deposits (lateral moraines) do exist on theE side 40, 80, 140 m (Kuhle 1997 Fig.14, No.11, Stadia I, I', I") and 400–500 m above the valley bottom (Stadium 0; 38°38'N/75°30'E). These lateral moraines, erratic blocks on roches moutonnées and on the valley flanks are evidence of an ice margin position at 1800 m asl (near 38°50'N/75°31'E, Kuhle 1997 Fig.14; Kuhle 1995b). In places where no basal till is found, Late Glacial sandars (Table: Nos.4–1) may have covered the Main Glacial bed. On the Muztagh Ata and Qungur Massif the present ELA runs at 5000 m asl. Thus the present glacier tongues reach down to the East Pamir Plateau at about 4200–4400 m asl (Kuhle 1997 Fig.14, Nos.18, 23, 22, 24, 30). A plateau–covering ice cap requires an ELA depression of 700-900 m.

In the north/east facing Oytag Valley of the King Ata Tagh Mountains (Fig.1, No.37), which rise to 6634 m asl (Kara-Bak-Tor Peak), the lowest LGP margin, confirmed by ground– and end moraines, extended down to 1850 m asl (38°56'N/75°26'E; Kuhle 1997 Fig.14, No.4, Stadium 0), whereas the present margin lies at 2750 m asl (Oytag Glacier; Kuhle 1997 Fig.14 between No.6 and 10).

For the definition of the LGM ELA of the E–Pamir Plateau (37°50'–39°N/74°40'–75°40'E) the end moraines of marginal outlet- and mountain glaciers, such as the Oytag– and Gez Glaciers provide evidence. The Oytag Glacier terminated at 1850 m asl. For a mean altitude of the catchment area of 6000 m results an ELA at 3925 m asl. At present the Oytag Glacier terminates at 2750 m asl. Its medium catchment area is 800 m higher than that of the LGP. Thus the ELA depression amounts to 850 m. The present ELA is running at 4775 m asl. A depression of only 850 m is a small value for the LGM (Table). Further to the E, on the S edge of the Tarim Basin, ELA depressions of 1300 m have been determined (Fig.3b; Norin 1932; Kuhle 1994 Fig.138).

The lowest ice margin of the Gez Glacier was lying at 1800 m. The next recent glacier terminus at 2820 m asl (Kuhle 1997 Fig.14, No.3) indicates an ELA about 5000 m asl. The mean altitude of the catchment area of the Gez outlet glacier was lying at 5700 m. There for the LGM an ELA about 3750 m asl is calculated. Thus the ELA, having decreased by 1250 m, was tangent to the East Pamir High Plateau. The topographic arrangement on the 5568 m massif (Kuhle 1997 Fig.14, No.33) allows to check the LGM ELA on the south/west edge of the test area. At an altitude of the catchment area about 5160 m asl, the glacier came down to 3200 m asl, i.e the ELA was running at 4180 m asl. In case the ELA depression of only 820 m does not belong to Stadium I, but to the LGM (Table: Stadium I or 0), the Tahman Basin may have been free of ice.

Thus an ELA depression of ca. 820–1250 m is observed on the leeward side of the high mountain chains down to the Tarim Basin (Fig.1, Nos.37, 36). It rose to 1300 m to the south/west edge of the Tarim Basin, on the north slope of the Kuen Lun (Fig.1, No.16, see 2.3.11).

2.3.14. Glaciation of the Tian (Tien) Shan

The Tian Shan (Fig.3a) is the northern continuation of the East Pamir Highland with an area of ca. 75 000 km² between the mountain chains of the Kokshaal Tau and Terskey Alatau (Fig.1, Nos.38, 39). Having been interrupted by Lake Issyk Kul, the Tian Shan continues toward the Kungey Alatau (Fig.1, No.40), the most northerly chain, at 43°N (Kuhle 1998 Fig.19). The Tian Shan Plateau was covered by a ca. 800 m thick ice cap with outlet glaciers on its edge. Evidence of this is observed as a blanket of ground moraine that covers the plateau, and end moraines, which extend down into Lake Issyk Kul (Kuhle 1998 Fig.18). The lowest ice margin position was found at Bishkek at 980 m asl (Kirgisen Shan; 42°45'N/74°35'E) (ibid. Fig.17). The complex of terminal moraines is evidence of a very extensive foreland glaciation with an ice body of more than 400 m in thickness. Extending over tens of kilometres, this tongue basin emerges from the Ala Archa Valley with a catchment area at 4300 m asl (Fig.1, No.41). The present (1991) terminus of the Golubin Glacier lies at 3250 m asl, so that an ELA depression for the ca. 2250 m deeper LGP ice margin is calculated to exceed 1100 m. In the area of Lake Issyk Kul (Fig.1, No.39-40), which is 14C dated as belonging to the LGP (26.100 +/-600 YBP, JGAN-616 and 32.390 +/-1780 YBP, JGAN-971), the ELA depression was about 1000–1200 m (Kuhle 1998 Fig.19; Grosswald, Kuhle and Fastook 1994). On the 100 km wide Tian Shan Plateau, south of the lake (Fig.1, No.38–39), the present snow line runs at 3900–4200 m asl (e.g. Petrov Glacier, Ak Shirak Massif), so that it was beneath the plateau level at 2700–3200 m asl during the LGP. This resulted in a plateau–covering ice cap, which is also evidenced by smoothly abraded and polished mountain forms. An ice cap thickness of 800 m was reconstructed on the basis of ice scour limits. The Dankov Massif (Kokshaal Tau, 41°02'N/77°37'E; 5982 m asl) on the southern edge of the Tian Shan Plateau (Fig.1, No.39) was covered by an ice stream with outlet glaciers flowing down into the Tarim Basin. A similar situation occurred for the East Kokshaal Tau (Fig.1, No.38), with the 7439 m high Pik Pobedy Massif (Tuomur Feng; 42°N/80°E), which is presently glaciated by an ice stream net (Fig.3a, Tian Shan).

The LGP glaciation of the Tian Shan 520–700 km further east (Fig.1, No.42) toward Urumqi (43°15'N/86°10'E) and on the Bogdo Ul Massif north slope (43°50'N/88°20'E) reached down far below 2000 m asl, but without having extended much further into the mountain foreland. In the catchment area of the Urumqi–River–Valley (Daxi Gou Valley) five hanging glaciers and corrie glaciers currently (1992) reach down to 3750 m asl. The present ELA is around 4000 m asl. During the LGP there was an approximately 50 km long valley glacier with a few tributary branches. Evidence of this appears as a U–shaped valley profile which extends up the valley flanks 600–800 m above the thalweg at 2500 m asl. In addition abraded valley flanks and truncated spurs, glacigenically polished and rounded bar mountains down–valley as far as ca. 2000 m asl can be observed. Outside of the valley two lateral moraine terraces indicate the continuation of the ice stream into the mountain foreland over a distance of several kilometres. Classified by the state of perservation of the moraines and by geomorphological sequences they belong to the LGP. Lower lateral moraines and kame terraces cease at 1900 m asl, whereas the upper moraine, with a 100 m lower ELA and classified as of the pre-LGP (Riß), drops to 1680 m asl. These moraine terraces have also been preserved in the Daxi Gou Valley, though at higher relative altitudes. With a mean altitude of the catchment area of about 3600 m, an LGP snow line at 2750 m asl is calculated. This is the equivalent of a snow line depression of 1250 m. In the Bogdo Ul Massif(Bogda Shan), in the north sloping Sky Lake Valley, an eastern lateral moraine was found at 1050 m asl. This is a position outside the rock and moraine threshold at 2000 m asl, which dams up the Sky Lake in a glacigenically over–deepened trough basin. The interpretation of an ice marginal position at 1000 m asl extends the Schroeder–Lanz snow line (Schroeder–Lanz 1986 Fig.8; see also Chiao 1984, p.123) 200 m down–valley to 2550 m asl (ELA depression=1400 m). This value likely belongs to the glaciation around 120 ka (Table: Riß). Thus during the LGP an ELA depression of 1200 m is a realistic interpretation based on global relations between Pre-LGP and LGP glaciations.

2.3.15. Extent of Glaciers and Snow Line Depression in North and North/Eastern Tibet

Reverse analogous to the arid area discussed above, the northern fringe of Tibet (Kuhle 1982b, 1982c, 1987b, 1997, 2003) stands for the more humid north/eastern area (Fig.1, Nos. 43, 45-49). The highest levels of the equilibrium line exist in the south (Fig.1, Nos.13, 17-19, 20, 22-31, 53, 54). If the extreme aridity, of today occurred also during the Pleistocene and Upper Pliocene, the likelihood of an extensive Pleistocene or Pliocene glaciation on the western edge of the plateau is minimal (Fig.1 Nos.15, 16, 21, 32-34, 36-41, 52, 55, 56, 59-62; see 2.3.6–2.3.14).

Meters-thick laterites and several bornhards (Inselberge) as indicators of humidity and tropical climate have been observed between 2900 and 3000 m asl in this area (Kuhle 1982b, p.77). They occur on a plateau in the eastern Quilian Shan (Mountains; 37°50'N/101°37'E; Fig.1 No.47), which was glaciated later and latest during the LGP. The laterites remain under the plateau ice cover.

A 450x820 km large area in the NE completes the Tibetan glacial reconstruction (Fig.1, Nos.43-49). During an expedition in 1981, LGP glaciation was reconstructed on the basis of 35 end moraine complexes (Kuhle 1987b Fig.13, p.302, text Fig.9, p.275; 1998 Fig.15, Nos.I–XLIII; 1997, Photo 133–151; 2003 Fig.15). Figures 13 and 14 (Kuhle 1998; Kuhle 1986) show the details of this reconstruction (Fig.1, Nos.43, 44). In the extreme north on the edge of the Gobi Desert, at 39°50'N/97°33'E (Kuhle 1998 Fig.15, No.XXIII) and at 39°49'N/97°49'E (Kuhle 1998, Fig.15, No.XXIV; Fig.3, Nos.45, 46) glaciers flowed down to 2150 m asl. This corresponds to a depression of the ELA of 1225 m to 3375 m asl. The ELA depression of 1450 m to 3250 m asl, which also occurred during the LGP at 39°23'N/98°49'E (Kuhle 1998 Fig.15, No.XXV; Fig.3, No.47) represents an extreme value. An ELA depression of 1430 m (with a maximum value of 1575 m) in NE Tibet was established for an older glaciation (Riß, see Table) on the basis of five ice margins (Kuhle 1987b).

On the basis of these values the ELA from its highest altitudes in southern Tibet to its northern edge amounts to 1470 m (4720-3250 m asl, Kuhle 1998 Fig.3). The 1 to 1000 ratio of the ELA depression (1470 m) to distance (1500 km) is in accordance with the modern temperature slope of 100 vertical meters to 100 km of horizontal distance (Kuhle 1998, Fig.19). A map of the reconstructed contour lines of the snow line altitudes (Kuhle 1998 Fig.15, I2) in NE Tibet during the LGP showed one contour line running south of the Tsaidam Depression at 4100 m, about 100–400 m below the mean altitude of the plateau (Kuhle 1998 Fig.15, from No.XV to XIII and from No.XIII to III). This must have led to the formation of an inland ice (Fig.3a, I2). Polishing or abrasion limits are evidence of an ice thickness of at least 500–700 m in NE Tibet (Kuhle 1987b). The existence of an ice cap is proven by ground moraine covers (see above 2.2.3) with locally prominent rough granite erratics (Fig.1, Nos.3, 4, 44). Granite erratics occur in a valley on the northern edge of the plateau, extending over the 4540 m high Oh La Pass (Kuhle 1998 Fig.15 Nos.III and II) and rising another 200 m along the slopes, before dropping to 3950 m asl toward the south (35°25'N/99°25'E, Kuhle 1987b, p.256).

North of the Tsaidam Depression, at the approximate latitude of the Kukunor (36°30'–50'N/98°–100°E; Fig.1, No.48), the equilibrium line descended to 3750–3650 m asl (Kuhle 1998 Fig.15, Nos.XI, IX–IV). Toward the Quilian Shan's north slope it fell below 3400 m (Kuhle 1998 Fig.15, Nos.XXV, XXXIV). Using a median altitude of 4000 m for the northern plateau, interpreting the ELA depression apart from observations of ground moraines, allows important conclusions to be drawn concerning the most northerly inland ice complex, referred to as "I1" (Fig.3a; Kuhle 1998 Fig.15). In the west and east it was linked to ice complex "I2" by bridges of glaciated mountain chains (Kuhle 1998 Fig.15, Nos.XX and IX, XI, X). At its center was the 5704 m high Kakitu Mountain Range (Fig.1, No.43) (Kuhle 1998 Fig.13), the forelands of which are covered by a ground moraine measuring tens of metres in depth, containing five types of granite, sandstones and metamorphics (38°09'N/96°29'E; Kuhle 1998 Fig.15, No.XXXVII). A 150 m deep core (Fig.2) was drilled in the foreland of the Kukunor Shan (see above 2.2.1.). It shows alternating deposits of stratified advance scree, ground moraine and limnic sediments and accordingly 6 to 15 Pleistocene glaciations in the mountain foreland.

2.3.16. Lowest Past (LGP) Ice Margin Positions in East–Tibet Between Daocheng and E of the Minya Konka–Massif and From the Xuebao–Massif to the Minyang River (Fig.1, No.51, 57, 58)

According to field- and laboratory investigations carried out on the E-margin of Tibet, an LGP Dadu–He–valley (also Ta–tu–ho or Tung–ho) parent–glacier has been reconstructed. Due to the fresh forms preserved and in consideration of the extremely destructive and reshaping forces of the monsoon–specific morphodynamics, it has been classified as belonging to the LGM. This main glacier has flowed down from E–Tibet and been fed by numerous side glaciers, i.e., outlet glaciers from the E–margin of the Tibetan inland ice (Kuhle 1988d, 2001, 2003). It has also been supplied by the connection to the local ice stream network of the Minya Gonka–massif (Fig.3a,b). With the help of ground moraines and glacigenic lateral forms, as e.g. several hundred–meter–high kames, the LGP glacier infilling of the Che Chu (dshu, valley), reaching the town of Kangting (also Tatsienlu, 2620 m asl) from the N, has been evidenced. Between Kangting and the inflow of the Tatsienlu-ho (valley) into the Dadu He valley at 1650 m asl, flank abrasions have widened the Tatsienlu gorge to a "gorge-shaped trough". Up to 600 m above the river mouth, a 30-70 m-high moraine has been preserved at the Wassöko settlement, which can be approached as a lateral-, i.e. medial-moraine between the side- and parent glacier. But on the opposite side of the tributary valley ground moraine has also survived (30°05'N/102°12'E). In the course of the continuing 60 km down the Dadu valley (main valley), from Wassöko as far as the junction with the Taitho (valley), i.e. from 1620 m down to 1240 m asl, the author in many places has mapped further ground moraine covers and lateral forms in continuation of those moraines. Additional clear evidence of the LGM-Dadu valley parent glacier occurs e.g. on the orographic left side, immediately above the town of Luding (or Lu-ting shao). Here, too, the thickness of the valley glacier might have amounted to 500–600 m. Above the inflow of the Taitho, through which the LGP Hailuogou– (Hailoko–) and Yan–tsöko glaciers have flowed down from the Minya Konka (Gonka) SE– and NE flank into the Dadu main glacier, the main glacier had still a thickness of ca. 300 m. Accordingly, the lowest LGM–ice margin position of this eastern outlet glacier of the Tibetan inland ice must have been situated at ca. 1150 m asl at 29°30'N/102°11'30"E. Here, below the Wantung settlement, a markedly winding, narrow gorge valley stretch starts, which was probably free of ice (Kuhle 2001, 2004).

In the eastern marginal area of E–Tibet, situated 200–500 km more northward, S of the Xuebao–massif, the ca. 150 km–long Minyang trunk glacier flowed down deepest (Fig.1, No. 58). It is one of the greatest outlet glaciers of the eastern Tibetan inland ice margin, the tongue end of which reached down to ca. 1040 m asl. The glacier came to an end ca. 45 km S of the city of Wenchuan and 75 km down–valley from the city of Maowen (Mao Xian) (Kuhle 2004), which is built on its ground moraine.

2.3.17. The LGP Glaciation of the Eastern Sayan Mountains and the Southern Trans–Baikal Mountains (Kuhle 2004)

It has been possible to follow up the snow line depression from High Asia to Central Asia and to SE Siberia, about 1300–1400 km NE of the Tian Shan at 52°N/102°30'E. Grosswald (1987, Fig.38, p.154) has deduced theoretically a 60.000 km² ice cap glaciation in the eastern Sayan Mountains. The author was able to carry out field research in this area in February/March 1993, and confirms that Pleistocene glaciers have calved into Lake Baikal at 455 m asl (see Grosswald and Kuhle 1994 Fig.2). The ice may have flowed from the west, from the 3492 m high Munku Sardyk Massif via the Tunkinskaya Galina (Irkut Valley; see Grosswald and Kuhle 1994 Figs.3–5) and definitely from the south, from the more humid Khamar Daban Range (2323 m; ibid. Fig.6). The catchment area of the Tunkinskaya Galina Glacier includes the 12.100 km² granite–gneiss plateau of the eastern Sayan Mountains (2300 m asl; ibid. Fig.3) along with 2600–3110 m high peaks immediately north of the valley. In the Tunkinskaya Galina Valley glaciers flowed from these three areas and merged to form a 30 km wide trunk glacier (ibid. Fig.4) which sloped eastward toward Lake Baikal. The lowest ice margin position established by the author was in the Tunkinskaya Galina at 500 m asl (51°45'N/103°36'E) at a mean altitude of ca. 2400 m asl. This would lead to a LGM snow line around 1450 m asl (ibid. Fig.2). At present, only a small glacier exists on the NE slope of Munku Sardyk, providing evidence of an orographic ELA around 3000 m. Thus, the LGP snow line depression is assumed to be 1500 m. This considerable ELA depression, in an area with a present annual precipitation of only 500–700 mm, is attributed to the presence of giant ice–dammed lakes in Central Asia with evaporation surfaces of several million km² (Grosswald 1987 Fig.25, p.109).

In case the circulation during the LGP has similarities to that from 4–5 Ma winters advection of moist air appears possible. This however requires that during summer a potential ice–cover of these lakes reduced such that evaporation became possible.

Late Glacial ice margins at 1300 m asl occur just beyond the valley mouth near the Mundi settlement (51°40'N/101°05'E; see Grosswald and Kuhle 1994 Figs.4, 5). The largest glacier thickness in this valley, using moraine and erratic altitudes of 2000 m asl, exceeded 700 m. Because of the large altitude above sea–level, this glacier is classified as being from the later Late Glacial period (Table: III or IV). Observations of erratic blocks of granite at 1915 m asl, are evidence of an potentially enormous LGP glacial cover. If the present–day high humidity in the mountains south of the southern end of Lake Baikal (Khamar Daban Range, upper Slujanka Valley, 51°32'N/103°37'E), applied also partially to the LGP, an additional indicator for a large glaciation exists.

The author is not aware of studies that test glacial aridity / humidity for this region.

These elevated erratics occur at a distance of only 18 km from the southern shore of Lake Baikal. The area between the Mundi settlement and Lake Baikal is witnessed by flat–bottomed trough valleys with steep valley heads, that transported glacier ice directly from ca. 2000 m asl to the lake shore (Grosswald and Kuhle 1994 Fig.2).

Further south, in the Khamar Daban Mountains, there was a relief– covering ice cap, as shown by the U–shaped transfluence passes and smoothed abraded and polished mountain forms. Evidence that this ice cap calved into Lake Baikal with its larger outlet glaciers is found, for example, in the Murina tongue basin at the exit of the Snirsdaja Valley (51°27'N/104°51'E). Side and end moraines fringe both valley flanks and run into the lake. With a mean altitude of 2100 m asl of the catchment area, the LGP snow line in this northern area is calculated to be around 1250 m asl (ibid. Fig.2). The present snow line is estimated to be at 2750 m asl. Thus the ELA depression is about 1500 m. Precipitation inferred to be provided by Lake Baikal influenced this depression.

If this inferred mechanism is correct, Lake Baikal must have been during LGP summer ice–free. Otherwise other sources of moisture need to be considered.

From the Quilian Shan (Fig.1, No.45) 1500 km further south, with an ELA at 3250 m asl, to the Sayan Mountains in Central Asia, ELA at 1350 m asl, the LGP ELA decreased by 125 m per 100 km.

2.3.18. Dating of the Tibetan Ice Cover

The LGP age of the last complete glacier cover described above has been established by 14C dating of recent outwash plains that interfinger with LGP limnites of the Tsaidam Depression. These limnites were obtained in 1981 by coring 80–180 m below the present sediment surface (36°48'N/96°27'E, 2886 m asl; Fig.1, No.50; Kuhle 1998 Fig.15). The peat from the base was found to date from 35 120+/-625 to more than 47 270 YBP. The average of the values amounted to 35 530+/-540 YBP. (C14-analyses: M.A. Geyh, Hannover, Germany documented in Kuhle 1987b, p.307). LGM ages of 26 100+/-600 to 32 390+/-1780 C14-years BP (IGAN-616,-971) from organic detritus from the base of a 60 m glaciolacustrine terrace were obtained in the Chu River Valley (Tian Shan; Fig.1, No.40) at the junction with the

Kokpak Kyrkoo Valley (Kuhle 1998 Fig.18; Grosswald, Kuhle and Fastook 1994 pp.282-283). Further radiocarbon datings are of base peat with ages of 9400+/-185 and 8660+/-135 YBP in the Kakitu Massif (38°02'N/96°26'E; Fig.1, No.43, Kuhle 1986; 1987c, pp.456-457; 1998 Fig.13; samples taken by J. Hövermann) and of detritus with conchyliums–fossils in glaciolacustrine (ice–dammed) sediments in the Animachin with an age of 8640–785 to +910 YBP (3600 m asl, 34°43'N/100°12'E; Fig.1, No.44; Kuhle 1986, 1987b, p.303; 1998 Figs.14, 15). Thus the more wide–ranging Late Ice age glaciation in Tibet had been completed by approximately 9400 to 8600 YBP (Fig.1, No.44-50). This is also supported by a date of a trunk older than 8670 14C-years BP by Yamanaka (1982/83) for a moraine located by the author in southern Tibet (Jhong Khola 28°48'N/83°51'E; Fig.1, No.19) at 3250 m asl (Kuhle 1980; 1982a Fig.6,J; 1983).

In the area of the Tsangpo bend in SE Tibet (Fig.1, No.12; Kuhle 1991a Fig.43, No.38) a last strong glacier advance is documented about or after 9820+/-350 YBP (see 2.2.12). The author shows (Kuhle 1998 Tab.2) further seven datings of trees from the 80 m high opening of glaciolimnic sediments, which has been analysed in 1989 (Kuhle 1991a Potos 53, 54, 55). The together eight 14C-samples stem from the lower 32 m of above–mentioned opening. They are shown in descending order (Kuhle 1998 Tab.2). These limnic basal sands, in which the trunks were embedded, have been overlain by eight meter thick varved clays. They provide evidence of an ice–dammed lake in this lower section of the Tsangpo valley at about 3000 m asl. Accordingly, this ice-dammed lake is of the same age or younger than 48 580 to 9 820 YBP. Therefore it is classified as ranging between the LGM (0: Table) and the Late Glacial (I–IV: Table). Varve counting shows that the lake existed ca. 1000 years. It was situated between the ice complexes I2 and I3 (Fig.3a, W of Namcha Bawa). It was dammed up by the Nyang Qu Glacier. The Nyang Qu Glacier, which was an outlet glacier of the ice complex I2, followed the Nyang Qu (Valley), reaching the Tsangpo Valley (Kuhle 1991a Fig.43, No.38) 17,5 km down–valley from the opening of the glaciolimnic sediments (Kuhle 1998 Tab.2). Ground- and lateral moraines (Kuhle 1991a Fig.43, Nos.36, 39, 40, Photos 48–52, 56, 57) confirm that the glacier bended into the Tsangpo Valley. Thus the ice–dammed lake was formed (Kuhle 1991a; 1998a; 1997, p.127, 128). Up to now these are the best datings of the author which give evidence of an inland glaciation of Central Tibet during the LGP.

On the Karakorum N slope the glacier retreat already took place before 12 870 +/- 180 YBP. This is evidenced by the 14C age of a basal mud which overlays a ground moraine in the Muztagh Valley (36°03'N/76°25'20"E, 3990 m asl; Fig.1, No.15; Table; see Kuhle 1994 Tab.2 Fig.36, Fig.138 No.10).

A Holocene, Neoglacial advance was established for two stages, in accordance with the 14–16–step stadial scale (Table) developed in the Dhaulagiri Himal (Kuhle 1982a, 1983; Fig.1, Nos.19, 26): the first is the Nauri Stadium (V) 4490+/-95 (see 2.2.12 and 2.3.1) and 4165+/-150 C14-years ago. The second is the older Dhaulagiri Stadium (VI) 2050+/-105 to 2400+/-140 C14-years ago (Kuhle 1986, 1987c). In the Khumbu Himalaya (north of Cho Oyu; 27°52'N/86°42'E; Fig.1, No.14), a Nauri Stadium (V) ELA depression of 560 m was averaged from seven glaciers. North of the Dhaulagiri and Annapurna Himal (28°43'N/83°45'E) Nauri Stadium (V) was found to have experienced an average equilibrium line depression of 570 m for 17 glaciers, while on the south side the ELA descended slightly less than 400 m (28°35'N/ 83°45'E), calculated on the basis of eight glaciers (Kuhle 1982a).

A reconstruction of the ELA depression in the Animachin Massif in NE Tibet (34°48'N/ 99°33'E; Fig.1, No.44) during the Nauri Stadium (V) (Kuhle 1986, 1998 Fig.14) yields a value of only 240 m. This is an indication that the response of the equilibrium line to short–term climatic changes is less significant in northern than in southern Tibet. The younger, Late Glacial ice margin positions and perhaps the Neoglacial advances, may provide a conceptual outline of the extent of glaciation on the Tibetan Plateau during the early LGP. The author thinks that the Tibet glaciation played an important role for shaping Neogene and Pleistocene glaciations and deglaciations.

3. Synopsis of the Extent of the Inland–Ice in Tibet Since the Earliest LGP

Fig.3 shows the reconstruction of the maximum glaciation in Tibet, with an area of about 2.4 million km². In the central part it formed a compact inland ice sheet with outflows that descended through the surrounding mountains and terminated at the steep edges of the high plateau.

To err on the safe side, the 50–70 m high, glaciofluvial gravel terraces east of 84°–85°E (Fig.1, Nos.10, 11), which lie there on a valley floor at 3800–3900 m asl, are classified as High Glacial deposits (Kuhle 1988c Fig.2, No.9 and Figs.7, 8). They indicate a glacier free Tsangpo section, which separates glacier complexes I3 and I2 as far west as 84°E (Fig.3a).

Further in the west the complexes I2 and I3 join again during the maximum of the LGP (Stadium 0: Table; Fig.3a). With an equilibrium line as low as 600 m below the average plateau altitude (Fig.3b) glaciation seems likely, even in the case of the more easterly Tsangpo section. This would place the gravel terraces and varved clays of the deepest parallel valleys in the Late LGP. Thus until now, at least for one time–interval during the LGP this valley section is regarded as free of ice. A precise dating of the terraces is to be done (Fig.3a, from north of Shisha Pangma to near Kola Kangri, Fig.1 from No.11 to 12, Kuhle 1991a Fig.43, Nos.37–45). The second ice free area is the Tsaidam Depression. In the block diagram of Fig.3a it shows up as narrow strip below I1 (Fig.1, No.50, Kuhle 1987b, p.302; 1998 Fig.15).

The third large ice–free area, documented by lake sediments, was observed in Mongolia. Dates of lake sediments in the area of the former ice sheet are all younger than 13,000 years old because they developed only after the glaciers had melted (Gasse et al. 1996, Avouac et al. 1996, Van Campo & Gasse 1993, Kashiwaya et al. 1991). By contrast, lakes in nearby, nonglaciated areas such as the Qaidam basin and the Gobi desert display continuous sediment records going back more than 40,000 years (Chen Kezao & Bowler 1986, Pachur & Wünnemann 1995, Wünnemann et al. 1998, Rhodes et al. 1996).

In a north–westerly direction from Mt. Everest to K2 (Fig.1, from No.14 to 15; cf. Fig.3b left), and from Dhaulagiri to the western Kuen Lun (Fig.1, from No.19 to 16), the cross section shows that the LGP ELA runs parallel to the present ELA (Kuhle 1998 Figs.16, 17, 20, 21). Over southern Central Tibet the ELA attained an altitude of over 4700 m asl (see 2.3.2) (Kuhle 1988c Fig.2). Nonetheless an LGP ELA depression of at least 1200 m means that 83–86% of the plateau surface was above the ELA. The accumulating ice necessarily led to the filling of valleys that incised the Tibet plateau. The remainder accounts for the remaining 14–17%.

LGP glaciers attained an approximate thickness of 2700 m. Glacier thicknesses, ascertained by means od abrasions, polishings and erratics, reach 1600–2000 m in the Himalaya (Kuhle 1982a; 1991a), and 700–1200 m in Central and North Tibet. In the northern and western Karakorum thicknesses of more than 1750–2000 m have been observed (Kuhle 1994). However, these are probably minimum values. The ice may well have risen to 2000–3000 m in Central Tibet (Fig.3b) owing to a compact ground plan that extended over 1500 x 3000 km. The high viscosity of cold, continental glacier ice with annual temperatures of around –6 to –10°C at the ELA (Kuhle 1988e; 1994) supports the build–up of ice, provided that there is sufficient precipitation. An average thickness for all of the Tibet ice of approximately 1000 m implies that 2.2 million km³ of water was bound in the ice sheet of Tibet. This corresponds to a lowering of sea level of about 5.4 m (calculated on the basis of data provided by Flint 1971).

Earlier it has been shown (Kuhle 1998 Fig.22) how the glaciated area in Tibet and in the Karakorum relates to an ELA depression or an uplift of the plateau- and mountain relief of only 500 m. At the same time it permits an estimation of conditions if the equilibrium line drops by 1200 m, and makes the cupola–shaped build–up of the inland ice to a considerable thickness plausible. With the ice held back by mountain barriers, build–up was assisted by the low run-off, and initially by freezing to the subsurface. At a later stage, due to ice build–up, the pressure–induced melting point was reached. Run–off from the central inland ice sheet gradually increased until an equilibrium had been achieved. This was the end of ice build–up.

4. The Tibet Plateau – A Stabilizing Factor For the Climate System

4.1. The Radiation Balance

The preceding chapters presented a synopsis of data that describe warm and cold environments in Tibet and Central Asia. Important data are laterites. At the respective sites they either have been overridden by LGP glaciers or LGP glacial sediments overlie unconformably these laterites. Due to adiabatic cooling on the High Plateau north of the Himalayas no laterites have been found. Whether in older times Tibetan laterites could have formed in warm times before a glaciation commenced is not known. A drillsite at (36°48'N/99°04'E, 3170 m asl) showed 6-15 Quaternary ice advances and retreats (Fig.2). The location near the NE fringe of the inland ice in the foreland of the Kukunor Shan permits tentatively a correlation with the Vostok ice–core. A drillsite in Central Tibet, probing through the Pleistocene into the Pliocene, studying also palynological samples, is needed to solve this question. Recent warming shrank some existing valley glaciers on the Himalayan south side (Kuhle 2004b, p.18-20). It is therefore inferred that continued warming to Pleistocene greenhouse conditions, 2.5–3°C above preindustrial values, shrinks present valley glaciers further. Testing climate models for such times using a reduced mountain glaciation appears reasonable. Data about vegetation on the Tibet Plateau during Pliocene, Pleistocene and possible future warm times do not exist. Coupled GCMs such as the current version of CSM from NCAR might shed light onto this question. If the present vegetation and the present glaciers are modeled correctly the modeled vegetation of Pleistocene and Pliocene warm times can be tested against data from drillsites. This applies to the present long cores from Lake Baikal and to a necessary future drillsite on the Tibet Plateau. The ELA on the Himalayan South is expected to be at roughly 5300 m, on the Tibet Plateau at 5600 m asl. To provide milestones climate models have to meet hard data on the ice–coverage of the Tibet Plateau. The preceding sections provide them. This applies both to ice–thickness in valleys and to the extension of inland ice of considerable thickness on the Tibet Plateau.

These data are summarized (ice extension and thickness) as overview in Fig.3. This Figure, that is based on about 39 expeditions during the past decades shows that the preliminary maps of CLIMAP (1981), indicating a small mountain glaciation for the LGM, have now been improved by hard data that show a large inland ice for the LGP and LGM. The author suggests to use these up to date field data to drive or test climate models.

Today the Tibet Plateau has an altitude of 5200 m meters. The low latitude (about 30°N, corresponding angle of incidence) together with the high altitude implies that from 1360 W/m^2 about 1000 to 1300 W/m^2 reach the surface. To determine the actual radiation–balance between August and November in 1984 and 1986 climatic parameters were measured on Mt. Everest and Shisha Pangma in southern Tibet (28°N; Fig.1, No.14), as well as on K2 in NW Tibet (36°N; Fig.1, No.15).

Eight climatic stations were installed for that purpose at altitudes varying from 3800 to 6650 m asl. At the same time portable, handoperated instruments allowed comparative measurements in other places. Measurements of radiation, and radiation balance, on rock or scree, as well as on glaciers have been carried out. Approximately 25,000 data records on radiation, return radiation and albedo were obtained. The data are published in Kuhle & Jacobsen (1988) and Kuhle (1985, 1988b; 1989, p. 277-279; 1994; 1996c, p. 209-212; 2005a).

When weather conditions were such that radiation was not impeded by cloud cover, the values of incoming radiation were between 1000 and 1300 W/m^2. The latter is approximately the solar constant at the upper limit of the atmosphere in relation to the corresponding position of the sun at that time. Theoretical incoming radiation on September 21st, as a mean value, is about 1180 W/m^2 at the latitude of the area under investigation (30°N).

An unglaciated Tibet Plateau with its large fraction of dark matter absorbs most of this energy (albedo 15–20%), either directly heating rocks and disseminating the heat at night and by wind or indirectly through sublimination of snow, melting and evaporation. Lumping all processes together the Tibet Plateau is today one of the most effective radiative heating sources of the Earth. Thus a non–glaciated Tibet tends to stay in this situation. In other words: A non–glaciated Tibet–Plateau contributes to stabilizing the climate system in this state.

On glacier surfaces, especially on fresh surfaces in feeding areas, 85–90% of the short–wave radiation (0.3-3µm) is reflected. During the LGP about 97% of the highland area was covered by ice. Being glaciated, also at such high altitudes and due to geometric conditions (angle of incidence) the Tibet–Plateau reflected considerable amounts of energy. In addition at 6600–7000 m asl, the greenhouse–effect no longer applies since the glacial aridity reduces the moisture content of the atmosphere.

With evidence for a glaciated area of 2.4 Mio km^2 and 1180 W/m^2 as average value for September this is an energy input at the surface of about 2832 x 10^{12} W (1180 x 1 10^6 x 2,4 x

10^6) during non–glaciated times. This reduces by 85%, i.e. 24072 x 10^{11} W to only 4248 x 10^{11} W during glaciated times. This means that during glaciated times the Tibet–Plateau tends to keep the system in this state as well. Based on a rough approximation of the radiation energy–balance at the surface the Tibet–Plateau appears to have the potential to serve as one of several stabilizing factors of the climate system both during warm times and during glaciated times.

Also in the case of the scandinavian and northern hemisphere glaciers excluding the Tibet Plateau that means in both cases it must be noted that the preceding lines refer to energy input at the surface. In mid and mid–high latitudes the atmosphere itself absorbes also considerably. The processes discussed for high altitudes in Tibet apply also to high latitudes in Scandinavia, Eurasia and North America: Glacial aridity reduces also the moisture content of the atmosphere in high latitudes. Thus energy absorption by water vapor of the atmosphere might reduce. An optically more transparent atmosphere permits more shortwave radiation down to the surface. High albedo–values of the glaciers contribute to stabilization of the system in the glaciated state.

With an atmosphere approximately transparent to radiation, incoming radiation at Tibetan altitudes produces an energy input at the surface that is at least four times higher than that between 60°N and 70°N. This applies also to the Pleistocene North European inland ice centre (Bernhardt and Phillips 1958). Radiation loss through diffuse reflection of the atmosphere down to sea level (Lauscher 1956) is about 7%. Therefore the Northern Hemisphere glaciers excluding Tibet (32,5 Mio km²), scaled down to the approximated average radiation input (ca. 300 W/m² instead of average ca. 1180 W/m² in Tibet) contribute ca. 4 times to the stabliziation of the climate system either in a warm or in a cold state.

A quantitative test through long transient coupled GCM runs is a task for the future. Above approximatation might explain (data in the Vostok ice–core, Petit et al. 1999) why, after a glaciation was initiated (such as on the Tibet Plateau) the window of warm conditions appeared difficult to reach again.

A Tibetan inland ice surface of 2.4 x 10^6 km² implies a cooling effect equal to at least 9.6 x 10^6 km² (4 x 2.4 x 10^6) of Nordic inland ice. This latter value represents an ice sheet more than twice the size of the North European Weichsel ice.

A statistical test of the radiative impact of the Tibet inland ice can be found in Lautenschlager et al. (1987, pp. 8-40).

4.2. Isostatic Impact of the Tibetan Inland Ice

A glacio-isostatic drop of the plateau of about 600-700 m must have occurred (Kuhle, M. & Kuhle S. 1997 p.124-129). (Density difference of ice versus material of the earth's mantle (g=1: 2.4 at an ice thickness of 1000 to 2700 m=417 to 1125 m glacio-isostatic drop). Since deglaciation occurred about 9000 years ago, remnants of glacio-isostatic uplift with downward tendency should still be observable, as in Central Scandinavia (Mörner 1978) or the Pleistocene glaciation of North America (Andrews 1970)

The latest tectonic map to be published in China (Seismographic Map of China, 1987) shows rates of uplift of more than 10 mm/y for Central Tibet (Chen 1991; Fang 1991). This value is two to three times higher than what had been established for the much younger High

Himalayas, which were actually uplifted more rapidly than Central Tibet since the Late Tertiary as the evidence of antecedent valleys demonstrates (Kuhle 1982a).

Schneider (1957, pp.468, 475) puts the uplift of the NW-Karakorum at 12,000 m since the end of the Late Tertiary. This corresponds to 3–4 mm/year for Tibet and the Himalaya since the Late Tertiary. Gansser (1983, p.19) gives an integral value of 10–15 mm/year for Tibet and the Himalaya while for the Himalaya alone a value of 4–8 mm/year is given (personal communication, 1982). The Swedish triangulation by Norin (1932, 1982) showed that since the Survey of India in 1861 the profile by way of Leh and into SW Tibet, had undergone an uplift of 37 m. This implies a rate of uplift of 521 mm/yr. In the Scandinavian centre of uplift such extreme rates of uplift only took place around 10–8 ka (see Mörner 1978 Fig.1 and for North America: Andrews 1970). They decreased to the present rates of 10 mm/yr by about 4 ka.

One possible conclusion is that in the region of Na-K'ot Ts'o (Fig.1, No.59; Fig.3a NW of Kamet), which is characterised based on that Swedish triangulation profile by very strong uplift, the Tibetan ice persisted several thousand years longer than the Scandinavian inland ice.

On the other hand young subduction processes in the southern Himalayas might contribute might to strong uplift as well. Thus a melting history that is comparable to other areas of the world, is also possible. New drillsites on the Tibet Plateau characterising individual ice advances and retreats are thus needed.

As Fig.3b shows, the ice burden decreased in the direction of the outlet glaciers on the rim of the Tibetan highland as a function of its steep gradient curves. On the one hand this serves to reduce the uplift, owing to the reduction in pressure. On the other hand, mountains south of the inland ice, like the Himalayas, were affected merely by ice filling of the valleys. This leads to a comparatively linear ice burden, whereas the Central Plateau carried an extensive ice sheet.

4.3. Impact of Orbital Parameters in Low Latitudes

Data from the Vostok ice–core show for glacial times (Petit et al. 1999) temperatures of about 8,3-8.5°C below pre–industrial values. For LGP times an ELA depression of 1200 m was inferred (see above). Using an atmospheric temperature gradient of 0.007 C per meter this implies a temperature drop of 8.4°C. Thus the field data (ELA depression) and the isotope–data from the Vostok ice–core yield independently consistent results. Therefore, at a first glance, both for high–latitudes (Vostok ice–core) and the subtropics (Tibet–Plateau) a comparable drop of the temperature can be inferred. Using an atmospheric temperature gradient of 0.008°C/1 m (meaning extreme aridity) summer temperatures would drop 9.6°C.

An ELA depression is however a relative value documenting a balance of precipitation in the catchment area, altitude, temperature and ice–melting. In principle it can result from a temperature drop. It can also result from an increase in precipitation in the catchment area. This increase in precipitation may or may not be accompanied by a temperature drop. Conceptually the effect of an ELA depression by 1200 m can also be achieved by a respective uplift by the same value. Considering however the consistency with the Vostok ice–core data a temperature drop by 8.4°C is regarded as possible interpretation.

Progressive cooling, lowering of the equilibrium line, or uplift of the Tibetan plateau above the ELA must have led to an initially more extensive ice cover in some parts of the surrounding mountains, as well as on Tibet itself.

Due to geometric conditions the variation of orbital parameters is, regarding radiation differences, more important in the high latitudes than in the subtropics (Milankovic 1941; Schwarzbach 1974, pp.300, 303). Following the equations of the variation of orbital parameters contributes to temperature differences of about 3,5°C (Schwarzbach 1974). A global equilibrium line depression of about 500 m correlates with this 3,5°C drop in temperature. The ocurrence of an insolation maximum in connection with the earth being at perihelion is expected to have a higher effect than an insolation maximum in connection with aphelion. The same applies to a high insolation ocurring during NH summer versus SH summer (such as different albedos of the hemispheres). Apart from this, Milancovic radiation anomalies do not apply to the whole earth at the same time; they have an alternating cooling effect on the N and the S hemisphere.

Above it was shown that the Tibet Plateau receives radiative input that is close to the solar constant at this latitude. Thus radiative losses during glacial times are high. If an insolation maximum that results from a superposition of above–mentioned processes coincides with a glaciated Tibet–Plateau a stabilization of the glaciation by reflection is to be expected. If an insolation–maximum coincides with a non–glaciated Tibet–Plateau a warm time–interval is expected to be extended. If an insolation maximum coincides with a phase of either buildup or decay of glaciers, then depending on the degree of buildup/decay that already took place (amount of reflected radiation), decay or buildup will either be accelarated or stopped and reversed. Fluctuations of isotope values of the Vostok ice–core that tend to document a started but uncompleted deglaciation are consistent with this.

Thus, time–resolved, depending of the superposition of each of the involved factors the Tibet–Plateau might either have been a stabilizer of a warm age (scree effect), a pacemaker of a warm age, accelarating already ongoing decay of ice, a pacemaker of an ice–age, accelarating glaciation by backscattered radiation, a trigger of an ice–age or a stabilizer of an ice–age.

The isostatic effects mentioned above contributed also to this.

Even with an equilibrium line depression, or uplift of the plateau, of only 500 m, glaciers would have covered at least one third of Tibet. This is evidenced by the reconstruction of Late Glacial and Neoglacial ice margins and the corresponding equilibrium line depressions. The increase in the glacier surface area is a function of the depression of the equilibrium line (Figs.3b and 4c) and the average elevation of valley floors or the plateau. Owing to glacier tongues descending almost twice as far as what the ELA amounted to, an increase in surface area was achieved even before the equilibrium line has reached its lowest position. At that time large-scale foreland glaciations built up from the mountains towards the plateau, and in the shallow main valleys (Karakorum) this resulted in sudden gains in reflection surfaces as the pace at which they filled up with ice increased. These valley fillings and the build- up of glaciated mountain forelands extended the nourishment area of the glaciers. Such sudden increase in the glacier area could only originate from the high mountains above the plateau. The mountains acted, to a certain extent, as "crystallization centres" for the build-up of the ice. Glacier formation was promoted by mountains which towered high above the plateau. Initially, the volume to be filled was small due to the presence of shallow valleys. With a decreasing equilibrium line the proportion of supply areas increased. It follows that these shallow valleys were forced to fill up and that their glacier surfaces coalesced to form an ice cap. The glaciation of about one third of the Tibetan highlands, with subtropical radiation parameters, was more effective in triggering an autocyclic mechanism than all the remaining areas of the earth. This in turn led to the drop in temperature of about ca. 10°C, and thus to the High Ice Age proper.

4.4. Impact of ELA Depressions on Global Climate

The high altitude of the Tibet Plaeau implies that a large area can get glaciated with a low amount of ELA depression. About 70% of incoming solar radiation is then reflected into space after an ELA depression of 500 m. This supports further cooling. If the radiation–deficit equivalates a further temperature drop a cooling of 1–1.5°C can be inferred (Kuhle 1989, p.279). This implies that an initial temperature drop of 3.5°C based on orbital parameters might be amplified by further 1–1.5°C through the Tibet Plateau. For warm times the reverse magnitude applies.

An ELA depression on the Tibet–Plateau may or may not have counterparts in other regions of the world. If – which needs to be tested – various regions on earth are coupled such that an ELA depression in one region might, with or without a time–lag, appear in other regions as well, then respective ELA depressions in other regions, such as Scandinavia, might be expected as a consequence. There an ELA depression of 500–700m implies the coverage of a large region with ice and a subsequent further feedback through backscattered radiation. This is outlined in Kuhle 1998 Fig.25. If the link between various continents is not that close the glaciation of the Tibet–Plateau is one of several important factors that contributed to glacials/interglacials. This needs to be tested quantitatively by future long transient coupled GCM runs.

The earliest time for additional amplifications of glaciations through the Tibet–Plateau is due to the necessary uplift during the late Neogene.

In this context attention ought to be drawn to the fact that a 1000 to over 2000 m thick burden of inland ice has existed. The Vostok ice–core shows that Pleistocene interglacials have been characterized by temperatures between 2.5 and 3.5 C above pre–industrial values. As today the Tibet–Plateau is generally unglaciated it is inferred that during Pleistocene interglacials it was ice–free as well. Consequently the isostatic depression during glacials was replaced by isostatic uplift during the short times of Pleistocene interglacials. Whether during these short warm times the Tibet–Plateau was able to uplift considerable amounts (viscosity of the mantle) or whether it factually remained generally depressed needs to be tested. If during Pleistocene warm times it uplifted only slightly then since the Holocene considerable uplift can be expected.

4.5. Duration of Ice Buildup from Warm Conditions

After the polar regions the Tibet–Plateau, due to its high altitude, is the first to be widely glaciated. This raises the question of timing and magnitude, also with respect to a potential isotope signal this might produce.

Today in central Tibet the precipitation is 400 mm/yr (Lhasa 437 mm/yr: more in the east and less in the west). Assuming 50% evaporation, 200 mm water column remain to form ice. With a corresponding ice–column of 222 mm 5000 years are needed to form 1110 m ice.

With a potential weaker summer–monsoon (Kuhle 1982a, 1982c) and 167 mm/yr precipitation the observed 1000 to 2000 m thick inland ice could have formed within 10 000 years. This appears to be consistent with the steep flanks of the fluctuations of the Vostok ice–core. A cooling of 8°–10°C, with precipitation similar to current values, would provide the most simpel model of ice build–up. During warm times the precipitation was most likely higher. During cold times it might have been less. Thus it appears justified to use present–day values for the precipitation as first approximation.

On the other hand during glacial times (Flohn, pers. comm. November 1987) the aridity was globally 10-30% higher than at present. At a first glance an even more reduced precipitation might be expected.

While globally this mechanism is correct, increased cooling means a lowering of the snowline and increased precipitation at lower altitudes. In Tibet and in the Karakorum the situation is different. In the Karakorum and Hindukush, the most arid area under investigation (Fig.1, No.15, 16, 32-34, 36, 37, 55, 56), exponential increase in precipitation of 1000–2000 mm/yr presently occurs at altitudes above 4000 m on the southern side, and above 4700–5000 m on the northern side (Kuhle 2004c, p.96). On the other hand, on the valley floors between 1000 and 4000 m asl, precipitation fluctuates between 40 and 250 mm/yr (Kuhle 1994; Miehe et al. 2000). During glacials, the altitude of heavy precipitation is expected to have been lowered by 1200 m to 2800-3500 m asl to be in accordance with the observed ELA depression.

Thus during glacials a very substantial increase in precipitation in the supply area can be expected. Added to this is build-up of the ice cover, which led to an additional increase in the supply area surfaces, and thus to an improvement in the glacier nourishment. This auto–feedback supports the concept of a rapid glacier builtup once it is triggered.

4.6. The Relief–Induced Decay of Glaciers at the End of a Glacial

Elevating temperatures by 3.5°C through changes of the orbital parameters, rises in wide areas the equilibrium line by about 500 m.

The reaction of Nordic lowland regions to the rise of the equilibrium line could have been dramatically. A rise of the equilibrium line by 500 m must have led to vast changes in the catchment and ablation areas. The corresponding melting process resulted in large, rapid loss of glacier area (approximately 1.5 times in the newly added ablation areas; see v. Höfer 1879; Louis 1955) (Kuhle 1998 Fig.25 step from No.3 to No.2). Being freed from ice, the lowland areas reflect only 15% of the incoming radiation and reattain a 70% heat gain.

Due to the position of the equilibrium line on steep plateau edges, on the Tibet Plateau this rise only produces a small reduction in glaciated areas. By melting, the glacier tongues receded and ascended about 800 m. They still terminated at about 1800–3500 m asl, below the average level of the plateau and the inland ice (ibid. Fig.25 No.2). Induced by the lowland areas, global warming eventually reduces Tibetan ice by way of a feedback process of raising slowly the equilibrium line.

This demonstrates that Milankovic cycles are primarily interpreted to trigger short cycles of deglaciations. Overall the Tibet inland–ice serves as stabilizer for the climate system.

4.7. Climatically Steered Tectonics

Bridgeland and Westaway showed that climatic changes can indirectly through isostasy and subcrustal flow have a considerable impact on tectonic situations, well beyond the extension of the isostatic load.

For the Tibet–Plateau this has been calculated as well: Uplift of the plateau was slowed and later depressed by the progressive increase in ice thickness. Each depression was later reversed by interglacial deglaciation. According to this model the average Pleistocene level of Tibet is unlikely to have changed since the beginning of the late Neogene ice ages (Kuhle 1993).

This theoretical deduction is confirmed by field observations. There are ice margins from the glaciations older than the LGP at eleven localities in High Asia. These lie at approximately 150-300 m lower than the LGP–age moraines in NW and NE Tibet (Kuhle 1988e, 1987a) (Fig.1, Nos.16, 45-49). This equates to the 100–200 m difference between the equilibrium lines of the second but last glaciation (Riß ca. 150 000- 120 000 YBP) and the LGP glaciations (Würm ca. 60 000 – 18 000 YBP), known from tectonically less active areas as in Europe.

Two alternatives are possible to explain the observations:

4.7.1. Neogene and Pleistocene Uplift as in the Holocene

In a region of continuing uplift, as Tibet is suspected to be, these older ice marginal positions would have been uplifted at a rate of about 3 mm/yr or about 390–420 m during the last 130–140 ka prior to the LGP if the boundary conditions that established the respective ELA (moisture availability, temperature, firn accumulation) had been the same in different ice–ages.

The preservation of ice marginal positions from the second but last ice–age shows that either the uplift at that time was slower, or the boundary conditions (such as cold vs. warm glaciers) had been different. If the boundary–conditions would have been the same as during the LGP and if the uplift would also have been the same, the old moraines left over from the second but last glacial would have been overridden and destroyed. The reason is that the uplift of approximal 400 m exceeds the normal vertical distance from moraines of the seond but last glacial to those from the LGP. Even if the equilibrium line had run out at a greater altitude, the younger glaciers would have flowed at least 400 m further down than the glaciers of the second but last glacial.

4.7.2. Pronounced Uplift since the Holocene

The other possibility is that in connection with a high mantle–viscosity the short Pleistocene interglacials did not allow for isostatic adjustion through uplift. In this case the Tibet–Plateau would have stayed in a depressed low position. Different positions of ice–margins would reflect overall different glacial advances. Only since the Holocene long and uninterrupted uplift occurs. This interprets the current high uplift rate as partially or nearly fully isostatic.

4.8. Tropical Weathering in Tibet and Pre–LGP Interglacials

Remains of a widespread Hipparion fauna in Middle Pliocene sediments of central Tibet are indicators of a warm–tropical steppe climate and show that the altitude of the winter snow zone had not yet been reached at this time (Chen 1981, Ji et al. 1981).

From 2.5 Ma onwards did sediment begin to accumulate on the loess plateau of China. This is interpreted as indicator of the onset of winter monsoon circulation (Kukla and An 1989; An et al. 1990; Ding et al. 1992).

Tibet's climatic impact first became effective when uplift raised the plateau to the level of the seasonal snowline from ~2.5 MaB.P. onwards, and subsequently to the level of maximum glaciation starting at ~1 MaB.P. The progressive glaciation of the Tibetan plateau may thus have been the decisive terrestrial factor causing orbital variations to translate into global ice ages.

During warm interglacials in Tibet the conditions had most likely been considerable warmer and wetter than today. On the eastern edge of Tibet (30°45'N/103°20'–40'E; Fig.1, Nos. 51, 57, 58), west of the town of Chengdu, at approximately 600 m asl there are mud flow-, or debris flow sediments or even ground moraines with a thickness of several (2–15) meters. They stretch over an area of about 40x50 km. Their origin as older than the LGP is suggested by the intensive degree of tropical weathering. Li Tianchi from the Geographical Institute in Chengdu confirms the glacial origin of the sediments and interprets these material as even older than the second but last glacial period. Li Tianchi and coworkers suggest that the debris flows and moraines belong to the Early or Middle Pleistocene (ca. 500 ka, pers. comm. August 1991) without, however, an absolute date being available. This implies that in East Tibet tropical conditions prevailed at that time. Together with an ice–covered Arctic Ocean considerable latitudinal temperature gradients are interpreted to have existed in this time, particularly in winter. The extension of tropical Pleistocene and Pliocene sediments in Tibet needs to be determined quantitatively by drillsites. Near the older ground moraines and a few kilometres from the edge of the plateau is the Chung Leh Shan (moutain massif) at about 4500 m asl. There is a moraine and linked debris flow indication, dated by weathering and position, of Early to Middle Pleistocene age. It documents a piedmont glaciation. Because it extends down into present warm subtropical climatic conditions, the glacier must have flown from an already highly uplifted Tibetan Plateau in order to have been able to reach down so far. The lowest of the Last Glacial glacier terminal moraines in this area reached down to 1300–1500 m asl (see v. Loczy 1893; Li Tianchi 1988). This means that at the time of that older piedmont glaciation the ELA must have been 350–450 m lower than during the LGP assuming a similar altitude for the Tibetan Plateau and surrounding mountains.

As the moraines appear in an area that is today characterized by subtropical conditions, comparable conditions can be inferred for pleistocene interglacials. Most likely (see Vostok ice–core) the conditions had been warmer. This is consistent with considerable latitudinal temperature gradients in times preceding a glaciation.

Judging from a world-wide comparison of ELA differences between the LGP and older glacial periods, the snowline of these earlier ice ages was only about 100 m lower than during the LGP. This means that the height of the base of the catchment area (the height of the plateau) must have been some 300 m greater than during the Last Glacial period. The lowest ice margin, related to the LGP, must have been 800 m higher than at the time of the older piedmont glaciation (see above).

The preservation of old moraines implies a stagnation in the uplift, if not subsidence of Tibet, at least during the Late Pleistocene. This points to a glacio–isostatic compensation or overcompensation of tectonic uplift in Tibet (Kuhle 1993). This supports the above–mentioned alternative 4.7.2.: reduced or nearly non–existing uplift in the Pleistocene, pleistocene interglacials too short to allow for large uplift (mantle viscosity), pronounced uplift since the Holocene.

This has implications for the boundary conditions for warm climates, both in the Pliocene and in the future. In the Pliocene, after about 2.75 Ma, large scale northern hemisphere glaciations began after the Tibet–Plateau reached a high–altitude. As various other factors changed also around this time, the uplift of the Tibet–Plateau is regarded as one important co–factor. The mechanism to terminate a warm climate are explained above. During the Pleistocene a considerable isostatic depression is inferred. An inland–ice existed. During the Holocene no reglaciation occurred. The uplift that is observed today is interpreted to be caused partially by isostatic uplift. Depending on its surface characteristics, particularly glaciated versus non–glaciated (measured incoming radiation close to the solar constant for that latitude), the Tibet–Plateau contributes to stabilizing the climate system: Either in glacial conditions or in non–glacial conditions. It is thus inferred that the Tibet–Plateau will contribute to stabilizing the climate system in the present non–glacial conditions until an elevation of ca. 5600 m will be reached. Other conditions, such as pronounced latitudinal temperature gradients are expected to modify this value.

5. Start and End of Warm Climates

The preceding chapters showed data on the Tibetan inland–ice and its impact on climate. The nature of the data prescribe that the preceding chapters had to focus on the LGP. Now the data are put into a broader context. The onset of the large northern Hemisphere glaciations is as well discussed as boundary conditions for the future. The following chapter discusses the role of important subfactors that govern climate change. The purpose is to outline mechanisms that either extend, terminate or stabilize present, warmer or colder climates.

5.1. The Panama Seaway

$\delta^{18}O$ records of planktonic foraminifera (Tiedemann et al. 1994; Shackleton et al. 1988; Morley and Dworetzky 1991) have shown that large northern Hemisphere glaciations began about 2.75 MaB.P.; from ~1 MaB.P. onwards, its intensity (ice volume) and the length of its glacial phases approximately doubled (Fig.4b).

Sometimes the closure of the Panamanian seaway and the resulting impact on North Atlantic deep water formation are mentioned in connection with the onset of the ice ages (Haug and Tiedemann 1998). The closure occurred 4.6 to 3.6 Ma ago. This is 1 Ma before the large Northern Hemipshere glaciations. The argument that increasing obliquity amplitudes between 3.1 and 2.5 MaB.P. caused the ice to build up, contradicts the fact that the decrease in obliquity amplitudes between 1 and 0.8 Ma does not coincide with ice retreat; on the contrary, it corresponds in time with an intensification of global glaciation (Fig.4a,b).

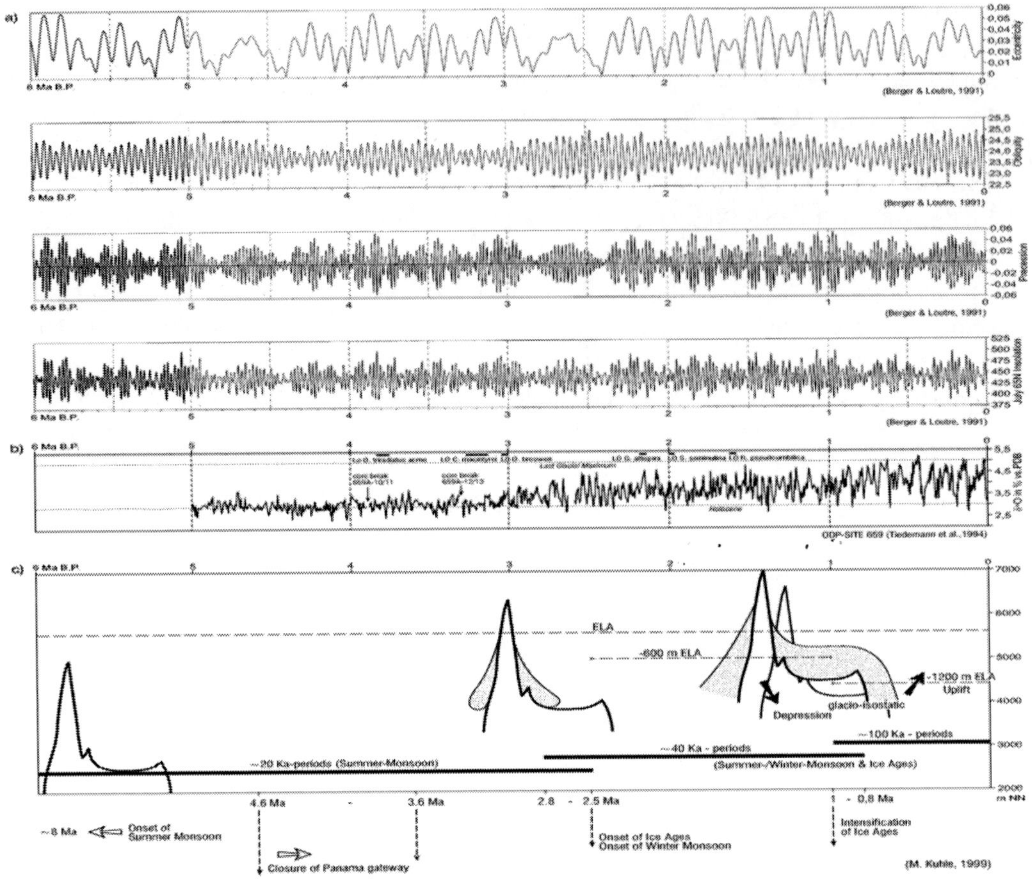

Fig. 4: (a) Orbital parameters of the earth and corresponding insolation values for 65°N for the last 6 million years according to Berger & Loutre (1991).(b) Benthic oxygen isotope records from Ocean Drilling Program Site 659 according to Tiedemann et al. (1994). The fluctuations in the $\delta^{18}O$ content of the foraminifera reflect the fluctuations of the global ice volume, with high values corresponding to the glacials and low values to the interglacials. Neither the beginning nor the intensification of the Quaternary glaciation period is correlated with the insolation (a).(c) Synopsis of the uplift and glaciation of the Tibetan plateau in their relation to other geoecological events. Comparison between (a) and (b) shows that an additional factor apart from orbital variations is required to explain both the start of the ice ages about 2.8 Ma and their increasing intensity from 1 Ma onwards. The closure of the Panama gateway occurred too early to be the terrestrial cause. The uplift of the Tibetan plateau, as far as it can be reconstructed from the onset of the summer- and winter monsoons, and, derived from this, the start of an autochthonous glaciation of Tibet from ~2.5 MaB.P. onwards, were synchronous with the onset of the global ice ages. Evidence that variations of the summer- and winter monsoon intensity documented by marine dust flux records and loess-palaeosol sequences on the Chinese loess plateau occurred in phase not with the insolation variation but with glacial-interglacial cycles (40 ky and ~100 ky periods), is a strong pointer to the existence of a Tibetan glaciation. Gradual uplift of the Tibetan Plateau towards the ELA (equilibrium line) level enabled an ice sheet of 2.4 million km² to grow from ~1 Ma B.P. onwards. The resulting cooling effect supported a maximum expansion of the Nordic lowland inland–ice.

5.2. Orbital Parameters

Various factors contribute to glacial interglacial changes. One of these factors are changes of the orbital parameters (Berger 1999; Berger and Loutre 1991; Hays et al. 1976; and Fig.4a.). The known orbital parameters can be expected to operate over long times. Fundamental changes of environmental conditions, such as glacial/interglacial changes after 1 Ma point to either a change of other parameters, including boundary conditions or to a change of the orbital parameters.

5.3. Carbon Dioxide

In recent years, the continuous decline in levels of the greenhouse gas CO_2 in the atmosphere and a concomitant global cooling have been considered to be one likely cause for glaciations (Ruddiman 1997, Broecker 1995). Computer simulations by Berger et al. (1999) show that, to trigger ice ages in this way, atmospheric CO_2 must have decreased from more than 320 parts per million by volume (ppmv) to 200 ppmv during the past 3 Ma.

Several authors such as Pearson and Palmer 1999, Pagani et al. 1999 aim at estimating past atmospheric CO_2 contents from chemical (alkenones) and isotope data. A crucial point is the transfer between measured data and atmospheric gas concentrations. As long as one step involves greenhouse experiments the applicability is restricted to the Quaternary. The reason is that biological systems discriminate well in the data–range for which they are adopted. This is for plants often between 100 and 1200 ppm. Data well beyond this range, such as 10%, CO_2 can normally not be resolved well by Holocene biological systems. Such high percentages occur if for example as first lower estimate the known Tertiary coal and oil deposits are recalculated to moles and then to respective CO_2 contents.

Thus, at present, the precise value of the CO_2 levels during the Teriary is, until new methods appear, regarded as unknown. Given the amount of Tertiary coal deposits it is expected to be higher than at present. The author thinks therefore that cooling towards the Quaternary may or may not be linked to a reduction of CO_2 concentrations. Oceanographic, orographic (Tibet Plateau) and factors generated by the system itself, such as high temperature gradients, contributed to climate change.

For the Eem (MIS11) CO2 concentrations might have been slightly above the pre–industrial value of 270 ppm around > 300 ppm (Fischer et al. 1999). For selected times of the Tertiary temperatures might have been even lower than at present.

The fine-scale records of the GRIP ice core confirm that during the last three glaciations the CO_2 values lagged climatic change by as much as several thousands of years (Fischer et al. 1999, Mudelsee 2001). This reverses the previously accepted causal nexus:

Based on principles of physics elevated CO_2 concentration contribute to elevated temperatures until upper limits are reached. It is however also possible that, depending on the boundary conditions that apply, atmospheric changes can be followed by elevated atmospheric CO_2 concentrations.

5.4. Effects of Cenozoic High Plateaus

Long-term global climatic changes may also have been caused by Cenozoic plateau uplift (especially of the Himalaya-Tibetan- and the N-American plateaus), inducing a change of zonal wind and precipitation patterns and a steepening of the climatic south–north gradient (Ruddiman and Kutzbach 1992). However, uplift of the Tibetan plateau started 20 MaB.P. ago (Harrison et al. 1992; Copeland 1997). For the summer monsoon to occur, a plateau elevation of 2000-2500 m is required. This was attained around 8 Ma B.P. (Prell and Kutzbach 1992; Tiedemann et al. 1994). The winter monsoon additionally needs the albedo effect of a seasonal snow cover (Flohn 1981; Ding et al. 1995; Xiao et al. 1995). This requires plateau elevations of 4000-4300 m. Data on the onset of the summer monsoon suggest, the Tibet–Plateau began to act as a climate–effective barrier some 8 Ma ago already (Manabe and Broccoli 1985; Quade et al. 1989; Prell and Kutzbach 1992; Tiedemann et al. 1994; De Menocal 1995). This is well before the large northern hemisphere glaciations.

In addition, the monsoon chronology reveals the start of the Tibetan plateau's impact on the climate (Fig.4c).

This implies that the uplift of the Tibet Plateau is one of several co–factors for climate change. It supports boundary conditions, such as for ice–formation and melting.

Data on the existence of a Tibetan inland ice are presented in detail above. These data show that once the Tibet–Plateau reached LGP altitudes around 4600 m glaciations are possible and proven by data. At this altitude even small temperature changes, such as 2–3°C can lead to extensive ice–sheets. Other studies show that the Tibetan plateau produces similar effects on wind patterns to those observed at the present time when average elevations of 2000–2500 m are attained (Manabe and Broccoli 1985; Prell and Kutzbach 1992). Thus the above–mentioned stabilizing effect on global climate can be expected at least since the Pliocene. In non–glaciated times the Tibet–Plateau delays a glaciation; in glaciated times it delays a deglaciation.

5.5. Impact on the Global Energy Budget and Ice Volume

At subtropical latitudes and high altitudes, insolation levels on the Tibetan plateau are close to the solar constant (Kuhle and Jacobsen 1988). They are four times higher than in the areas formerly occupied by the Nordic ice sheets. An ice–free plateau surface absorbs 80% of incoming solar radiation, converts it into longwave radiation, and thus contributes substantially to warming the Earth's atmosphere. Our measurements have shown that snow–covered glaciers directly reflect 75-95% of insolation, which is therefore lost to the global heat budget (Kuhle and Jacobsen 1988). Calculations indicate that during the LGM as much as 32% of the albedo-induced energy loss of the Earth's atmosphere was due to the ice sheet of the Tibetan plateau (Bielefeld 1997).

As a result of the albedo of a winter snow cover at high subtropical insolation levels, the Tibetan plateau was able to influence the absolute heat balance of the Earth. The respective altitude of the Tibet–Plateau and the expansion of the Nordic ice–sheets appear to be contemporaneous. Because its elevation was lower than it is today, the plateau had only a 25–50% ice cover at most, even during cold phases. Sensitivity experiments conducted by Marsiat (1994) have shown that under conditions of a reduced glaciation of Tibet the Nordic lowland glaciers also remained rudimentary. This correlates with marine $\delta^{18}O$ records according to which global ice volumes between 2.5 and 1 MaB.P. were only half those during the late Pleistocene (Shackleton et al. 1988; Morley and Dworetsky 1991; Tiedemann et al. 1994).

Glaciation of the Tibetan plateau could only reach its proven LGM extension of 2.4 million km² when today's average elevation of at least 4600 m had been attained, i.e. from ~1 MaB.P. onwards, then inducing maximum global ice volumes.

This effect must have been a strong promoter of ice build–up on a global scale. Thus if for example an extended and/or long–term snow cover occurs today on the Tibet–Plateau it is expected to have a measurable effect on the temperatures of the region and possibly a large part of the northern hemisphere.

5.6. Impact on the Regional Wind Circulation

The consequences are even more far-reaching. Under present conditions, summer warming of the Tibetan plateau leads to areas of low surface pressure and hence to a marked atmospheric pressure gradient in relation to the relatively cold adjacent oceans. Strong winds are the result: the East Asian and the Indian summer monsoons (Findlater 1974). The pattern is reversed in winter. Then the high albedo of the winter snow cover of the Tibet Plateau leads to cold–induced high surface pressure and thus to a compensatory flow of air to the low surface pressure areas. This occurs over the now relatively warm oceans (i.e. the winter monsoon circulation, Flohn 1981; Ding et al.1995; Xiao et al. 1995). Inevitably, the existence of a perennial Tibetan ice sheet must have modified this seasonally alternating large-scale pattern of monsoonal circulation. Whereas the summer monsoon would have been either weaker or non–existent (Sirocko et al. 1993), the winter monsoon must have been much stronger. Data from deep–sea cores from the Arabian Sea are consistent with this. They allow the reconstruction of changes in the upwelling system off Arabia due to variations in strength of the SW Indian monsoon circulation. For the last 500,000 years the data show that the summer monsoon was substantially weaker during the glacial phases, but the winter monsoon was stronger (Anderson and Prell 1993; Emeis et al. 1995). From 2.5 MaB.P. onwards loess accumulated on the loess plateau of China. It is interpreted as record of the onset of winter monsoon circulation (Kukla and An 1989; An et al. 1990; Ding et al. 1992).

High–resolution loess–paleosol sequences from China spanning the last 2.5 Ma permit a reconstruction of the intensity fluctuations of the East Asian summer monsoon. They supply further confirmation that the summer monsoon had been dramatically weaker during glacial times (Rutter and Ding 1993). For the East Asian winter monsoon, however, the same sequences record a marked increase in intensity during glacial phases (Ding et al. 1995; Xiao et al. 1995). Both marine and terrestrial sediment records document a highly significant correlation between variations in global ice volumes and corresponding counterfluctuations of monsoon circulation throughout the entire ice age era.

Thus two interpretations for neogene and future climates are possible:

The first is that warm tropical oceans in connection with an ice–covered Arctic Ocean and the polar night cause pronounced latitudinal temperature gradients. These contribute to high wind speeds during northern hemisphere winter. This includes also, as GCM runs that had been driven by reconstructed sea surface temperatures from the Pliocene showed a strong winter monsoon. Whether this contributes to the Tibetan inland–ice needs to be studied.

The second is that the Tibetan inland–ice itself generates respective regional temperature gradients.

If both mechanisms apply both strong latitudinal temperature gradients, particularly warm tropical oceans and the Tibetan inland–ice can stabilize winter conditions that support an ice–age.

6. Open Questions, Simulation Design and Possible Solutions

6.1. Linking Models and Data

Glacio–isostatic subsidence of the Tibetan plateau below the ELA under ice pressure is interpreted to be one important factor for deglaciations. The resulting albedo drop is a plausible mechanism to pace a self–organizing path towards ice–free conditions. The high uplift rates of 12 mm/a measured in the region of the Tibetan plateau (Hsu et al. 1998) may be a first indication of glacio–isostatic recovery. The author's own observations of moraine deposits on the northern slope of Shisha Pangma, in the border area between the Himalayas and the Tibetan plateau, are consistent with this (Kuhle 1988c). These extensive pedestal moraines were left by the ice during the late Late Glacial.

Kaufmann and Lambeck (1997), Kaufmann (2003) have shown that, on the basis of secular changes in geoid anomaly and free–air gravity anomaly, it is possible to distinguish the amount of glacio-isostatic uplift from uplift caused by tectonic movements. The predictable effects of the melting of an up–to–2 km thick incland–ice on the Tibetan plateau are so profound that the current satellite missions CHAMP and GRACE would be able to identify them.

This needs to be tested quantitatively.

A further issue is the extent of the influence of the Tibetan ice sheet on the heat balance of the atmosphere, particularly whether this was marked enough to have a decisive initial impact on the pattern of global ice ages. Only modeling results can provide an answer.

Using an atmospheric general circulation model (GCM), Verbitsky and Oglesby (1992) studied the evolution of ice sheets in relation to the atmospheric CO_2 concentration (i.e. temperature change) using a computed "glaciation sensitivity" index. Their results showed that Tibet and Siberia were more likely to develop an ice sheet than Canada or Fenno–Scandinavia.

Similar conclusions are drawn by Marsiat (1994), who used a 3-D climate model to simulate global ice growth generated by temperature changes caused by orbital variations. Here too, the first large ice sheet formed on the Tibetan plateau, followed by Siberia. In contrast to Verbitsky and Oglesby, Marsiat sees a strong glaciation tendency in Alaska too.

Both models thus confirm the hypothesis that the Tibetan ice sheet played a foremost role in terms of time.

The latest modeling results presented by Kutzbach et al. (1998) point into the same direction. However, these climate and biome simulations carried out using the Community

Climate Model, Version 1 (CCM1) do not investigate the pattern of ice build-up but are based on the reconstructed glaciation conditions described in CLIMAP (1981), i.e. the Tibetan plateau is assumed to be non–glaciated.

As model results show, there are still "small areas of permanent snow–cover over non-glaciated areas of western Canada–Alaska, northern Eurasia between the Eastern Siberian and Western Eurasian ice sheets and over Tibet at 21 ka. This result indicates that the model would develop a larger area of glaciated land than actually appeared to have been present at 21 ka. The simulation of permanent snow cover over Tibet contributed to the reduction of the Asian monsoon at 21 ka." (Kutzbach et al. 1998, p.496).

That a Tibetan ice sheet not only influenced the monsoon but also had a global impact, is demonstrated by a sensitivity experiment conducted by Marsiat (1994). By reducing the global albedo of snow–covered mountain areas Marsiat attempts to prevent what he considers to be the unrealistic formation of ice sheets on the Tibetan plateau and the Rocky Mountains: "although the mountainous areas were covered by ice after some period, the perturbation occuring at the beginning of the glacial cycle influences the remainder of the simulation, showing lower ice volumes during the entire glacial cycle." This suggests that the albedo effect of the Tibetan ice sheet has a global impact. Thus, unintended, the results of Marsiat (1994) are consistent with the field data: Without tuning the global albedo the model produces the inland–ice on the Tibet–Plateau. Exactly this was observed.

In order to investigate this impact in greater depth, it is appropriate to use sensitivity experiments with circulation models that consider the Tibetan ice sheet to be a realistic possibility rather than a simulation error. It would make sense to place alternative simulations side by side, showing, on the one hand, the climatic effects of a glaciated Tibetan plateau – as reconstructed by the author on the basis of his empirical fieldwork – and, on the other, simulations based on the old view (Wissmann 1959) of a non– or only slightly glaciated plateau, which is still being advocated by Derbyshire and Shi Yafeng (1991), among others. Sensitivity experiments with atmospheric circulation models could also be used to quantify the differential influence of the albedo of the various ice sheets on the global cooling balance.

The crucial role played by the Tibetan plateau for our understanding of Quaternary climatic change is undisputed (Hughes 1998). Due consideration of the Tibetan ice sheet is expected to significantly increase the realism of climatic simulations.

6.2. Future Perspectives

Above it was shown that the Tibet–Plateau contributes to stabilizing global climate. Glaciated it serves as stabilizer for glacial conditions; non–glaciated the low albedo together with the high, close to the solar–constant, insolation, prevents a glaciation. Large volcanic eruptions can cause the global climate to react accordingly. Following the eruptions of Laki in Iceland around 1783 the next year was in Europe known as the "year without summer".

If short–term extreme conditions lead to a permanent snow–cover of the Tibet–Plateau, it is not excluded that this snow–cover contributes to stabilizing these conditions.

At present the Tibet–Plateau is not glaciated. It contributes thus to stabilizing the non–glaciated state.

The following lines focus on the duration of potential future warmer climates.

As has been discussed, Tibet's uplift above the snowline will continue as long as the plate–tectonic boundary conditions and isostatic uplift are not overcompensated by the erosion of the plateau. Due to the minor relief energy on the high plateau, the fluvial erosion during interglacial periods and the glacial denudation during high glacial periods is very insignificant. For this reason, the insignificant erosion, the level of the high plateau will exceed 5000 m in average for several hundred ka. The center will exceed an altitude of 5400–5600 m.

Conservatively calculated, at present an uplift of 400 m would be necessary to reach altitudes that permit a permanent snow–cover. This corresponds to a cooling of ca. 2.4°C at a gradient of 0.6°C per 100 m of uplift measured in High Asia (Kuhle 1994). At the current uplift velocity of 12 mm/a (Hsu et al. 1998) the surface of Tibet needs ca. 33 ka to reach 5400–5600 m asl..

The predicted time interval to trigger an ice age may be delayed by an unforseeable global warming. A future warming of the atmosphere by 1°C requires an additional uplift by 160 m up to 5160, i.e. 5560–5760 m. At the assumed constant uplift velocity, about 13.35 (13.333) ka would be necessary to reach this altitude. Accordingly, the next Quaternary High Glacial would start ca. 46.35 ka from today, i.e. delayed by 13.35 ka.

These predictions apply to the impact of the Tibet–Plateau on global climate. If other parameters change, different time–scales apply. Such parameters are for example the deep ocean circulation or the formation of pronounced latitudinal temperature gradients. In case such changes will not be observed the Tibet–Plateau is expected to contribute to stabilizing the current climate in its non–glaciated state.

ACKNOWLEDGEMENT

The author thanks PD. Dr. Peter Paul Smolka for a very careful review of the manuscript.

REFERENCES

An, Z. S., Liu, T.S., Lu, Y.C., Porter, S.C., Kukla, G., Wu, X.H., Hua, Y.M., 1990. The long term paleomonsoon variation recorded by the loess-paleosol sequence in central China. Quaternary International 7/8, 91-95.

Anderson, D.M., Prell, W.L., 1993. A 300 kyr record of upwelling off Oman during the late Quaternary: Evidence of the Asian southwest monsoon. Paleoceanography 8, 193-208.

Andrews, J.T., 1970. A Geomorphological Study of Post–Glacial Uplift with Particular Reference to Arctic Canada. Can.J. of Earth Sciences 7/2, 703-715.

Avouac, J.-Ph., Dobremez, J.-F., Bourjot, L., 1996. Palaeoclimatic interpretation of a topographic profile across middle Holocene regressive shorelines of Longmu Co (Western Tibet). Palaeogeography, Palaeoclimatology, Palaeoecology 120, 93-104.

Berger, A., Li, X.S., Loutre, M.F., 1999. Modelling northern hemisphere ice volume over the last 3 Ma. Quaternary Science Reviews 18, 1-11.

Berger, A., Loutre, M.F., 1991. Insolation values for the climate of the last 10 million years. Quaternary Science Reviews 10, 297-317.

Bernhardt, F., Phillipps, H., 1958. Die räumliche und zeitliche Verteilung der Einstrahlung, der Ausstrahlung und der Strahlungsbilanz im Meeresniveau. Die Einstrahlung. Abh. Meteor. Hydrol. Dienst 15, 1-227.

Bielefeld, B., 1997. Investigation into albedo–controlled energy loss during the last glaciation. GeoJournal 42 (2-3), Tibet and High Asia IV, 329-336.

Broecker, W.S., 1995. Chaotic Climate. Scientific American 11, 44-50.

Broecker, W.S., Denton, G.H., 1990. What drives glacial cycles? Scientific American 1, 48-56.

Chen Kezao, Bowler, J.M., 1986. Late Pleistocene evolution of salt lakes in the Qaidam basin, Qinghai province, China. Palaeogeography, Palaeoclimatology, Palaeoecology 54, 87–104.

Chen, W.Y., 1981. Natural environment of the Pliocene basin in Gyirong, Xizang. In: Liu, D.S., (Ed.), Geological and Ecological Studies of Qinghai-Xizang Plateau", Volume 1. Science Press, Beijing, pp. 343-352.

Chen, X.,, 1991. The low velocity upper mantle and the thick-bedded lower crust as an important indicator of deep material state in the Tibetan Plateau. In: Bulletin INQUA, N.C. 14, 85.

Chiao, Chao Sung, 1984. Analysis of desert terrain in China using Landsat-imagery. In: F.El-Baz · (Ed.), Deserts and arid lands. Martinus Nijhoff Publishers, The Hague/Boston/Lancaster, pp. 115-132.

Climap Project Members, 1981. Seasonal Reconstructions of the Earth's Surface at the Last Glacial Maximum. Geological Society of America Map and Chart Series MC-36.

Cline, R., (Ed.), 1981. Climap Project: Seasonal Reconstructions of the Earth's Surface at the Last Glacial Maximum. Geol.Soc.Amer., New York, 1-18.

Copeland, P., 1997. The When and Where of the Growth of the Himalaya and the Tibetan Plateau. In: Ruddiman, W.F. (Ed.), Tectonic Uplift and Climate Change., Plenum Press, New York, pp. 20-41.

Dainelli, G., Marinelli, O., 1928. Le condizioni fisiche attuali. Risultati Geologicie Geografici, Relazioni scientifiche della Spedizione Italiana De Filippi nell'Himalaya, Caracorum e Turchestan Cinese (1913-1914), Ser. II, Bd. IV, Tavole, Bologna. pp. 1-68.

Derbyshire, E., Shi, Y., Li, J. Zheng, B., Li, S., Wang, J., 1991. Quaternary glaciation of Tibet: the geological evidence. Quaternary Science Reviews 10, 485-510.

Derbyshire, E., Li Jijun, Perrot, F.A., Xu Shuying, Waters, R.S., 1984. Quaternary glacial history of the Hunza Valley, Karakoram, Pakistan. In: Miller, K.K. (Ed.), Internat. Karakoram Project 1, pp. 457-495.

De Terra, H., 1932. Geologische Forschungen im westlichen K'un Lun und Karakorum-Himalaya. Wiss. Ergeb. d. Dr. Trinklerschen Zentralasien Expedition, Bd.II, Dietrich Reimer/Ernst Vohsen, Berlin, pp. 1-196.

Ding Zhongli, Rutter, N., Han Jingtai, Liu Tungsheng, 1992. A coupled environmental system formed at about 2.5 Ma in East Asia. Palaeogeography, Palaeoclimatology, Palaeoecology 94, 223-242.

Ding Zhongli, Liu Tungsheng, Rutter, N., Zhiwei Yu, Zhengtang Guo, Rixiang Zhu, 1995. Ice-Volume Forcing of East Asian Winter Monsoon Variations in the Past 800,000 Years. Quaternary Research 44, 149-159.

Ehlers, J., Gibbard, P.L., 2003. Extent and chronology of glaciations. – Quarternary Science Reviews 22, 1561-1568.

Emeis, K.-Ch., Anderson, D.M., Doose, H., Kroon, D., Schulz-Bull, D., 1995. Sea-Surface Temperatures and the History of Monsoon Upwelling in the Northwest Arabian Sea during the Last 500,000 Years. Quaternary Research 43, 355-361.

Fang, Q., 1991. The age of Hipparion guiznongensis and the upheavel of Qinghai-Xisang plateau. In: Bulletin INQUA, N.C. 14, 87.

Findlater, J,. 1974. The low-level cross-equatorial air current of the western Indian Ocean during the Northern Summer. Weather 29, 411-416.

Fischer, H., Wahlen, M., Smith, J., Mastroianni, D., Deck, B., 1999. Ice core records of atmospheric CO_2 around the last three glacial terminations. Science 283, 1712-1714.

Flint, R.F., 1971. Glacial and Quaternary Geology. John Wiley and Sons Inc., New York, pp.1-892.

Flohn, H., 1981. The elevated heat source of the Tibetan highlands and its role for the large scale atmospheric circulation. In: Geological and Ecological Studies of the Qinghai-Xizang Plateau, Science Press, Beijing, Gordon and Breach, New York 2, pp.1463-1469.

Gansser, A., 1983. The Wider Himalaya. A Model for Scientific Research. Mat. Fys. Medd. Dan. Vid. Selsk. 40 (14), 3-30.

Gasse, F., Fontes, J. Ch., van Campo, E., Wei, K., 1996. Holocene environmental changes in Bangong Co basin (Western Tibet). Part 4: Discussion and conclusions. Palaeogeography, Palaeoclimatology, Palaeoecology 120, 79-92.

Grosswald, M.G., 1983. Ice Sheets of the Continental Shelves. Nauka, Moscow, pp. 1-215.

Grosswald, M.G., 1987. Last Glaciation of the Sayano- Tuvan Highland: The Morphology, Rate of Nourishment, and Ice-Dammed Lakes. In: Interaction of Glaciers with Atmosphere and the Ocean. Nauka, Moscow, pp. 152-170.

Grosswald, M.G.; Kuhle, M., 1994. Impact of glaciation on Lake Baikal. In: Shoji Horie (Ed.), Intern. Project on Paleo-Limnology and Late Cenozoic Climate, No. 8., Innsbruck, pp. 48-60.

Grosswald, M.G., Kuhle, M., Fastook, J. L., 1994. Würm Glaciation of Lake Issyk-Kul Area, Tian Shan Mts.: A Case Study in Glacial History of Central Asia. Geo Journal 33, 2-3, Tibet and High-Asia III, Dordrecht/Boston/London, 273-310.

Harrison, T.M., Copeland, P., Kidd, W. S. F., An Yin, 1992. Raising Tibet. Science 255, 1663-1670.

Haserodt, K., 1989. Zur pleistozänen und postglazialen Vergletscherung zwischen Hindukusch, Karakorum und Westhimalaya. Beitr. u. Mat. z. Reg. Geogr. H.2, 181-233.

Haug, G.H., Tiedemann, R., 1998. Effect of the formation of the Isthmus of Panama on Atlantic Ocean thermohaline circulation. Nature 393, 673-676.

Hays, J. D., Imbrie, J., Shackleton, N. J., 1976. Variations in the Earth's Orbit: Pacemaker of the Ice Ages. Science 194, 1121-1132.

Hess, H., 1904. Die Gletscher. Friedrich Viehweg und Sohn. Braunschweig, pp. 1-426.

Heuberger, H., 1986. Untersuchungen über die eiszeitl. Vergletscherung des Mt.Everest-Gebietes, Südseite, Nepal. In: Kuhle, M. (Ed.), Intern.Symp. üb. Tibet u. Hochasien vom 8.-11.10.1985, Göttinger Geogr.Abh. 81, 29-30.

Höfer, H.v., 1879. Gletscher und Eiszeitstudien. Sitz. Ber. Akad. d. Wiss. Wien. Math.-Phys. Kl.I, Vol. 79. 331-367.

Hsu, H., Zhang, C. Wang, Y., 1998. Study on crustal movements of Tibetan plateau and ist mechanism by geodetic methods. Intern. Symp. on the Qinghai-Tibet Plateau (Xining, China), 4.

Hughes, T.J., 1998. Ice Sheets. Oxford University Press, New York. pp. 343.

Iwata, S., Yamanaka, H., Yoshida, M., 1982. Glacial Landforms and River Terraces in the Thakkhola Region, Central Nepal. Journal of Nepal Geol. Soc. 2, 81-94.

Ji, H.X., Huang, W.P., Chen, W.Y., Wu, Q.Q., Zheng, S.H., 1981. The discovery of the *Hipparion* faunas in Tibet and its significance on the problem of the uplift of the Plateau. Studies on the Period, Amplitude and Type of the Uplift of the Tibetan Plateau. Science Press, Beijing, 19-25.

Kamp, U., 1999. Jungquartäre Geomorphologie und Vergletscherung im östlichen Hindukusch, Chitral, Nordpakistan. Berliner Geographische Studien 50, 254.

Kashiwaya, K., Yaskawa, K., Yuan, B., Liu, J., Gu, Z., Cong, S., Masuzawa, T., 1991. Paleohydrological processes in Siling-Co (Lake) in the Qing-Zang (Tibetan) Plateau based on the physical properties of its bottom sediments. Geophysical Research Letters 18 (9), 1779-1781.

Kaufmann, G., Lambeck, K., 1997. Implications of Late Pleistocene Glaciation of the Tibetan Plateau for Present-Day Uplift Rates and Gravity Anomalies. Quaternary Research 48, 267-279.

Kaufmann, G., 2003. Satellite gravimetry and the Late Pleistocene Tibetan Ice Sheet. Geophys. Research Abstracts, Vol. 5, 04058.

Kotljakow, W. M., Grosswald, M. G., 1987. Wsaimodejstwie oledenenia s atmorsferoi i okeanom. Akademia Nauk SSSR, Institut Geografii, Moskwa. pp. 1-155.

Kuhle, M., 1974. Vorläufige Ausführungen morphologischer Feldarbeitsergebnisse aus den SE-Iranischen Hochgebirgen am Beispiel des Kuh-i-Jupar. Zeitschrift für Geomorphologie N.F. 18 (4), 472-483.

Kuhle, M., 1976. Beiträge zur Quartärgeomorphologie SE-Iranischer Hochgebirge. Die quartäre Vergletscherung des Kuh-i-Jupar. Göttinger Geographische Abhandlungen 67, Vol. I, 1-209; Vol. II: 1-105.

Kuhle, M., 1980. Klimageomorphologische Untersuchungen i.d. Dhaulagiri- u. Annapurna-Gruppe (Zentraler Himalaya). Tagungsber. u. wiss. Abh. 42. Dt. Geographentag 1979, 244-247.

Kuhle, M., 1982a. Der Dhaulagiri- und Annapurna-Himalaya. Zeitschrift für Geomorphologie, Suppl. 41, Vol. I, Vol. II, Abb. 1-184, Stuttgart. 1-229.

Kuhle, M., 1982b. Erste Deutsch-Chinesische Gemeinschaftsexped. nach Tibet u.i.d. Massive des Kuen-Lun-Gebirges 1981. Ein Expeditions- u. vorl. Forschungsber., Tagungsber. u. wiss. Abh., 43. Dt. Geographentag 1981, 63-82.

Kuhle, M., 1982c. Was spricht f. eine pleistozäne Inlandvereisung Hochtibets? Sitzber. Braunschweig. Wiss. Ges. Sonderh. 6, Göttingen, 68-77.

Kuhle, M., 1983. Der Dhaulagiri- und Annapurna-Himalaya. Empirische Grundlage. Z. f. G., Suppl. Bd. 41, 1-383.

Kuhle, M., 1984/85. DFG-Abschlußbericht über d. Ergebn. d. 2. Chines./Deutsch. Gemeinschaftsexped. nach S-Tibet u. i.d. N-Flanken von Shisha Pangma u. Mt. Everest (Chomolungma) 1984, 1-52.

Kuhle, M., 1985. Ein subtropisches Inlandeis als Eiszeitauslöser. Südtibet- u. Mt. Everest-Expedition 1984. Georgia Augusta, 35-51.

Kuhle, M., 1986. Former Glacial Stades in the Mountain Areas surrounding Tibet - in the Himalayas (27°-29°N: Dhaulagiri, Annapurna, Cho Oyu and Gyachung Kang areas) in the south and in the Kuen Lun and Quilian Shan (34° - 38°N: Animachin, Kakitu) in the north. In: Joshi, S.C. (Ed.), Nepal, Himalaya. Geo-Ecological Perspectives, HR Publishers, Kay Kay Printers, New Delhi, 437-473.

Kuhle, M., 1987. Glacial, Nival and Periglacial Environments in Northeastern Quinghai-Xizang-Plateau. Hövermann, J., Wang Wenjing (Eds.), Reports on the North-Eastern part of the Quinghai-Xizang (Tibet) Plateau by the Sino-German Scientific Expedition. Beijing, China, 176-244.

Kuhle, M., 1987a. Subtropical mountain- and highland-glaciation as ice age triggers and the waning of the glacial periods in the Pleistocene. GeoJournal 14 (4), 393-421.

Kuhle, M., 1987b. The Problem of a Pleistocene Inland Glaciation of the North-Eastern Quinghai-Xizang-Plateau (Tibet). In: Hövermann, J., Wang Wenjing (Eds.), Reports on the North-Eastern part of the Quinghai-Xizang (Tibet) Plateau by the Sino-German Scientific Expedition. Beijing, China, 250-315.

Kuhle, M., 1987c. Absolute Datierungen zur jüngeren Gletschergeschichte im Mt. Everest-Gebiet u. die mathem. Korrektur von Schneegrenzberechnungen. Tagungsber. 45. Dt. Geographentag, Berlin 1985, Stuttg., 200-208.

Kuhle, M., 1987d. Letzteiszeitliche Gletscherausdehnung vom NW-Karakorum bis zum Nanga Parbat (Hunza-Gilgit- und Indusgletschersystem). Tagungsber. Wiss. Abh., 46. Dt. Geographentag 1987, München, Vol. 46, Stuttgart, 606-607.

Kuhle, M., 1988a. Topography as a Fundamental Element of Glacial Systems. A new Approach to ELA-Calculation and Typological Classification of Paleo- and Recent Glaciations. GeoJournal 17 (4), 545-568.

Kuhle, M., 1988b. Die eiszeitl. Vergletscherung W-Tibets zwischen Karakorum und Tarim-Becken und ihr Einfluß auf die globale Energiebilanz. Geogr. Z., 76 (3), 135-148.

Kuhle, M., 1988c. Geomorphological findings on the build-up of Pleistocene glaciation in southern Tibet and on the problem of inland ice. Results of the Shisha Pangma and Mt.Everest Expedition 1984. GeoJournal 17 (4), 457-511.

Kuhle, M., 1988d. The Pleistocene Glaciation of Tibet and the Onset of Ice Ages- An Autocycle Hypothesis. GeoJournal 17 (4), 581-596.

Kuhle, M., 1988e. Zur Geomorphologie der nivalen und subnivalen Höhenstufe in der Karakorum-N-Abdachung zwischen Shaksgam-Tal u. K2-N-Sporn: Die quartäre Vergletscherung und ihre geoökologische Konsequenz. In: Tagungsber. Wiss. Abh., 46. Dt. Geographentag München 1987 Stuttgart. 413-419.

Kuhle, M., 1988f. Letzteiszeitliche Gletscherausdehnung vom NW-Karakorum bis zum Nanga Parbat (Hunza-Gilgit- und Indusgletschersystem).- In: Tagungsber. Wiss. Abh., 46. Dt. Geographentag München 1987, Stuttgart. AK- Sitzung "Neueste physisch-geographische Forschungsergebnisse aus Hochtibet und angrenzenden Gebieten". Stuttgart (Steiner), 606-607.

Kuhle, M., 1989. Die Inlandvereisung Tibets als Basis einer in der Globalstrahlungsgeometrie fußenden, reliefspezifischen Eiszeittheorie. Petermanns Geographische Mitteilungen 133 (4), 265-285.

Kuhle, M., 1990. New data on the Pleistocene glacial cover of the southern border of Tibet: the glaciation of the Kangchendzönga Massif (8585 m, E-Himalaya). GeoJournal 20 (4), 415-421.

Kuhle, M., 1991a. Observations supporting the Pleistocene inland glaciation of High Asia. GeoJournal 25 (2-3), 133-231.

Kuhle, M., 1991b. Die Vergletscherung Tibets und ihre Bedeutung für die Geschichte des nordhemisphärischen Inlandeises. Sitzungsberichte der Akademie der Wissenschaften und der Literatur Mainz, Paläoklimaforschung 1. Klimageschichtliche Probleme der letzten 130000 Jahre. Frenzel, B. (Ed.), 293-306.

Kuhle, M., 1993. Eine Autozyklentheorie zur Entstehung und Abfolge der quartären Kalt- und Warmzeiten auf der Grundlage epirogener und glazialisostatischer Bewegungsinterferenzen im Bereich des tibetischen Hochlandes. Peterm. Geogr. Mittlg. 1993 (3), 137, 133-152.

Kuhle, M., 1994. Present and Pleistocene glaciation on the north-western margin of Tibet between the Karakoram Main Ridge and the Tarim Basin supporting the evidence of a Pleistocene inland glaciation in Tibet. GeoJournal 33 (2-3), 133-272.

Kuhle, M., 1995a. Glacial Isostatic Uplift of Tibet as a Consequence of a Former Ice Sheet. GeoJournal 37 (4), 431-449.

Kuhle, M., 1995b. New Results Concerning the Age Glaciation of Pamir, Karakorum and Himalaya: Findings of the Expeditions 1992-1994/95. Abstracts of the International Symposium on Karakorum-Hindukush-Himalaya: Dynamics of Change, Sept. 29th- Oct. 2nd, 1995, Islamabad, Pakistan. 27.

Kuhle, M., 1996a. Rekonstruktion der maximalen eiszeitlichen Gletscherbedeckung im Nanga Parbat Massiv (35°05'-40°N/74°20'-75°E). Kick, W. (Ed.), Beiträge und Materialien zur Regionalen Geographie 8, Forschung am Nanga Parbat. Geschichte und Ergebnisse. Geographisches Institut der T.U. Berlin, 135-156.

Kuhle, M., 1996c. Glazial-geomorphologische Expedition in die Karakorum-Nordseite 1986. In: Publikationen zu wissenschaftlichen Filmen des Institutes für den Wissenschaftlichen Film (IWF), Technische Wissenschaft/Naturwissenschaft 12, Film C 1856; Göttingen: 185-256.

Kuhle, M., 1997. New findings concerning the Ice Age (LGM) glacier cover of the East Pamir, of the Nanga Parbat up to the Central Himalaya and of Tibet, as well as the age of the Tibetan inland ice. GeoJournal 42 (2-3), 87-257.

Kuhle, M., 1998. Reconstruction of the 2.4 Million km^2 Late Pleistocene ice sheet on the Tibetan Plateau and its impact on the global climate. Quaternary International 45/46, 71-108 (additional Figures in: 47/48, 173-182).

Kuhle, M., 1998a. Neue Ergebnisse zur Eiszeitforschung Hochasiens in Zusammenschau mit den Untersuchungen der letzten zwanzig Jahre. Petermanns Geogr. Mitteilungen, 142, 3+4, 219-226.

Kuhle, M., 1999. Reconstruction of an approximately complete Quaternary Tibetan inland glaciation between the Mt.Everest- and Cho Oyu massifs and the Aksai Chin.- A new glaciogeomorphological SE-NW diagonal profile through Tibet and its consequences for the glacial isostasy and ice age cycle. GeoJournal 47 (1-2), 3-276.

Kuhle, M., 2001. The maximum Ice Age (LGM) glaciation of the Central- and South Karakoram: an investigation of the heights of its glacier levels and ice thickness as well as lowest prehistoric ice margin positions in the Hindukush, Himalaya and in East-Tibet on the Minya Konka-massif. GeoJournal 54 (1-4) and 55 (1), 109-396.

Kuhle, M., (2003): New geomorphological indicators of a former Tibetan ice sheet in the central and northeastern part of the high plateau. Zeitschrift für Geomorphologie N.F. Suppl. 130, 75-97.

Kuhle, M., 2004. The High Glacial (Last Ice Age and LGM) glacier cover in High- and Central Asia. Accompanying text to the mapwork in hand with detailed references to the literature of the underlying empirical investigations. In: Ehlers, J., Gibbard, P.L. (Eds.), Extent and Chronology of Glaciations, Vol. 3 (Latin America, Asia, Africa, Australia, Antarctica), INQUA Commission on Glaciation, 175-199.

Kuhle, M., 2004a. The High Glacial (pre-LGM and LGM) glaciation of SE-Iranian mountains exemplified by Kuh-i-Jupar massif in The Zagros. Quarterly Geographical Journal of Territory 1 (2-3), 1-15, 1-8.

Kuhle, M., 2004b. Aktuelle Gletscherdynamik in Hochasien - Fallstudien. In: Geographie und Schule. Aulis-Verlag Köln., 26 (148), 16-20.

Kuhle, M., 2004c. Past glacier (Würmian) ice thickness in the Karakoram and on the Deosai Plateau in the catchment area of the Indus river. Eiszeitalter und Gegenwart 54, 95- 123.

Kuhle, M., 2005. The maximum Ice Age (Würmian, Last Ice Age, LGM) glaciation of the Himalaya – a glaciogeomorphological investigation of glacier trim-lines, ice thicknesses and lowest former ice margin positions in the Mt. Everest-Makalu-Cho Oyu massifs (Khumbu- and Khumbakarna Himal) including informations on late-glacial-, neoglacial-, and historical glacier stages, their snow-line depressions and ages. In: Tibet and High Asia (VII) Glaciogeomorphology and Former Glaciation in the Himalaya and Karakoram. GeoJournal Volume 62 (no. 3-4); Kluwer, Dordrecht/ Boston/ London, 191-650.

Kuhle, M., 2005a: The Maximum Ice Age Glaciation between Karakoram Main Ridge (K2) and the Tarim Basin and its Influence on the global Energy Balance. Journal of Mountain Science 2 (1), 5-22.

Kuhle, M., Herterich, K., Calov, R., 1989: On the Ice Age Glaciation of the Tibetan Highlands and its Transformation into a 3-D Model. In: GeoJournal 19 (2), 201-206.

Kuhle, M., Jacobsen, J.–P., 1988. On the geoecology of Southern Tibet. GeoJournal 17 (4), 597-613.

Kuhle, M., Kuhle, S., 1997. Der quartäre Klimawandel - System oder geschichtliches Ereignis? Überlegungen zur geographischen Methode am Beispiel von Eiszeittheorien. Erdkunde 51, 114-130.

Kuhn, M., 1981. Climate and Glaciers. Sea Level, Ice and Climatic Change. Proceedings of the Canberra Symposium, Dec. 1979. IAHS Publ.No.131, 3-19.

Kuhn, M., 1983.: Die Höhe der Schneegrenze in Tirol, berechnet aus Fliris klimatischen Profilen. Arbeiten zur Quartär- und Klimaforschung. Innsbrucker Geogr. Stud. 8 (Fliri-Festschr.), 85-91.

Kukla, G., An, Z. S., 1989. Loess stratigraphy in central China. Palaeogeography, Palaeoclimatology, Palaeoecology 72, 203-225.

Kutzbach, J., Gallimore, R., Harrison, S., Behling, P., Selin, R., Laarif, F., 1998. Climate and Biome Simulations for the Past 21,000 Years. Quaternary Science Reviews 17, 473-506.

Lauscher, F., 1956. Strahlungs- und Wärmehaushalt. Ber. Dt. Wetterdienst, 4 (22), 21-29.

Lautenschlager, M., Herterich, K., Schleese, U., Kirk, E., 1987. Simulation of the January Climate 18 000 YBP. Report No.11, MPG-Inst. Meteor., Sept. 87, Hamburg, 1-42.

Lichtenecker, N., 1938. Die gegenwärtige und die eiszeitliche Schneegrenze in den Ostalpen. Verhandl. d. III. Internat. Quartärkonferenz, Wien 1936, 141-147.

Li Tianchi, 1988. A Preliminary Study on the Climatic and Environmental Changes at the Turn from the Pleistocene to Holocene in East Asia., GeoJournal 17 (4), 649-657.

Loczy, L.v., 1893. Die wiss. Ergebn. d. Reise des Grafen Bla Széchéniy in Ostasien 1877-1880. 3. Abschn., Geologie 1, Wien, 307-836.

Louis, H., 1955. Schneegrenze und Schneegrenzbestimmung. Geogr. Taschenbuch 1954/55, Wiesbaden, 414-418.

Manabe, S., Broccoli, A. J., 1985. The influence of continental ice sheets on the climate of an ice age. Journal of Geophysical Research 90, 2167-2190.

Marcinek, J., 1985. Gletscher der Erde. Thun u. Frankfurt/Main, 1-214.

Marsiat, I., 1994. Simulation of the northern hemisphere continental ice sheets over the last glacial-interglacial cycle: experiments with a latitude-longitude vertically integrated ice sheet model coupled to a zonally averaged climate model. Palaeoclimates 1, 59-98.

Meinardus, W., 1938. Das Klima der Antarktis. Handbuch der Klimatologie, Teil U., Borntraeger Berlin, pp. 1-133.

Milankovic, M., 1941. Kanon der Erdbestrahlung. Belgrad, pp. 1-132.

Miller, A. A., 1943. Climatology. Methuen, New York, pp. 1-288.

Morley, J. J., Dworetzky, B. A., 1991. Evolving Pliocene-Pleistocene climate: a North Pacific perspective. Quaternary Science Reviews 10, 225-237.

Mörner, N.A., 1978. Faulting, fracturing and seismicity as functions of glacio-isostasy in Fennoscandia. Geology 6, 41-45.

Mudelsee, M., 2001. The phase relations among atmospheric CO2 content, temperature and global ice volume over the past 420 ka. Quaternary Science Reviews 20, 583-589.

Norin, E., 1932. Quaternary Climatic Changes within the Tarim Basin. Geograph. Rev. 22, 591–598.

Norin, E., 1982. The Pamir's, K'un Lun, Karakoram and Chang T'ang regions. Sino-Swedish Expedition, Publ. 54/1, Geography, 3 (1), Stockholm, 1–61.

Pachur, H.-J., Wünnemann, B., 1995. Lake evolution in the Tengger Desert, northwestern China, during the last 40,000 years. Quaternary Research 44, 171-180.

Pagani, M., Freeman, K. H., Arthur, M. A., 1999. Late Miocene atmospheric CO_2 concentrations and the expansion of C4 Grasses. Science 285, 876-879.

Pearson, P.N., Palmer, M.R., 1999. Middle Eocene seawater pH and atmospheric carbon dioxide concentrations. Science 284, 1824-1826.

Petit, J. R., Jouzel, J., Raynaud, D., Barkov, N. I., Barnola, J.-M., Basile, I., Bender, M., Chappellaz, J., Davis, M., Delaygue, G., Delmotte, M., Kotlyakov, V. M., Legrand, M., Lipenkov, V. Y., Lorius, C., Pépin, L., Ritz, C., Saltzman, E. & Stievenard, M., 1999. Climate and atmospheric history of the past 420,000 years from the Vostok ice core, Antarctica. Nature 399, 429-436.

Porter, S. C., 1970. Quaternary glacial record in Swat Kohistan, West Pakistan. Geol. Soc. Amer. Bull. 81, 1421-1446.

Prell, W.L., Kutzbach, J. E., 1992. Sensitivity of the Indian monsoon to forcing parameters and implications for its evolution. Nature 360, 647-652.

Prinz G., 1927: Beiträge zur Glaziologie Zentralasiens. Mitt. Jahrb. Kgl. Ungar. Geolog. Anst. 25, 127-335.

Quade, J., Cerling, T. E., Bowman, J. R., 1989. Development of Asian monsoon revealed by marked ecological shift during the latest Miocene in northern Pakistan. Nature 342, 163-166.

Rhodes, T. E., Gasse, F., Lin Ruifen, Fontes, J. Ch., Wei Keqin, Bertrand, P., Gibert, E., Melieres, F., Tucholka, P., Wang Zhixiang and Cheng Zhi Yuan, 1996. A Late

Pleistocene- Holocene lacustrine record from Lake Manas, Zunggar (Northern Xinjiang, Western China). Palaeogeography, Palaeoclimatology, Palaeoecology 120, 105-118.

Ruddiman, W. F., Kutzbach, J. E., 1991. Plateau Uplift and Climatic Change. Scientific American 3, 42-50.

Ruddiman, W. F., Raymo, M. E., Prell, W. L., Kutzbach J. E., 1997. The Uplift-Climate Connection: A Synthesis. In: Ruddiman W.F. (Ed.), Tectonic Uplift and Climate Change, Plenum Press, New York, 471-515.

Rutter, N., Ding, Z., 1993. Paleoclimates and monsoon variations interpreted from micromorphogenic features of the Baoji Paleosols, China. Quaternary Science Reviews 12, 853-862.

Schneider, H. J., 1957. Tektonik und Magmatismus im NW-Karakorum. Geol. Rdsch. 46, 426-476.

Schneider, H. J., 1959. Zur diluvialen Geschichte des NW-Karakorum. Mitt. Geogr. Ges. München 44, 201-216.

Schroeder-Lanz, H., 1986. Beobachtungen zur rezenten und eiszeitlichen Vergletscherung am Kongurshan/Pamir, Himmelssee/Tienshan und Koko Nor-Nanshan Qaidam., Internat. Sympos. über Tibet und Hochasien, Okt.1985, Göttinger Geogr.Abh. 81, 39-58.

Schwarzbach, M., 1974. Das Klima der Vorzeit. 3.Aufl. F. Enke. Stuttgart, 1-380.

Schwerdtfeger, W., 1970. The climate of the Antarctic. In: Orvig, S. (Ed.), Climates of the polar regions. World Surv. of Climatology 14, 253-355.

Seismographic Map of China, 1987. Beijing.

Shackleton, N. J., Imbrie, J., Pisias, N. G., 1988. The evolution of oxygen-isotope variability in the North Atlantic over the past three million years. Phil. Trans. R. Soc. Lond. B 318, 679-688.

Shi Yafeng, Wang Jing-Tai, 1979. The fluctuations of climate, glaciers and sea level since Late Pleistocene in China. Sea level, ice and climatic change. Proc. Canberra Symp. 1979, 1.

Shi Yafeng et al., 1982. Inquiry on Ice Ages in Mount Xixabangma Region. Reports of Scientific Expedition in Mount Xixabangma Region. Science press, Beijing, 155-176.

Sirocko, F., Sarnthein, M., Erlenkeuser, H., 1993. Century-scale events in monsoonal climate over the past 24,000 years. Nature 364, 322-324.

Tafel, A., 1914. Meine Tibetreise. Eine Studienreise durch das nordwestliche China und durch die innere Mongolei in das östliche Tibet. 2 Bde., Karte, Stuttgart, 1-346.

Tiedemann, R., Sarnthein, M., Shackleton, N. J., 1994. Astronomic timescale for the Pliocene Atlantic O18 and dust flux records of Ocean Drilling Program Site 659. Paleoceanography 9 (4), 619-638.

Trinkler, E., 1932. Geographische Forschungen im westlichen Zentralasien und Karakorum-Himalaya. Wiss. Ergeb. d. Dr. Trinklerschen Zentralasien Exped. Berlin, pp. 1-133.

Van Campo, E., Gasse, F., 1993. Pollen- and diatom-inferred climatic and hydrological changes in Sumxi Co Basin (Western Tibet) since 13,000 yr B.P. Quaternary Research 39, 300-313.

Verbitsky, M.Y., Oglesby, R. J., 1992. The effect of atmospheric carbon dioxide concentration on continental glaciation of the northern hemisphere. Journal of Geophysical Research 97 D5, 5895-5909.

Ward, F. K., 1926. Explorations in South-Eastern Tibet. Geographical Journal 67, 97-123.

Wissmann, H.v., 1959. Die heutige Vergletscherung und Schneegrenze in Hochasien mit Hinweisen auf die Vergletscherung der letzten Eiszeit. Akad. D. Wiss. U. d. Lit., Abh. D. Math.-Nat.wiss. Kl. 14, 1103-1407.

Wünnemann, B., Pachur, H.-J., Li Jijun, Zhang Hucai, 1998. Chronologie der pleistozänen und holozänen Seespiegelschwankungen des Gaxun Nur/Sogo Nur und Baijian Hu, Innere Mongolei, Nordwestchina. Petermanns Geograph. Mitt. 142 (3+4), 191-206.

Xiao, J., Porter, S.C., An, Z., Kumai, H., and Yoshikawa, S., 1995. Grain size of quartz as an indicator of winter monsoon strength on the loess plateau of central China during the last 130,000 yr. Quaternary Research 43, 22-29.

Xie Zichu, 1984. Quinghai-Glacier Map (ELA). Science Press, Beijing.

Yamanaka, H., 1982/83. Radiocarbon Ages of Upper Quaternary Deposits in Central Nepal. Sci. Rep. Tohoku Univ., 7th ser. (Geogr.), 32 (1). 46-60.

Zabirov, R. D., 1955: Oledenenie Pamira. Moskau. pp. 1-372.

In: New Permafrost and Glacier Research
Editors: Max I. Krugger and Harry P.Stern

ISBN: 978-1-60692-616-1

Chapter 8

FUNGI IN HIGH ARCTIC GLACIERS

***Lorena Butinar*[*1], *Silva Sonjak*[2] *and Nina Gunde-Cimerman*[2]**
[1]University of Nova Gorica, Laboratory for Environmental Research, P.O.Box 301, SI-5001 Nova Gorica, Slovenia[2] University of Ljubljana, Biotechnical Faculty, Department of Biology, Večna pot 111, SI-1000 Ljubljana, Slovenia

ABSTRACT

Glacial ice was for a long time considered only as an extremely stable, frigid and static environment. However, recent investigations showed that glacial ice and glaciers are much more dynamic at the microscale, as well as at the geomorphological level, than previously assumed. Particularly in polythermal glaciers, characterized by a warm core, quick seismic shifts can occur and cause displacements of cryokarst formations and glacial ice masses. Increased pressure at the base of the glaciers can generate subglacial ice melting. These waters, supplemented with supraglacial melt-waters and groundwater, soak rocks and sediments below the glacier and become enriched with local solutes and suspended sediments. When frozen together with the base of the glacier, they constitute the subglacial environment.

Until recently, subglacial environments were thought to be abiotic. However, recent studies revealed the existence of aerobic heterotrophic bacterial communities able to survive the dynamic processes of thawing and freezing. To our knowledge, until our investigations, there were no reports on the presence of fungi in subglacial ice. Given the known adaptive behaviour of many fungi to low water activity (a_w) and a wide range of temperatures, we assumed that various types of ice can represent potential natural habitats for diverse halotolerant fungi. To evaluate this hypothesis, media with lowered a_w and incubations at low and "normal" temperatures were chosen to provide a selective advantage for the recovery of culturable fungi from supra- and subglacial environments of four different polythermal Arctic glaciers (Svalbard, Norway).

The dominant taxons isolated were basidiomycetous and ascomycetous yeasts, melanized yeast-like fungi, mainly represented by the genera *Cladosporium* and

* Correspondence to: Lorena Butinar University of Nova Gorica Laboratory for Environmental Research P.O.Box 301 SI-5001 Nova Gorica Slovenia E-mail: lorena.butinar@p-ng.si Fax: +386-5-3315232 Tel: +386-5-3315296

Aureobasidium and different species of the genus *Penicillium*. The fungal counts detected in the subglacial samples were two orders of magnitude greater when compared with those recovered from supraglacial samples, mainly due to yeasts (with counts reaching 4×10^6 CFU L^{-1}). Five ascomycetous and twenty-two basidiomycetous yeast species were isolated, including three new species. According to species diversity and abundance, the majority of species were assigned to the hymenomycetous yeasts (*Filobasidium*/*Cryptococcus albidus* taxa of the Tremellales). The stable core of the subglacial yeast communities were represented by *Cr. liquefaciens*, *Rhodotorula mucilaginosa*, *Debaryomyces hansenii* and *Pichia guillermondii*.

Among the isolated filamentous fungi the prevailing genus was *Penicillium*, with twenty-four different species being identified and a new species, *P. svalbardense* being described. The dominant species was *P. crustosum*, representing on the average half of all isolated strains from the studied glaciers. In contrast to yeasts, primarily associated with the clear subglacial ice, the highest counts for penicillia were obtained for debris-rich subglacial ice.

Enriched fungal populations in subglacial environments may represent a significant reservoir of biological activity with the potential to influence glacial melt-water composition, release of nitrogen and carbon in the polar environment and seeding of oceans with microbial life.

Introduction

Certain species of fungi, one of the ecologically most successful eukaryotic lineages, can be considered as rare examples of eukaryotic extremophiles. Although they display different adaptive strategies from those encountered in prokaryotic extremophiles (Gunde-Cimerman et al., 2005), they have been isolated from similar extreme environments as bacteria. Fungi have been isolated from hypersaline waters (Gunde-Cimerman et al., 2000), at 10 km depth below the surface of the oceans (Nagahama et al., 2001), from extremely acidic mine waters (Hölker et al., 2004), and from the surface of rocks in arid and cold climates (Gorbushina et al., 1996; Sterflinger and Krumbein, 1997; Sterflinger et al., 1999) as well as in extremely cold polar environments. Cold-tolerant species have been primarily reported in connection with sub-Arctic vegetation (Fisher et al., 1995; Babjeva and Reshetova, 1998; Tosil et al., 2002), in snow and below snow-covered tundra (Vishniac, 1993; Babjeva and Reshetova, 1998; Pennisi, 2003; Schadt et al., 2003), and in permafrost (Dmitriev et al., 1997; Babjeva and Reshetova, 1998; Golubev, 1998; Soinam et al., 2000; Tosil et al., 2002) and offshore polar waters (Broady and Weinstein, 1998). Very few studies describe their presence in Arctic glaciers. Viable fungi have, however, been isolated from Arctic and Antarctic ice, ranging in age from 10,000 and up to 140,000 years (Abyzov, 1993; Ma et al., 1999; Christner et al., 2000; Ma et al., 2000; Poglazova et al., 2001; Christner et al., 2003; Ma et al., 2005).

Glacial ice in polar regions has been thus regarded for long only as a life-preserving medium, chronologically entrapping deposited microbes, transported with atmospheric circulation. A new glacial habitat for active microbial life has been only recently discovered at the base of polythermal glaciers, where ice melting occurs due to changes in pressure. These environments were previously considered abiotic, but recent investigations have shown that they provide a habitat for nonphotosynthetic microorganisms and may constitute a significant global reservoir of biological activity (Skidmore et al., 2000; Foght et al., 2004). Subglacial bacterial populations were found in Northern and Southern hemispheres and beneath high and low altitude glaciers (Skidmore et al., 2000; Foght et al., 2004). Until our investigations, there were no reports on the presence of fungi or any other eukaryotic organism in subglacial ice.

We undertook a study involving the isolation of fungi from an Arctic coastal environment in Spitzbergen, Norway. By choosing appropriate selective isolation conditions our preliminary investigations of randomly sampled glacier ice revealed that fungi are occasionally present in high numbers (Gunde-Cimerman et al., 2003). These results prompted us to focus on the systematic isolation and identification of fungi from different layers of polythermal glaciers.

Herein we present the review of fungal diversity and spatial distribution along glacier layers from four different high Arctic polythermal glaciers (Svalbard, Norway), with emphasis on fungal presence in the unique subglacial environment.

1. Fungi in Polar Regions

Cold polar regions are extreme environments, in which the majority of studies have been oriented towards psychrotolerant/ psychrophilic bacteria, while the occurrence and diversity of psychrotolerant/ psychrophilic fungi remained largely unknown.

So far, fungi have been reported primarily in connection with sub-Arctic vegetation and soil in polar regions. Mainly basidiomycetous yeasts were isolated from berries, flowers, vegetation of the littoral zone, soils, forest trees, grasses [Babjeva and Reshetova, 1998] and Antarctic mosses [Tosil et al., 2002]. Recently, fungi belonging to the Ascomycota and Basidiomycota, many of them new, were discovered in abundance below snow-covered tundra [Pennisi, 2003; Schadt et al., 2003]. Although much of the water in tundra regions for most of the year is not biologically available, the peak in fungal activity was detected during winter, while in spring and summer bacteria prevailed [Hodkinson et al., 1999; Schadt et al., 2003].

Few studies exist on the biodiversity of fungi in Antarctic soils [Vishniac and Onofri, 2003]. Such soils represent an interesting habitat for xerophilic fungi, since they exhibit extraordinary aridity, with a correspondingly low a_w, as well as a relatively high salt content [Vishniac and Onofri, 2003]. The major soluble salts in Antarctic soils are sulfates, chlorides, nitrates of Na, K, Mg and Ca. Besides, microbial life is exposed to low temperatures, low nutrient availability, seasonally increased UV radiation, and geographic isolation [Onofri et al., 2004]. In contrast to the mycobiota present in mesophilic soils, dominated primarily by diverse ascomycetous filamentous fungi [Domsch et al., 1980], in Antarctic soils diverse basidiomycetous yeasts prevail. The dominant yeast genera were *Candida*, *Cryptococcus* and *Leucosporidium* [Vishniac and Klinger, 1986; Vishniac and Onofri, 2003]. In all cases their diversity was low and dominated by a few highly specialised and often endemic taxa [Abyzov, 1993; De Wit et al., 2003]. The highest halotolerance for yeasts isolated from Antarctic soil was recorded for *Cr. albidus* and *Cr. himalayensis* (9% NaCl) [Onofri et al., 2004], although basidiomycetous yeasts in general show low salt tolerance and inability to grow on media with low a_w.

Viable yeast and fungi were isolated sporadically also from Siberian permafrost sediments, firmly fixed by ice. They were maintained in a frozen state for extended periods, but upon thawing they were nevertheless able to resume their metabolic activity [Takano, 2004]. The most common yeast genera were again *Cryptococcus* with the most frequently encountered species being halotolerant *Cr. albidus*, followed by *Sporobolomyces*, *Rhodotorula* and *Cystofilobasidium*. They were found with the highest frequency of occurrence in the youngest layers, less than 10,000 years old, although they were also detected in three millions years old Pliocene samples. In all cases the share of the yeasts represented 25% of all aerobic heterotrophs, independent of the organic matter content [Dmitriev et al., 1997; Rivkina et al., 2000].

The occurrence of fungi in polar aquatic habitats has been even less investigated. Yeasts and fungi were isolated from fresh water samples, benthic microbial mats and biofilms on pebbles beneath the ice of Antarctic lakes [Baublis et al., 1991; De Wit et al., 2003].

From polar offshore sea waters mainly basidiomycetous yeasts of the genera *Leucosporidium*, *Rhodosporidium* and *Sporobolomyces* were isolated [Jones, 1976]. Sequences belonging to Eumycota were detected in up to 3,000 m deep Antarctic polar front waters [Lopez-Garcia et al., 2001]. There is no report on the isolation of fungi from sea ice, although their characteristic small subunit rRNA gene sequences were present in DNA extracted from diverse Antarctic and one Arctic sea ice sample [Brown and Bowman, 2001]. Fungi were detected as well in the hypersaline Antarctic Lake Wanda [Kriss et al., 1976].

The presence of fungi was least investigated in polar glaciers: Filamentous fungi and yeasts were found in the microbial cryoconite holes that probably serve as biological refuges during extreme cold [Margesin et al., 2002; Reeve et al., 2002]. Viable filamentous fungi and yeasts have been isolated from 10,000-13,000 years old Greenland ice [Ma et al., 1999, 2000], 12,000 years old Antarctic Vostok ice core sections [Christner et al., 2000, 2002], and even from Antarctic ice layers up to 38,600 years old [Abyzov, 1993]. In all these cases the isolated fungi were filamentous and their numbers were low, while viable yeasts of the genera *Cryptococcus* and *Rhodotorula* have been found only in the upper, younger ice-sheet horizons and surface layers of ice and snow. The oldest yeasts were isolated from horizons 700-3,250 years old [Abyzov, 1993].

By PCR amplification of fragments of the eukaryotic 18S rRNA gene, a diversity of fungi was identified in 2,000-4,000 years old ice-core samples from North Greenland. They were not tested for viability [Price, 2000]. All findings of fungi in glacier ice were interpreted as the result of coincidental Aeolian deposits of spores or mycelium into the ice during its geological history.

2. Glacial Ice as Potential Ecological Niches for Xerotolerant/ Halotolerant Fungi

Due to the adaptive fungal behaviour at low a_w, we have assumed that coastal Arctic environments, in particular diverse type of ice, could represent a potential ecological habitat for xerotolerant/ halotolerant fungi. High concentrations of NaCl create both ionic and osmotic stress, while high concentrations of sugars and drought cause osmotic stress. There is great similarity between osmotic stress and matric water stress since in both cases the water activity is low. As water becomes ice it is not biologically available, and diverse types of ice can therefore be characterized as well as low a_w environments. Additionally, freezing leads to cellular dehydration due to reduced water absorption, while high salinity causes the same effects due to osmotic imbalances.

Atmospheric circulation over polar regions provides air-mass exchange with lower latitudes. As a result, microorganisms from air-borne terrestrial dust may become embedded in ice formed from snow. Glacial microbial diversity is thus represented by taxa that are probably endemic to the polar regions as well as exotic species from temperate and tropical regions, which can originate from ocean mist, wind-borne pollen and soil particles, infected plant surfaces, and many other sources. They may have been transported and deposited by the action of waves, wind, rain, snow, animals, or by other means [Abyzov, 1993; Ma et al., 1999, 2000]. Glacial ice thus provides a unique global source of microorganisms, enabling the study of both contemporary and ancient microbial diversity. Viable microorganisms, randomly entrapped in ice even for thousands of years, are destined to be released during glacial melts or after the calving of icebergs into the ocean [Ma et al., 1999, 2000].

Glacial ice is known as an extremely stable, frigid and static environment. However, recent investigations have shown that glaciers are much more dynamic than previously assumed on the micro scale as well as on the geomorphological level. Ice in temperate glaciers is permeated by a continuous network of aqueous veins, formed at the linear junctions of three ice crystals. They are formed due to sea salts deposited as aerosols, that are essentially insoluble in ice crystals. These liquid veins, with diameters range from ~1 μm at -50 °C to ~10 μm at -4 °C [Rohde and Price, 2007], can have high ionic strength. Due to the percolation of salts from the top of the glacier to its bottom, salts can be accumulated to relatively high concentrations in the bottom parts of polythermal glaciers [Price, 2000]. Besides, due to quick seismic shifts [Ekström et al., 2003; Fahnestock, 2003] and cryokarst phenomena in connection with massive surface ablations, liquid water can temporarily appear as ponds or streamlets on the surface of the glacier and as caves or interglacial lakes, artesian fountains and moulins within the glaciers [Christner et al., 2000]. These supraglacial waters can also reach the glacier bed and mix with groundwater and basal meltwater generated by frictional and geothermal melting of ice at the glacier base. These liquid waters interact with rocks and sediments, and hence contain high solute and suspended sediment concentrations. When frozen onto the basal glacier ice, they can be transported to the glacier margins, where subglacial ice can be aseptically sampled.

These processes create subglacial environments, until recently considered abiotic. However, recent studies have revealed prokaryotic microbial communities dominated by aerobic heterotrophic Betaproteobacteria [Foght et al., 2004]. They have been mainly associated with sediment particles [Foght et al., 2004; Gaidos et al., 2004; Skidmore et al., 2005; Siegert et al., 2001; Skidmore et al., 2000], since thin films of liquid water around embedded mineral grains act as potential microniches [Rohde and Price, 2007]. In all these cases, there were no reports on the presence of fungi.

3. Collection of Samples, Isolation and Identification of Fungi from High Arctic Glaciers

A study involving isolation of xerotolerant/ halotolerant fungi from an Arctic coastal environment was performed in Kongsfjorden. The fjord is located at 79°N, 12°E and is one of the larger fjords on the western coast of Spitsbergen, in the Svalbard Archipelago. It is 26 km long and 8 km wide and stretches from ESE to WNW from the Greenland Sea. The majority of the drainage basin is covered by glaciers, which calve pieces of glacier ice into the fjord throughout the year. The annual mean temperature is around -5 °C, although the water is warmer and less salty than the open sea during the summer. On average, the fjord water temperature is ≥0 °C by the end of May and 3.8 °C at the end of August. The mean salinity ranges from 34.00 to 35.00 PSU. Lowering of salinity can occur in summer and near the surface [Ito and Koduh, 1997].

The glaciers studied were Conwaybreen, Kongsvegen, austre Lovénbreen and austre Brøggerbreen. They have polythermal characteristics, and therefore they mainly consist of ice at subfreezing temperatures [Copland and Sharp, 2001]. Melting in the temperate cores of the glaciers and seasonal inputs of meltwater from the glacier surfaces provide liquid water at their base [Skidmore et al., 2000]. The unfrozen sediments beneath the glaciers are entrained into the basal ice where the meltwaters refreeze beneath the cold-based marginal regions of the glacier. The ice flow then transports them to the glacier margins, where they can be easily accessed and aseptically sampled [Butinar et al., 2007; Skidmore et al., 2000]. Although all of the glaciers studied are polythermal, austre Brøggerbreen is almost entirely cold-based and thus no prolonged interactions between the meltwaters and the glacier bed occur [Hodson et al., 2005].

Samples from the supra- and subglacial environments were collected aseptically during the two melt seasons (in 2001 and 2003), as previously described [Butinar et al., 2007; Gunde-Cimerman et al., 2003]. Samples were collected from Conwaybreen, Kongsvegen and austre Lovénbreen coastal glaciers and inland glacier austre Brøggerbreen. Twenty-six subglacial samples included sediment-rich, overlying clear basal ice and subglacial meltwater. Additionally, two samples of subsurface ice from cryokarst formations were collected. The supraglacial samples comprised four samples of snow/ ice mixtures and nine samples of seasonal meltwaters on the glacier surfaces.

Physico-chemical parameters (pH, Na^+, Mg^{2+} and K^+ concentrations, and total phosphorus content) were determined for five basal ice samples (originating from Kongsvegen), and a sample of subglacial meltwater, as described by Gunde-Cimerman et al. [2003].

Isolation conditions were designed to accommodate xerotolerant/ halotolerant fungi by using media with high concentrations of salt or sugar, and thus low a_w, such as previously used for the isolation of xerotolerant/ xerophilic and halotolerant/ halophilic fungi from temperate hypersaline environments [Gunde-Cimerman et al., 2000] (Table 1). These media should give a selective advantage to cultivable microorganisms adapted to ice, thereby possibly enabling the isolation of higher fungal colony forming units (CFU) numbers than previously reported [Gunde-Cimerman et al., 2000, 2003].

Melanized yeast-like fungi were identified by their morphology, physiology, and by multi-loci sequencing to the species level [Zalar et al., 1999; data not published]. Isolates of filamentous fungi were identified to the species level by morphology, physiology, and in most cases also by secondary metabolite profiles using HPLC-DAD [Smedsgaard, 1997; Sonjak et al., 2005, 2006]. The identification methods for non-melanized yeasts followed those described by Yarrow [1998], Fonseca [1992] and Sampaio [1999], and the molecular characterisation included analyses of the electrophoretic band patterns following minisatellite-primed PCR (MSP-PCR) [Gadanho et al., 2003], and their determining the

D1/D2 domain sequences of the 26S rDNA and/ or ITS sequences [Kurtzman and Robnett, 1998; Fell et al., 2000; Fonseca et al., 2000].

Table 1. Enumeration and selective isolation media, and respective water activity (a_w) values, used in this study

Medium	a_w	Reference
Dichloran rose bengal chloramphenicol agar (DRBC)	~1	[King et al., 1979]
Dichloran 18% glycerol agar (DG18)	0.946	[Hocking and Pitt, 1980]
Malt yeast 10% glucose and 12% NaCl agar (MY10-12)	0.916	[Samson et al., 2004]
Malt yeast 20% glucose agar (MY20G)	0.941	[Gunde-Cimerman et al., 2000; 2003]
Malt yeast 35% glucose agar (MY35G)	0.915	[Gunde-Cimerman et al., 2000; 2003]
Malt yeast 50% glucose agar (MY50G)	0.890	[Samson et al., 2004]
Malt extract agar (MEA)	~1	[Samson et al., 2004]
Malt extract 5% NaCl agar (MEA5NaCl)	0.951	[Gunde-Cimerman et al., 2000; 2003]
Malt extract 10% NaCl agar (MEA10NaCl)	0.924	[Gunde-Cimerman et al., 2000; 2003]
Malt extract 15% NaCl agar (MEA15NaCl)	0.881	[Gunde-Cimerman et al., 2000; 2003]
Malt extract 17% NaCl agar (MEA17NaCl)	0.861	[Gunde-Cimerman et al., 2000; 2003]
Malt extract 24% NaCl agar (MEA24NaCl)	0.828	[Gunde-Cimerman et al., 2000; 2003]
Malt extract 30% NaCl agar (MEA30NaCl)	0.782	[Gunde-Cimerman et al., 2000; 2003]

All isolates are maintained in a genetically stable way in the Culture Collection of the National Institute of Chemistry (MZKI) (Slovenia) and in the EXF Culture Collection of the Department of Biology, Biotechnical Faculty, University of Ljubljana (Slovenia). Most penicillia are preserved as well in the fungal collection (IBT) at the Centre for Microbial Biotechnology (BioCentrum-DTU), Denmark, while yeast-like melanized fungi are preserved in CBS, Utrecht, The Netherlands.

4. Physico-Chemical Parameters at the Isolation Sites

pH of the melted glacial ice within individual glaciers varied between 7.1 and 7.4. The cation concentrations in glacial ice ranged from 5 to 340 mg kg^{-1} for sodium, from 20 to 310 mg kg^{-1} for potassium, and from 70 to 550 mg kg^{-1} for magnesium. The highest phosphorus content was determined in Kongsvegen glacier ice sample, 9 mg kg^{-1}; otherwise the content was low, <1-3 mg kg^{-1}. The Kjeldahl nitrogen content ranged from 3.45 to 7.1 mg L^{-1}, whereas ammonium nitrogen content ranged from 0.8 to 1.57 mg L^{-1}. The water activity of melted ice ranged between 0.95 and 0.98, indicating a salt concentration of less than 5% [Gunde-Cimerman et al., 2003; Sonjak et al., 2006; Butinar et al., 2007].

5. Frequency of Occurrence of Fungi in the Glaciers

There are no reports in the literature on the frequency of occurrence of fungi in glacier ice, apart the infrequent sporadic isolations of individual fungal species, deposited by wind or snowfall into the ice [Abyzov, 1993; Ma et al., 1999, 2000; Price, 2000; Christner et al., 2003].

Table 2. Frequencies of occurrence of isolated *Penicillium* species in Conwaybreen (C), Kongsvegen (K), and Austre Lovénbreen (A) glaciers

Penicillium spp.	Filtration method (CFU L^{-1})						Direct spreading of the sediment (CFU g^{-1})		
	C [a]	C [b]	K [a]	K [b]	K [c]	A [a]	C	K	A
P. bialowiezense [d]			88	161	11,280	410		160	
P. brasilianum							2		
P. brevicompactum [d]			3	14				2	
P. chrysogenum [d]	2		48	65		828		26	
P. commune [d]			7			548	25		
P. corylophilum [d]	2		7						
P. crustosum [d]	783	9000	552	1180	1720	1600	169	36	1
P. discolor [d]			10			248			2
P. echinulatum [d]	100						3		
P. expansum [d]						105			
P. glabrum [d]			42	20		10		1	
P. lanosum [d]	26						1		
P. nordicum [d]		10	5				1		
P. olsonii [d]			42	40					
P. palitans [d]	106		2						
P. polonicum [d]			53	10		600		1	
P. roqueforti [d]				40					
P. solitum [d]	35	102	28	200			1	1	
P. thomii [d]				550				58	
P. tulipae	5					20	1	1	
P. svalbardense [d]			43	200		60		15	
Total	1059	9112	930	2480	13,000	4429	203	301	3

[a]Clear glacial ice samples.
[b]Sediment rich glacial ice samples.
[c]Outflow water samples
[d]Penicillia isolated also on media with a_w from 0.9 to 0.78.

During our investigation ice from four different glaciers was sampled at the surface (supraglacial part), subglacial cavities and moulins, but primarily basal glacier ice (subglacial part) was sampled. The difference in spatial distribution along the sampled glaciers was observed as counts increased when supraglacial layers were compared with subglacial layers. In concordance, the non-melanized yeasts counts from the subglacial samples were two orders of magnitude greater when compared with those recovered from supraglacial samples (below 25×10^3 CFU L^{-1}) [Butinar et al., 2007] (Fig. 1). In general non-melanized yeasts predominated in clear glacial ice, although in samples of glacier ice with high gypsum inclusions, melanized yeast-like fungi prevailed (with CFU as high as 6×10^5 L^{-1}) [Gunde-Cimerman et al., 2003]. A similar trend was observed also for penicillia [Sonjak et al., 2006]. Penicillia were in contrast with non-melanized and melanized yeast-like fungi primarily detected in sediment-rich subglacial ice (up to $9{,}2 \times 10^3$ CFU L^{-1}) or subglacial meltwater (up to 13×10^3 CFU L^{-1}) with mineral inclusions deriving from the glacier bed (Table 2) [Sonjak et al., 2006]. The counts of penicillia in the supraglacial samples were also considerably lower, only up to 50 CFU L^{-1} (Table 2) [Sonjak et al., 2006].

The highest fungal CFUs were in all cases obtained on media containing 5% NaCl (a_w = 0.951), and 20% glucose (a_w = 0.941) respectively [Gunde-Cimerman et al., 2003; Sonjak et al., 2006; Butinar et al., 2007]. Without salt and with salinity higher than 5%, the number of fungal CFU decreased. The upper salinity range for the detection of fungi was 24% NaCl with CFU values up to 5 CFU L^{-1} only. At lower salinities non-melanized yeasts dominated, but with increasing salinity the proportions changed in favour of melanized fungi a

6. Diversity of Fungi

The main taxa of fungi isolated from the high Arctic glaciers include melanized fungi that were mainly represented by the oligotrophic genus *Cladosporium*, taxonomically and phylogenetically closely related to the black yeast-like halotolerant genus *Aureobasidium*. Among melanized fungi the genera *Alternaria* and *Phoma* occured with low frequencies. Filamentous fungi, isolated with high frequency were represented by the cosmopolitan anamorphic genera *Aspergillus* and *Penicillium*, together with the teleomorphic form *Eurotium*, with the prevalence of g. *Penicillium* over *Aspergillus* and *Eurotium* [Gunde-Cimerman et al., 2003; Gunde-Cimerman et al., 2005b; Sonjak et al., 2006]. Other filamentous genera that appeared with low frequency were *Mucor* and *Trichoderma* [Gunde-Cimerman et al., 2005b].

Non-melanized yeasts that were detected in the glaciers include Bulleromyces, Candida, Cryptococcus, Cystofilobasidium, Debaryomyces, Filobasidium, Leucosporidiella, Pichia, Protomyces, Rhodosporidium, Rhodotorula and Trichosporon [Butinar et al., 2007; sub.).

Species diversity in glacier ice depended considerably on the occurrence of mineral inclusions. In samples with mineral sediments penicillia prevailed, and their species diversity was high, while in samples of clear ice, overlying the ice reach in sediment inclusions, the species diversity was low, although individual species of non-melanized yeasts occurred with high frequency [Gunde-Cimerman et al., 2003; Gunde-Cimerman et al., 2005b; Sonjak et al., 2006].

1.1 Aureobasidium

Aureobasidium is a genus of xerotolerant fungi, primarily inhabiting oligotrophic environments. The most ubiquitous species is the halotolerant *A. pullulans*, which is regularly detected in salterns, where it represents one of the core species of saltern mycobiota at lower salinities (up to 10% NaCl). Besides salterns, sea and fresh water its ecological niches are oligotrophic inert surfaces such as glass or stone or it can be frequently isolated from xeric phylloplane of plants. It is also known for its ability to grow at high levels of radioactive gamma contamination, since some strains have been isolated even from the walls of the Chernobyl reactor. In polar environments *A. pullulans* has been previously isolated from continental Antarctic soil samples [Onofri et al., 2004].

The dominant genus of black yeast-like fungi isolated from subglacial samples containing abundant crystal gypsum inclusions was *Aureobasidum*. Based on multi-loci sequencing, isolates were identified as diverse genotypes of *A. pullulans* [data not published]. *A. pullulans* strains were primarily isolated on media with 5% NaCl after one week of incubation at 22°C, or two weeks at 10°C [Gunde-Cimerman et al., 2003; Gunde-Cimerman et al., 2005b].

1.2 Cladosporium

The genus *Cladosporium* is taxonomically and phylogenetically closely related to black yeasts belonging to the order Capnodiales [Crous et al., 2007]. It comprises about 500 plant-pathogenic and saprophytic species, many of which are among the most airborne species in the fungal kingdom.

The species *C. cladosporioides*, *C. herbarum*, *C. sphaerospermum*, *C. tenuissimum*, and *C. oxysporum* are often isolated from salty/ sugary food [Samson et al., 2004] and other environments with low water activity such as saline (coastal) soils and salt marshes, the phylloplane of Mediterranean plants, and the rhizosphere of halophytic plants [Abdel-Hafez et al., 1978]. During the study on the occurrence of fungi in hypersaline environmnts, strains from this genus were isolated from hypersaline waters worldwide. They were in general among the most abundant and most consistently detected fungi in all natural hypersaline waters sampled, at environmental salinities between 15-25% [Zalar et al., 2001; Butinar et al., 2005a; Gunde-Cimerman et al., 2005a).

The majority of cladosporia isolated from Kongsfjorden comprise the same species as most frequently isolated from Antarctic soil and hypersaline waters in the salterns: *C. cladosporioides* complex of species and *C. herbarum* complex of species, although several yet unidentified strains were isolated as well [Abyzov, 1993; Ma et al., 1999, 2000; Gunde-Cimerman et al., 2005b]. These *Cladosporium* species were primarily isolated from glacial ice samples and considerably less frequently from sea water and sea ice [Gunde-Cimerman et al., 2003; Gunde-Cimerman et al., 2005b]. The highest frequency of occurrence was obtained on media with 50% glucose added at 10 °C, although all tested strains grew well on 17% NaCl media [Gunde-Cimerman et al., 2003; Gunde-Cimerman et al., 2005b].

The genus *Cladosporium* is known for its psychrotolerance. The reported temperature range for growth of *C. cladosporoides* is from 0-32 °C, with slight growth possible at -3 °C and even at -10 °C. Some isolates of *C. herbarum* were shown to grow even at -6 °C. All three species can grow at high levels of radioactive gamma irradiation [Onofri et al., 2004].

1.3. Non-Melanized Yeasts

1.3.1 Yeast Diversity

Around 500 yeast isolates were obtained from the four glaciers sampled. Most of these were of basidiomycetous origin, while ascomycetous yeasts represented ~15% of all of the isolates. The ratio of basidiomycetous towards ascomycetous yeasts changed with the a_w and the solute used to increase the osmotic pressure of the media. On media with added NaCl, the ratio between ascomycetous and basidiomycetous yeasts increased above 50% (Fig. 2). However, on media with a_w between ~1 and 0.95, basidiomycetous yeasts prevailed (Fig. 2), they constituted ~85% of all identified strains from subglacial ice samples [Butinar et al., 2007; Butinar et al., sub.]. Although basidiomycetous yeasts are generally recognized as being more nutritionally versatile and tolerant to low temperatures than ascomycetous yeasts [Sampaio, 2004], these results also reflect the biased approach of most isolation procedures.

Overall, 22 different basidiomycetous species of the classes Hymenomycetes and Urediniomycetes were found [Butinar et al., 2007] (Table 3). With respect to their abundance and species diversity, the dominant group corresponded to the hymenomycetous yeasts and amongst them the non-pigmented *Filobasidium/ Cryptococcus albidus* taxa of the Tremellales predominated [Butinar et al., 2007]. Our isolates fell into the Albidus clade [Fonseca et al., 2000], where *Cr. adeliënsis*, *Cr. albidosimilis*, *Cr. albidus*, *Cr. liquefaciens* and *Cr. saitoi* were identified, and into the Magnus clade [Fonseca et al., 2000], where *Cr. gilvescens* and *Cr. magnus*, *Cr. oeirensis* and *F. uniguttulatum* were recognised. Differentiation between *Cr. magnus* and *F. floriforme* required additional ITS analysis. No representatives of the Aerius clade were found [Butinar et al., 2007].

The remaining species of the Tremellales belong to the Bulleromyces (*Bulleromyces albus*, species from *Cr. laurentii* complex) and the Dimennae/Victoriae clades (*Cr. carnescens*, *Cr. victoriae*) [Scorzetti et al., 2002; Takashima et al., 2003]. *Cryptococcus* sp. MZKI K-282 from *Cr. laurentii* complex differed by 10 bp when compared with *Cr. laurentii* CBS 139^T (Butinar et al., 2007). The remaining strains from this complex were grouped into the phylogenetic group II [Takashima et al., 2003].

Two species assigned to the Cystofilobasidium clade of the Cystofilobasidiales [Scorzetti et al., 2002], *Cr. macerans* and *Cystofilobasidium* sp., were also found [Butinar et al., 2007]. Sequence of the strain EX-F 1547 differed significantly (2%) from the type strain of *Cystofilobasidium bisporidii* [Butinar et al., 2007].

In the clade Cutaneum of the Trichosporonales [Scorzetti et al., 2002] *Trichosporon mucoides*, which is characterized by the presence of septate hyphae and the production of arthroconidia, was identified [Butinar et al., 2007].

The urediniomycetous yeasts were represented by four carotenoid-pigmented and one non-pigmented species and were distributed in three lineages: the Sporidiobolus lineage, with *Rhodosporidium diobovatum* from the clade Glutinis and *Rhodotorula mucilaginosa* belonging to the clade Sphaerocarpum (Scorzetti et al., 2002); the Erythrobasidium lineage, with *Rh. laryngis* and *Rh. lysiniphila* from the clade Occultifur, and the Microbotryum lineage, with *Leucosporidiella fragaria* from the clade Leucosporidium [Scorzetti et al., 2002; Sampaio et al., 2003; Butinar et al., 2007].

Table 3. Occurrence of asco- and basidiomycetous yeast species in the glaciers studied.

	Conwaybreen	Kongsvegen				Austre Lovénbreen		Austre Brøggerbreen	
	Subglacial samples	Supraglacial samples	Ice from cryokarst formations	Subglacial samples	Glacier melt-water	Supraglacial samples	Subglacial samples	Supraglacial samples	Subglacial samples
Bulleromyces albus		+							
Candida parapsilosis	+								
C. pseudorugosa-like				+					
Cryptococcus adeliensis	+								
Cr. albidosimilis				+					
Cr. albidus	+	+		+					+
Cr. carnescens	+			+					
Cr. gilvescens		+							
Cr. laurentii complex	+			+					
Cr. liquefaciens		+	+	+		+	+	+	+
Cr. macerans				+					
Cr. magnus				+		+	+		+
Cr. oeirensis				+					
Cr. saitoi				+					
Cr. victoriae		+		+		+			+
Cystofilobasidium sp.				+					
Debaryomyces hansenii	+			+			+		
Filobasidium uniguttulatum	+			+					
Leucosporidiella fragaria				+				+	
Pichia guilliermondii	+			+					
Protomyces inouyei			+						
Rhodotorula laryngis							+		
Rh. lysiniphila				+					
Rh. mucilaginosa	+	+	+	+	+	+	+	+	+
Rhodosporidium diobovatum			+						+
Trichosporon mucoides				+					

The majority of ascomycetous yeasts identified to the species or genus levels were assigned to the Hemiascomycetes lineage, Order Saccharomycetales (Table 3) [Kurtzman and Robnett, 1998]. Within the *Debaryomyces/ Lodderomyces* clade [Kurtzman and Robnett, 1998], *Candida parapsilosis*, *Pichia guilliermondii* and *Debaryomyces hansenii* were identified [Butinar et al., sub.].

Sequence analysis of the D1/D2 domains of the 26S rDNA revealed that the sequences of two strains (MZKI K-259 and K-269) differed (3 and 4 nucleotides, respectively) from the type strain of *C. pseudorugosa*. Only one representative, *Protomyces inouyei*, was found to belong to the "Archiascomycetes" lineage (Protomycetales) [Butinar et al., sub.].

1.3.2 Yeast Distribution in the Glaciers

Basidiomycetous yeasts were always recovered in low numbers (up to 25×10^3 CFU L^{-1}) from supraglacial samples, whereas their counts increased approximately 15-fold (up to 4×10^5 CFU L^{-1}) in samples from diverse cryokarst formations (Fig. 1) [Butinar et al., 2007]. In comparison, the ascomycetous yeasts were absent in supraglacial samples, however they were detected in subsurface samples (up to 10^3 CFU L^{-1}) (Fig. 1) [Butinar et al., sub.].

The counts of basidomycetous yeasts from the subglacial samples were two orders of magnitude greater when compared with those recovered from supraglacial samples (Fig. 1) [Butinar et al., 2007]. A similar trend was observed for ascomycetous yeasts [Butinar et al., sub.]. The ratio shifted in favour of ascomycetous yeasts in particular when the a_w of the media was lowered with NaCl (Fig. 2) [Butinar et al., sub.]. In subglacial ice, the highest mean values recorded for ascomycetous yeasts (with a maximum around 10^4 CFU L^{-1}; Fig. 2) were obtained on media with 5% NaCl and 20% glucose ($\geq$300 CFU L^{-1}) [Butinar et al., sub.]. On the remaining media, with few exceptions, the mean counts remained below 10^2 CFU L^{-1} (Fig. 1) [Butinar et al., sub.].

In this study, yeasts were isolated primarily from the subglacial ice at the bottom of the glaciers. Their frequency of occurrence was surprisingly high, reaching 4×10^3 CFU ml^{-1} [Butinar et al., 2007; sub.], while total bacterial counts in thawed subglacial ice was within the range 10^4-10^7 cells ml^{-1} [Foght et al., 2004]. It is worth noting that yeast counts in freshwater rivers and lakes of glacial origin are normally in the range of 0-0.25 cells ml^{-1} [Botha and Wolfaardt, 2000; Libkind et al., 2003). In contrast to subglacial bacteria, primarily associated with debris-rich subglacial ice [Skidmore et al., 2000], the highest yeast counts were obtained for the clear glacier ice, just above the ice with sediments [Butinar et al., 2007, sub.].

Taxonomic characterisation of the isolates showed that the dominant yeasts are of basidiomycetous affinity [Butinar et al., 2007, sub.] and belong to the same genera that occur frequently on Antarctic soils, Siberian permafrost and polar offshore sea-waters, where they represent the most important eukaryotic microorganisms [Jones, 1976; Vishniac and Klinger, 1986; Abyzov, 1993; Dmitriev et al., 1997; Vishniac, 2002; Onofri et al., 2004]. *Cr. liquefaciens* was the predominant yeast species and it was isolated primarily from sediment-free subglacial ice, reaching counts as high as 4×10^6 CFU L^{-1} [Butinar et al., 2007].

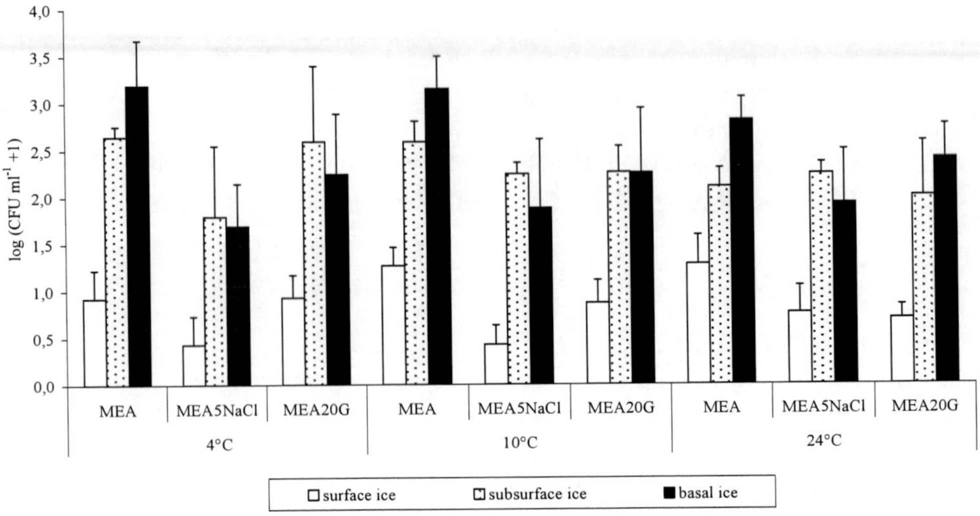

Figure. 1 Yeast abundance and distribution in the surface ice, subsurface cryokarst formations ice and basal ice of four different glaciers. Yeasts were isolated using malt extract medium (MEA) as such and supplemented with either 5% NaCl (MEA5NaCl) or 20% glucose (MEA20G), after incubation at 4, 10 and 24°C. Abundance data for yeasts were transformed by taking logarithm log (CFU ml^{-1} + 1) and mean value and 95% confidence interval were calculated for each layer.

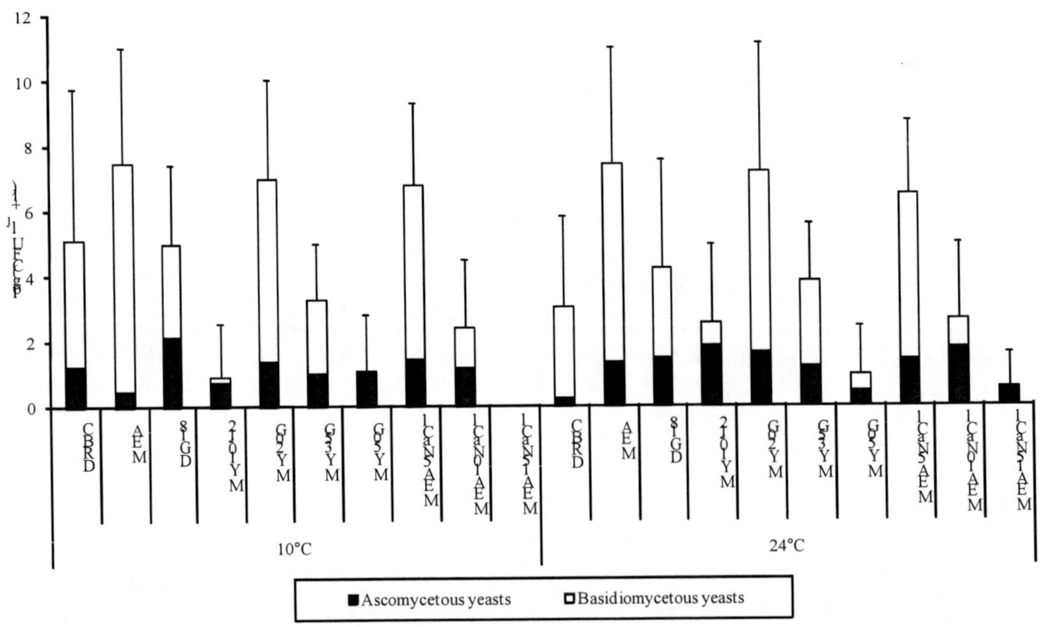

Figure 2. Total number of yeasts and the relative ascomycetous/ basidiomycetous yeasts ratio in subglacial ice at 10 °C and 24 °C, on enumeration media, on media with 0-15% NaCl or 0-50% glucose added. The abundance data were log-transformed by taking log (CFU L^{-1} + 1), and mean ± SD were calculated.

Although strains in the Albidus clade are widely distributed in nature, both in natural and man-made environments, they have been isolated only sporadically from soil and plants in the polar regions [Middelhoven, 1997; Vishniac and Onofri, 2002; Onofri et al., 2004]. Since this species was reinstated not long ago [Fonseca et al., 2000], the ecology of *Cr. liquefaciens* (Fig. 3), apart its dominant presence in subglacial environments, is not well known. The second most frequent species, *Rh. mucilaginosa*, was always isolated together with *Cr. liquefaciens*, although it was the prevailing yeast in subglacial ice with sediment inclusions (up to 10^5 CFU L^{-1}) [Butinar et al., 2007]. Interestingly, both *Cryptococcus* sp. and *Rh. mucilaginosa* were the only microorganisms that has been isolated from Antarctic Vostok ice core [Christner, pers. comm.] and at various depths of the Greenland ice sheet [Starmer et al., 2005]. *Rh. mucilaginosa* has been associated to human infections, besides it is rather common in diverse aquatic environments, including polar waters of glacial origin [Libkind et al., 2004]. Interestingly, it has been isolated from hypersaline waste water ponds in Israel as one of the two dominant microorganisms [Lahav et al., 2002] as well as from deep sea [Nagahama et al., 2001]. *Cr. liquefaciens* was detected with the highest numbers on MEA medium, while the highest counts for *Rh. mucilaginosa* were obtained on MEA with the addition of 5% NaCl [Butinar et al., 2007].

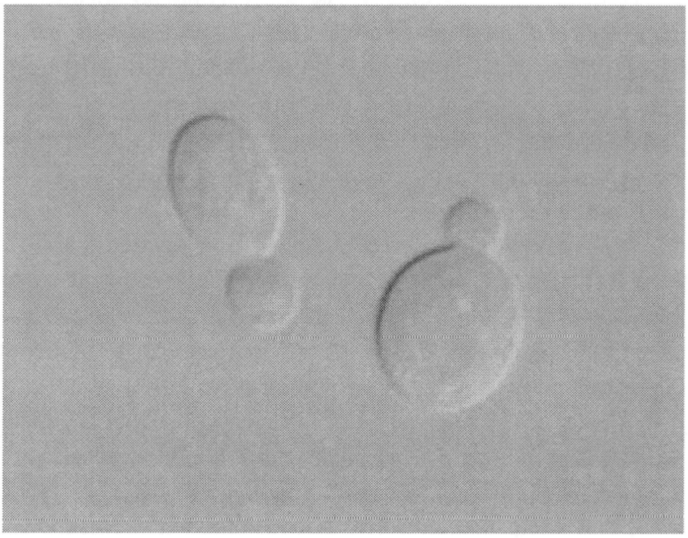

Figure 3. *Cryptococcus liquefaciens*, the yeast species most frequently isolated from subglacial samples

Among the remaining basidiomycetous yeast species detected in lower numbers, *Cr. albidus* was the most abundant (up to 5×10^5 CFU L^{-1}) [Butinar et al., 2007]. *T. mucoides* is considered an opportunistic pathogenic species [Kurtzman and Fell, 1998] and it was unexpectedly obtained in high counts (8×10^5 CFU L^{-1}), though only from one glacier [Butinar et al., 2007]. Our group isolated this species also from waters of the Dead Sea [Butinar et al., 2005b]. Surprisingly, half of all the isolated species are repeatedly isolated from humans and only around one quarter of the species (*Cr. gilvescens*, *Cr. macerans*, *Cr. victoriae*, *Cystofilobasidium* sp., *L. fragaria*) can be considered psychrophilic, with maximum temperatures for growth not exceeding 25 °C [van Uden, 1984; Vishniac, 1987]. In addition, two not yet described species of the genera *Cryptococcus* and *Cystofilobasidium* were discovered in our study [Butinar et al., 2007].

To our knowledge, ascomycetous yeasts have only been exceptionally isolated from cold polar regions [Di Menna, 1966; Atlas et al., 1978; Vishniac, 1987]. This is surprising, as ascomycetous yeasts represent the main spoilage agents of chilled or frozen foods [Davenport, 1980; Schmidt-Lorenz, 1982] and of food preserved with low water activity (a_w) [Deak and Beuchat, 1996; Samson et al., 2004]. It is possible that this deficit has been a result of inappropriate selective conditions used to recover the yeasts present in those extreme ecosystems.

In our studies by using low a_w media also ascomycetous species *D. hansenii* and *P. guillermondii* were frequently found and together with the two dominant basidiomycetous yeasts, *Cr. liquefaciens* and *Rh. mucilaginosa*, can probably be considered autochthonous subglacial species. The highest CFU values for *D. hansenii* were obtained on media with 20% glucose and 5% NaCl (up to 10^4 CFU L^{-1}), whereas the *P. guilliermondii* numbers were approximately half (up to 7×10^3 CFU L^{-1}) on the medium with 10% NaCl added, at 24 °C [Butinar et al., sub.]. To our knowledge this counts were the highest ever reported for this two species. However differences in distribution was observed as the two dominant ascomycetous species primarily occurred in Kongsvegen samples (Fig. 4), while the two dominant basidiomycetous species prevailed in samples originating from austre Lovénbreen and Brøggerbreen glaciers (Fig. 4) [Butinar et al., 2007, sub.]. In Brøggerbreen glacier no ascomycetous yeasts were obtained, probably due to its almost entirely cold base.

D. hansenii, with its anamorphic state *Candida famata*, is known as a salt- and cold-tolerant species, and it is usually isolated from materials of plant and animal (including clinical) origin, and from soil and air, in temperate climates. The species is ubiquitous in the oceans of the World [Jones, 1976]; however, it can also be isolated from hypersaline water in solar salterns [Butinar et al., 2005b]. It is a common spoilage yeast of frozen food, brine-preserved food and other low a_w products [Deak and Beuchat, 1996; Boekhout and Robnet, 2003]. Despite this wide distribution, reports of its occurrence in polar regions are scarce. The presence of *D. hansenii* has been recorded in Antarctic soils [Atlas et al., 1978], overcooled brine cryopegs in permafrost [Gilichinsky et al., 2005], polar waters [Vishniac, 1987; Ma et al., 1999, 2000; Bridge et al., 2008], and ice from Antarctic glaciers [Di Menna, 1966]. In all cases, the isolates were present at very low densities

P. guilliermondii was the second most frequent ascomycetous yeasts species in the subglacial ice. It was always found in association with *D. hansenii*, both forming a distinct phylogenetic group [Kurtzman and Robnett, 1998]. *P. guilliermondii* is also widely distributed in nature, as strains of this species are routinely isolated from exudates of various trees, and from insects [de Araújo et al., 1995; Sibirny, 1996], soil, plants [Capriotti and Ranieri, 1964], the atmosphere and sea water [Diriye et al., 1993], and it was reported as the most frequently occurring species in the Adriatic salterns [Butinar et al., 2005b]. Moreover, *P. guilliermondii* is among the yeasts that are most commonly related to human disease [Kurtzman and Fell, 1998].

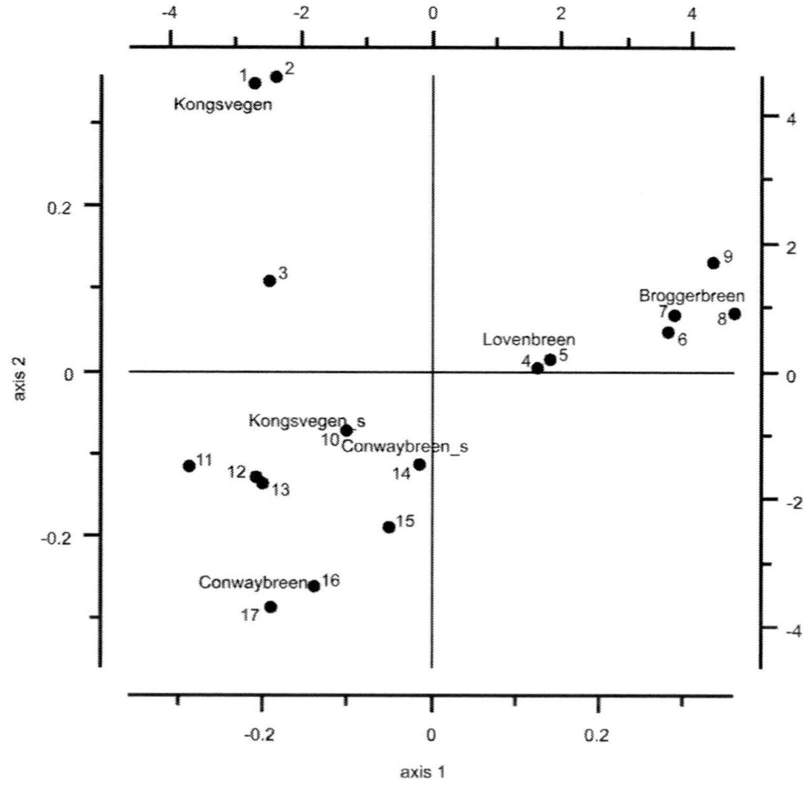

Figure 4. Principal component analysis of yeast species (1-Candida rugosa-like; 2-Cryptococcus oeirensis, Cr. saitoi, Leucosporidiella fragaria, Trichosporon mucoides; 3-Rhodotorula minuta; 4-Cr. magnus; 5-Rh. laryngis; 6-Cr. albidus; 7-Cr. victoriae, Rhodosporidium diobovatum; 8-Rh. mucilaginosa; 9-Cr. liquefaciens; 10-Cystofilobasidium sp.; 11-Pichia guilliermondii; 12-Filobasidium uniguttulatum; 13-Debaryomyces hansenii; 14-Cr. albidosimilis; 15-Cr. adeliensis; 16-C. parapsilosis, Cr. laurentii; 17-Cr. carnescens) isolated from different samples originating from four glaciers. Letter s, following the name of glacier, indicates subglacial ice with sediment. The first two axes explained 67.3% of the variation in the species data.

Although it is also known as a food-spoilage yeast of processed and refrigerated food [Diriye et al., 1993], in the polar regions *P. guilliermondii* has only been reported from permafrost cryopegs [Gilichinsky et al., 2005]. *C. parapsilosis*, which was isolated from the subglacial ice of a single glacier, is also a food-borne halotolerant yeast [Davenport, 1980; Schmidt-Lorenz, 1982]. Again, it has been frequently isolated from clinical specimens [Kurtzman and Fell, 1998], and it has, somewhat surprisingly, been isolated from solar salterns and from the Dead Sea [Butinar et al., 2005b]. This species has been isolated from the ice tunnel samples, collected at the Amundsen-Scott IGY South Pole Station [Jacobs et al., 1964].

The other ascomycetous yeast species appeared less consistently and with lower counts. *C. parapsilosis* occurred only in the Conwaybreen glacier, while the *C. pseudorugosa*-like species was detected only in the Kongsvegen glacier (Table 3) (Fig. 4) [Butinar et al., sub.].

Sampling of ice from a glacier cave resulted in the isolation of a rare species, *P. inoueyi* (Table 3) [Butinar et al., sub.]. The genus *Protomyces* may include up to 60 species [Reddy and Kramer, 1975], and as it is poorly studied and so far only six strains are available in culture collections, our finding is of considerable interest for future ecological investigations.

1.4 Penicillium

The cosmopolitan genus *Penicillium* comprises more than 225, mainly food-, soil-, or airborne species [Pitt et al., 2000]. The genus shows tolerance for cold environments as demonstrated by the fact that many species grow on food preserved in refrigerators [Pitt and Hocking, 1999] or are isolated from alpine, tundra [Domsch et al., 1980], and even polar soil [Frisvad, 2004; McRae et al., 1999]. Being often psychrotolerant and sturdy as well as having prolific production of conidia, it is not surprising that penicillia are among the few viable fungi isolated from glacial ice cores even up to 38,600 years old [Abyzov, 1993]. Spores and fragments of mycelia trapped within ice are exposed to several stressful factors such as freezing, desiccation, and starvation. However, ice also protects microorganisms from UV irradiation, oxidation, and chemical damage [Rogers et al., 2005].

Our study revealed that supraglacial samples contained only up to 50 CFU L^{-1} of penicillia, whereas subglacial samples harbored a surprisingly rich diversity and high occurrence of penicillia (up to 13×10^3 CFU L^{-1}) [Sonjak et al., 2006], equivalent to subglacial bacteria isolated primarily from subglacial debris-rich ice [Skidmore et al., 2000]. It seems that penicillia were present in the soils and sediments that glaciers overrode and became selectively enriched through the processes of melting and freezing that occurred at the glacier bed. The dominant species, both in sediment-rich and in overlying clear glacial ice, is the cosmopolitan *P. crustosum* [Sonjak et al., 2006], which is typically reported as being food-borne [Samson et al., 2004]. This species was isolated from all sampled glaciers (up to $9{,}2 \times 10^3$ CFU L^{-1}) and represents one of the two dominant species in the glacier outflow water (up to 1,720 CFU L^{-1}) [Sonjak et al., 2006]. It is interesting to note, that *P. crustosum* strains isolated from the Conwaybreen glacier, differed from all other Arctic strains and also from all tested strains isolated worldwide [Sonjak et al., 2005, 2007a]. Although growth on CREA medium is one of the basic characteristics of this species [Frisvad and Samson, 2004], a group of Arctic isolates showed surprisingly weak growth [Sonjak et al., 2005, 2007a]. Although AFLP analysis distinguished as a new *P. crustosum* genotype, they did not differ in other tested chemical, physiological, and morphological characteristics [Sonjak et al., 2005, 2007a].

From all isolated penicillia only *P. commune*, *P. discolor*, and *P. polonicum*, were detected with significantly higher counts in clear glacial ice (up to 600 CFU L^{-1}) than in the sediment-rich ice (below 10 CFU L^{-1}) [Sonjak et al., 2006]. These species might have originated in the sediment but they may also have been washed into the subglacial environment from the glacial surface. The only species besides *P. crustosum* that was detected with high counts in the glacial outflow water was *P. bialowiezense* (up to $11{,}3 \times 10^3$ CFU L^{-1}) [Sonjak et al., 2006].

A group of *Penicillium* strains were isolated that did not belong to any known *Penicillium* species [Sonjak et al., 2007b]. This species was isolated in high numbers from the Kongsvegen subglacial ice and was not detected in the surrounding environment [Sonjak et al., 2007b]. A detailed analysis of secondary metabolite profiles, physiological and morphological characteristics, and partial β-tubulin gene sequences showed that the proposed new species *P. svalbardense* is closely related but not identical to P. piscarium and *P. simplicissimum* [Sonjak et al., 2007b].

Of all the species of *Penicillium* recovered, most species found in subglacial ice are also among the very common foodborne penicillia, including *P. crustosum*, *P. polonicum*, *P. discolor*, *P. commune*, *P. palitans*, *P. nordicum*, *P. solitum*, *P. echinulatum*, *P. expansum*, *P. brevicompactum*, *P. chrysogenum*, and *P. tulipae* [Sonjak et al., 2006]. In contrast, these species are very rare in soil [Frisvad and Samson, 2004]. The only soilborne forms of *Penicillium* found in the subglacial ice were *P. brasilianum* from series *Simplisissima*, *P. lanosum* from series *Lanosa*, *P. corylophilum* from series *Citrina*, (all from *Penicillium* subgenus *Furcatum*) and *P. glabrum* and *P. thomii* from series *Glabra*, subgenus *Aspergilloides* [Sonjak et al., 2006]. Thus, penicillia that are primarily foodborne are clearly much more prevalent in subglacial ice than soilborne penicillia. A similar observation was made for penicillia found in the Antarctic, which were isolated from soil or bird nest material on Antarctica. McRae et al. [1999] reported on *P. aurantiogriseum*, *P. brevicompactum*, *P. chrysogenum*, *P. commune*, *P. echinulatum*, *P. expansum*, *P. palitans*, and *P. solitum* from subgenus *Penicillium*, whereas they found *P. antarcticum*, *P. corylophilum*, *P. fellutanum*, *P. glabrum*, *P. janthinellum*, *P. jensenii*, and *P. waksmanii* of the more soilborne type of penicillia from the subgenera *Aspergilloides* and *Furcatum*. This mycobiota is actually rather similar to that found by us in the Arctic subglacial ice. The foodborne penicillia from subgenus *Penicillium* are extremely rarely found in isolations from soil regardless of climatic zone [Christensen et al., 2000; Frisvad et al., 2000], but apparently they are very common in the polar regions both in soil (Antarctica) and in ice samples (Arctic). Several of the species from subgenus *Aspergilloides* and *Furcatum*, including *P. glabrum*, *P. corylophilum*, and *P. antarcticum*, have also been found regularly on foods, whereas some of the fungi reported from Antarctica, such as *P. janthinellum*, *P. jensenii*, and *P. waksmanii*, are common fungi in soil worldwide. These common soilborne types were not found in the subglacial samples. In addition to these species, some new species from soil in Greenland or alpine habitats were also found in the subglacial samples.

Penicillia strains that populate glacial ice must be physiologically adaptable and able to retain their viability throughout the dynamic processes of ice melting and freezing and extremes in pressure. Such enrichment would select for *Penicillium* populations best adapted to dark, cold, oligotrophic environments with shifting osmotic pressures. These predictions were confirmed by our isolation of the highest *Penicillium* counts on the enumeration medium used for the isolation of moderate xerophiles at 10°C, whereas almost all penicillia were isolated additionally also on media with even lower water activity. The dominant species, *P. crustosum*, and the other frequently occurring species apparently can propagate in this extreme habitat due to their sturdy nature, ability to withstand osmotic imbalances, and to adjust to nutritionally poor habitats. Besides, they are characterized by a high production of small-sized conidia, able to spread and colonize pockets and microchannels of the ice. In cold and oligotrophic conditions, small size could be advantageous for more efficient nutrient uptake and for the occupation of microenvironments [Sonjak et al., 2006].

1.5 Aspergillus

The genus Aspergillus and its teleomorphic states contains 254 broadly accepted species [Pitt et al., 2000]. Many of these are known for their xerotolerance and frequent occurrence as contaminants of food preserved with high concentrations of salt or sugar, primarily in warmer climates.

In spite of their ubiquitious nature and preference for low water activity, only few halotolerant *Aspergillus* species were isolated from Kongsfjorden: *A. niger*, *A. tubigensis*, *A. sydowii* and *A. versicolor*. *A. niger* and *A. sydowii* have been repored world-wide from sand dunes, salt marshes, estuaries, mangrove mud and other marine environments [Gunde-Cimerman et al., 2005a]. Extremotolerant nature of *A. sydowii* spores enabled it to survive prolonged suspension in Dead Sea water [Kis-Papo et al., 2003], and to survive in the Antarctic ice sheet for 10,000 years or more. *A. versicolor*, isolated most frequently from diverse ice samples in Kongsfjorden, was repeatedly isolated as well from Antarctic soils, and is also common in diverse low a_w environments in temperate zones. It belongs to the most xerophilic Antarctic species, able also to grow at high levels of radioactive contamination [Onofri et al., 2004].

1.6 Eurotium

The genus *Eurotium* is the holomorphic ascomycete genus for *Aspergillus* sections *Aspergillus* and *Restricti*. Species in this genus are particularly xero- and halophilic (Pitt and Hocking, 1985a, 1985b). Its representatives have been reported to live in concentrated salt or sugar solutions at a_w as low as 0.7 (Wheeler and Hocking, 1993; Martín et al., 1998). Different species of the teleomorphic genera *Eurotium* were often isolated from natural hypersaline water environments such as Dead Sea water and hypersaline waters of salterns around the globe. *Eurotium amstelodami* was isolated most frequently, followed by *E. herbariorum* and *E. repens*, *E. rubrum* and *E. chevalieri* were isolated with lower frequency [Butinar et al., 2005c].

In Kongsfjorden *Eurotium* species were rarely isolated, although more frequently than *Aspergillus* spp. They were primarily derived from glacial and sea ice. The most frequent isolate was *E. repens*, followed by *E. amstelodami* and *E. rubrum*. *E. repens* and *E. rubrum* were previously isolated from 8,150 years old ice and Antarctic soil [Onofri et al., 2004].

Conclusion

It is possible that the ability of fungi to adapt to low and changing water activities, whether due to frigid temperatures or high salinities, is crucial for their successful survival in some of the harshest extreme environments on Earth. Since the majority of microbiological studies in the polar regions are oriented on the investigation of psychrophilic bacteria, the occurrence of fungi, one of the ecologically most successful eukaryotic lineages, has received little attention so far. However, fungi have been reported in connection with sub-Arctic vegetation, in soil and permafrost in polar regions, and recently found to occur abundantly below snow-covered tundra.

Subglacial environments constitute quite a unique ecosystem, since they enable on the geological scale, occasional enrichment of selected genotypes of the most robust species, able to tolerate a broad range of temperatures and water activities. Surprisingly, the deserts, glaciers and salterns may result in similar environmental and evolutionary pressure on microorganisms. Results of this study favour the hypothesis that fungal adaptation to low a_w can be related to low temperature stress. Freezing, drying and hypersaline stress lead to cellular dehydration, and can therefore activate common responses. Cold-, salt- and drought-tolerant fungi may therefore belong to a limited group of extremophilic species that share more features and inhabit more extreme environments than we have imagined so far.

It seems that the occurrence of fungi in polar and other extreme natural environments has been largely underestimated, although these micro-eukaryotes can take up and transform nutrients very efficiently, and thus serve as a carbon sink and participate in the short food webs present. Our results indicate that subglacial environments may represent a significant, previously unrecognized reservoir not only of prokaryotic, but also of eukaryotic diversity. Fungi enclosed in ice will be released during periods of glacial melts and thus contribute to the biogeochemical processes and biodiversity in polar regions.

ACKNOWLEDGEMENTS

The work in Kongsfjorden was funded by EU Large Scale Facility Fund. The work on identification of yeasts was supported by the Slovenian Ministry of Education, Science and Sport and Slovenia-Portugal bilateral collaboration. The authors would like to thank to Nick Cox (NERC Station) for his logistic support and to Martin Grube for help with sequencing of yeasts.

REFERENCES

Abdel-Hafez, S.I.I., Moubasher, A.H. & Abdel-Fattah, H.M. (1978). Cellulose-decomposing fungi of salt marshes in Egypt. *Folia Microbiol., 23,* 37-44.

Abyzov, S.S. (1993) Microorganisms in the Antarctic Ice. In: Friedmann, E.I. (ed.), *Antarctic Microbiology*. Wiley-Liss, New York, pp. 265-285.

Atlas, R.M., Di Menna, M.E. & Cameron, R.E. (1978). Ecological investigations of yeasts in Antarctic soils. *Antarct. Res. Ser., 30,* 27-34.

Babjeva, I. & Reshetova, I. (1998). Yeast resources in natural habitats at polar circle latitude. *Food Technol. Biotechnol., 36*, 1-5.

Baublis, J.A., Wharton, R.A. & Volz, P.A. (1991). Diversity of micro-fungi in an Antarctic dry valley. *J. Basic Microbiol., 31,* 3-12.

Boekhout, T. & Robert, V. (2003). *Yeasts in food: Beneficial and detrimental aspects.* Woodhead Publishing, CRC Press and Behr's Verlag, Hamburg.

Botha, A. & Wolfaardt, G.M. (2000). Yeast biofilms: an unexplored natural phenomenon. In: Abstracts of the 10th International Symposium on Yeasts, Arnhem, The Netherlands, 27th August–1st September 2000.

Bridge et al., (2008). http://www.antarctica.ac.uk/Resources/BSD/Fungi/Speciespublic2.html#Lich.

Broady, P.A. & Weinstein, R.N. (1998). Algae, lichens and fungi in La Gorce Mountains, Antarctica. *Antarct Sci, 10,* 376-385.

Brown, M.V. & Bowman, J.P. (2001). A molecular phylogenetic survey of sea - ice microbial communities (SIMCO). *FEMS Microbiol. Ecol., 35,* 267-275.

Butinar, L., Spencer-Martins, I. & Gunde-Cimerman, N. (2007). Yeasts in high Arctic glaciers: the discovery of a new habitat for eukaryotic microorganisms. *Antonie Van Leeuwenhoek, 91,* 277-289.

Butinar, L., Strmole, T., Spencer-Martins, I. & Gunde-Cimerman, N. Relative incidence of ascomycetous yeasts in Arctic coastal environments. *Microb. Ecol.* (submitted)

Butinar, L., Santos, S., Spencer-Martins, I., Oren, A. & Gunde-Cimerman, N. (2005b). Yeast diversity in hypersaline habitats. *FEMS Microbiol. Lett. ,244,* 229-234.

Butinar, L., Sonjak, S., Zalar, P., Plemenitaš, A. & Gunde-Cimerman, N. (2005a). Population dynamics of melanized halophilic fungi in solar salterns. *Bot. Mar., 48,* 73-79.

Butinar, L., Zalar, P., Frisvad, J.C. & Gunde-Cimerman, N. (2005c). The genus *Eurotium* – members of indigenous fungal community in hypersaline waters of salterns. *FEMS Microbiol. Ecol., 51,* 155-166.

Capriotti, A. & Ranieri, L. (1964). Yeasts in the green-house environment I. From soil, flowers and animals. *Arch. Microbiol., 48,* 325–331.

Christner, B.C. (2002). Incorporation of DNA and protein precursors into macromolecules by bacteria at -15oC. *Appl. Environ. Microbiol., 68,* 6435-6438.

Christner, B.C., Mosley-Thompson, E., Thompson, L.G. & Reeve, J. (2003). Bacterial recovery from ancient glacial ice. *Environ. Microbiol., 5,* 433-436.

Christner, B.C., Mosley-Thompson, E., Thompson, L.G., Zagorodnov, V., Sandman, K. & Reeve, J.N. (2000). Recovery and identification of viable bacteria immured in glacial ice. *Icarus, 144,* 479-485.

Copland, L. & Sharp, M. (2001). Mapping thermal and hydrological conditions beneath a polythermal glacier with radio-echo sounding. *J. Glaciol., 47,* 232-242.

Crous, P.W., Braun, U., Schubert, K. & Groenewald, J.Z. (2007). Delimiting *Cladosporium* from morphologically similar genera. *Stud. Mycol., 58,* 33-56.

Davenport, R.R. (1980). Cold-tolerant yeasts and yeast-like organisms. In: Skinner FA, Passmore SM, Davenport RR (eds) Biology and Activities of Yeasts, Academic Press, London, pp 215–230

de Araújo, F.V., Soares, C.A., Hagler, A.N. & Mendonça-Hagler, L.C. (1995). Ascomycetous yeast communities of marine invertebrates in a southeast Brazilian mangrove ecosystem. *Antonie van Leeuwenhoek, 68,* 91-99.

De Wit, R., Dyer, P., Genilloud, O., Goetlich, E., Hodgson, D., de Hoog, S., Jones, B., Laybourn-Parry, J., Marinelli, F., Stackebrandt, E., Swings, J., Tindall, B.J., Vyverman, W. & Wilmotte, A. (2003). Antarctic lakes – 'hot spots' for microbial diversity and biotecnological screening, In: 1st FEMS Congress, June 29- July 3, Abstract book. Ljubljana, Slovenia, p. 228.

Deak, T. & Beuchat, L. (1996). Handbook of Food Spoilage Yeasts, CRC Series in Contemporary Food Science, CRC Press, Boca Raton.

Di Menna, M.E. (1966). Yeasts in Antarctic soils. *Antonie van Leeuwenhoek, 32,* 29-38.

Diriye, F.U., Giansante, C., Scorzetti, G. & Zilli, R. (1993). Indagine preliminare sulla ecologia dei lieviti delle acque del litorale Abruzzese. *Ann Fac Agr Perugia, XLV*, 339-345.

Dmitriev, V.V., Gilichinsky, D.A., Faizutdinova, R.N., Shershunov, I.N., Golubev, W.I. & Duda, V.I. (1997). Occurrence of viable yeasts in 3-million-year-old permafrost in Siberia. *Mikrobiologiya, 66,* 655-660.

Domsch, K.H., Gams, W. & Anderson, T.H. (eds.) (1980). *Compendium of Soil Fungi*. Academic Press, London.

Ekström, G., Nettles, M. & Abers, G.A. (2003) Glacial earthquakes. *Science, 302,* 622-624.

Fahnestock, M. (2003) Geophysics: glacial flow goes seismic. *Science, 302,* 578-579.

Fell, J.W., Boekhout, T., Fonseca, A., Scorzetti, G. & Statzell-Tallman, A. (2000). Biodiversity and systematics of basidiomycetous yeasts as determined by large-subunit rDNA D1/D2 domain sequence analysis. *Int. J. Syst. Evol. Microbiol., 50,* 1351-1371.

Fisher, P.J., Graf, F., Petrini, L.E., Sutton, B.C. & Wookey, P.A. (1995). Fungal endophytes of Dryas octopetala from a high arctic polar semidesert and from the Swiss Alps. *Mycologia, 87,* 319-323.

Foght, J., Aislabie, J., Turner, S., Brown, C.E., Ryburn, J., Saul, D.J. & Lawson, W. (2004). Culturable bacteria in subglacial sediments and ice from two southern hemisphere glaciers. *Microb. Ecol., 47*, 329-340.

Fonseca, A. (1992). Utilization of tartaric acid and related compounds by yeasts: taxonomic implications. *Can. J. Microbiol., 38,* 1242-1251.

Fonseca, A., Scorzetti, G. & Fell, J.W. (2000). Diversity in the yeast *Cryptococcus albidus* and related species as revealed by ribosomal DNA sequence analysis. *Can. J. Microbiol., 46,* 7-27.

Frisvad, J.C. & Samson, R.A. (2004). Polyphasic taxonomy of *Penicillium* subgenus *Penicillium*. A guide to identification of food and air-borne terverticillate Penicillia and their mycotoxins. *Stud. Mycol., 49,* 1-173.

Frisvad, J.C. (2004). Chemical diversity of psychrotolerant fungi. In: Watanabe, MM, Suzuki, K, Seki, T (Eds.) Innovative Roles of Biological Resources Centers. Japan Society for Culture Collections and Worlds Federation of Culture Collections, Tsukuba, pp 165-167.

Frisvad, J.C., Filtenborg, O., Lund, F. & Samson, R.A. (2000). The homogeneous species and series in subgenus *Penicillium* are related to mammal nutrition and excretion. In: Samson, RA, Pitt, JI (Eds.) Integration of Modern Taxonomic Methods for *Penicillium* and *Aspergillus* Classification. Harwood Academic Publishers, Amsterdam, pp 265–283

Gadanho, M., Almeida, J.M. & Sampaio, J.P. (2003). Assessment of yeast diversity in a marine environment in the south of Portugal by microsatellite-primed PCR. *Antonie van Leeuwenhoek, 84,* 217-227.

Gaidos, E., Lanoil, B., Thorsteinsson, T., Graham, A., Skidmore, M., Han, S-K, Rust, T. & Popp, B. (2004). A viable microbial community in a subglacial volcanic crater lake, Iceland. *Astrobiology, 4,* 327-344.

Gilichinsky, D., Rivkina, E., Bakermans, C., Shcherbakova, V., Petrovskaya, L., Ozerskaya, S., Ivanushkina, N., Kochkina, G., Laurinavichuis, K., Pecheritsina, S., Fattakhova, R. & Tiedje, J.M. (2005). Biodiversity of cryopegs in permafrost. *FEMS Microbiol. Ecol., 53.,*117-128.

Golubev, W.I. (1998). New species of basidiomycetous yeasts, *Rhodotorula creatinovora* and *R. yakutica*, isolated from permafrost soils of Eastern-Siberian Arctic. *Mykologiya I Phytopathologiya, 32,* 8-13.

Gorbushina, A.A., Panina, L.K., Vlaslov, D.Y. & Krumbein, W.E. (1996). Fungi deteriorating marble in Chersonessus. *Mikol. Fitopatol., 30,* 23-27.

Gunde–Cimerman, N., Butinar, L., Sonjak, S., Turk, M., Uršič, V., Zalar, P. & Plemenitaš, A. (2005b). Halotolerant and halophilic fungi from coastal environmnt in the Arctics. In: Gunde–Cimerman N, Oren A and Plemenitaš A (eds) *Adaptation to life at high salt concentrations in Arcahea, Bacteria and Eucarya*. Springer, Dordrecht.

Gunde-Cimerman, N., Frisvad, J.C., Zalar, P. & Plemenitaš, A. (2005a). Halotolerant and halophilic fungi, In: S.K. Deshmukh and M.K. Rai (eds.), *The Biodiversity of Fungi: Their Role in Human Life*. Oxford & IBH Publishing Cp. Pvt. Ltd., New Delhi, pp. 69-127.

Gunde-Cimerman, N., Sonjak, S., Zalar, P., Frisvad, J.C., Diderichsen, B. & Plemenitaš, A. (2003). Extremophilic fungi in Arctic ice: a relationship between adaptation to low temperature and water activity. *Phys. Chem. Earth, 28,* 1273-1278.

Gunde-Cimerman, N., Zalar, P., de Hoog, G.S. & Plemenitaš, A. (2000). Hypersaline water in salterns – natural ecological niches for halophilic black yeasts. *FEMS Microbiol. Ecol., 32,* 235-240.

Hocking, A.D. & Pitt, J.I. (1980). Dichloran-glycerol medium for enumeration of xerophilic fungi from low moisture foods. *Appl. Environ. Microbiol., 39,* 488-492.

Hodkinson, D., Webb, N.R., Bale, J.S. & Block, W. (1999). Hydrology, water availability and tundra in a changing climate: the need for a closer integration of ideas. *Glob. Chang. Biol., 5,* 359-369.

Hodson, A.J., Mumford, P.N., Kohler, J. & Wynn, P.M. (2005). The High Arctic glacial ecosystem: new insights from nutrient budgets. *Biogeochemistry, 72,* 233-256.

Hölker, U., Bend, J., Pracht, R., Müller, T., Tetsch, L. & de Hoog, G.S. (2004) *Hortaea acidophila*, a new acidophilic black yeast from lignite. *Antonie van Leeuwenhoek, 86,* 287-294.

Ito, H. & Koduh, S. (1997). Characteristics of water in Kongsfjorden, Svalbard. Proc. NIPR Symp. *Polar Meteorol. Glaciol., 11,* 221-232.

Jacobs, M.P.H., Taylor, H.C. & Shafer, J.C. (1964). Studies of fungi at Amundsen-Scott IGY South Pole Base (1957). *Arch. Dermatol., 89,* 117-123.

Jones, G.E.B. (Ed.) (1976). *Recent Advances in Aquatic Mycology*. Surrey. The Gresham Press, Old Woking.

King, A.D., Hocking, A.D. &Pitt, J.I. (1979). Dichloran-rose bengal medium for enumeration of molds from foods. *Appl. Environ. Microbiol., 37,* 959-964.

Kis-Papo, T., Oren, A., Wasser, S.P. & Nevo, E. (2003). Survival of filamentous fungi in hypersaline Dead Sea water. *Microb. Ecol., 45,* 183-190.

Kriss, A.E., Mitskevich, I.N., Rozanova, E.P. & Osnitskaia, L.K. (1976). Microbiological studies of the Wanda Lake (Antarctica). *Mikrobiologiia, 45,* 1075-1081.

Kurtzman, C.P. & Robnett, C.J. (1998). Identification and phylogeny of ascomycetous yeasts from analysis of nuclear large subunit (26S) ribosomal DNA partial sequences. *Antonie van Leeuwenhoek, 73,* 331-371.

Kurtzman, C.P. & Fell, J.W. (Eds.) (1998). *The Yeasts, A Taxonomic Study*. Amsterdam: Elsevier.

Lahav, R., Fareleira, P., Nejidat, A. & Abeliovich, A. (2002). The identification and characterisation of osmotolerant yeast isolates from chemical wastewater evaporation ponds. *Microb. Ecol., 43,* 338-396.

Libkind, D., Brizzio, S. & van Broock, M. (2004). *Rhodotorula mucilaginosa*, a carotenoid producing yeast strain from a Patagonian high-altitude lake. *Folia Microbiol., 49,* 19-25.

Libkind, D., Brizzio, S., Ruffini, A., Gadanho, M., van Broock, M. & Sampaio, J.P. (2003). Molecular characterization of carotenogenic yeasts from aquatic environments in Patagonia, Argentina. *Antonie van Leeuwenhoek, 84,* 313-322.

López-García, P., Rodríguez-Valera, F., Pedrós-Alió, C. & Moreira, D. (2001). Unexpected diversity of small eukaryotes in deep-sea Antarctic plankton. *Nature, 409,* 603-607.

Ma, L.J., Catranis, C.M., Starmer, W.T. & Rogers, S.O. (2005). The significance and implications of the discovery of filamentous fungi in glacial ice. In: J.D. Castello & S.O. Rogers (Eds.), *Life in ancient ice*. Princeton University Press, Princeton and Oxford.

Ma, L.J., Catranis, C.M., Starmer, W.T. & Rogers, S.O. (1999). Revival and characterization of fungi from ancient polar ice. *Mycologist, 13,* 70-73.

Ma, L.J., Rogers, S.O., Catranis, C.M. & Starmer, W.T. (2000). Detection and characterization of ancient fungi entrapped in glacial ice. *Mycologia, 92,* 286-295.

Margesin, R., Zacke, G. & Schinner, F. (2002). Characterization of heterotrophic microorganisms in Alpine glacier cryoconite. *Arctic, Antarctic and Alpine Res., 34,* 88-93.

Martin, S., Companys, E., Sanchis, V. & Ramos, A.J. (1998). Effect of water activity and temperature on competing abilities of common maize fungi. *Mycol. Res., 120,* 959-964.

McRae, C.F., Hocking, A.D. & Seppelt, R.D. (1999). *Penicillium* species from terrestrial habitats in the Windmill Islands, East Antarctica, including a new species, *Penicillium antarcticum*. *Polar Biol., 21,* 97-111.

Middelhoven, W.J. (1997). Identity and biodegradative abilities of yeasts isolated from plants growing in arid climate. *Antonie van Leeuwenhoek, 72,* 81-89.

Nagahama, T., Hamamoto, M., Nakase, T., Takami, H. & Horikoshi, K. (2001). Distribution and identification of red yeasts in deep-sea environments around the northwest Pacific ocean. *Antonie van Leewenhoek, 80,* 101-110.

Onofri, S., Selbmann, L., Zucconi, L. & Pagano, S. (2004). Antarctic microfungi as models for exobiology. *Planet. Space Sci., 52,* 229-237.

Pennisi, E. (2003). Microbiology. Neither cold nor snow stops tundra fungi. *Science, 301,* 1307.

Pitt, J.I. & Hocking, A.D. (1985a). The ecology of fungal food spoilage, In: J.I. Pitt and A.D. Hocking (eds.), Fungi and Food Spoilage. Academic Press, Sydney, pp. 5-18.

Pitt, J.I. & Hocking, A.D. (1985b). Xerophiles, In: J.I. Pitt and A.D. Hocking (eds.), Fungi and Food Spoilage. Academic Press, Sydney, pp. 313-333.

Pitt, J.I., Samson, R.A. & Frisvad, J.C. (2000). List of accepted species and their synonyms in the family *Trichocomaceae*, In: R.A. Samson and J.I. Pitt (eds.), Integration of Modern Taxonomic Methods for *Penicillium* and *Aspergillus* Classification. Harwood Academic Publishers, Amsterdam, pp. 9-47.

Pitt, J.I. & Hocking, A.D. (1999). Fungi and Food Spoilage, 2nd edn. Aspen Publishers, Inc., Gaithersburg.

Poglazova, M.N., Mitskevich, I.N., Abyzov, S.S. & Ivanov, M.V. (2001). Microbiological characterization of the accreted ice of subglacial Lake Vostok, Antarctica. *Mikrobiologiia, 70,* 838-846.

Price, P.B. (2000). A habitat for psychrophiles in deep Antarctic ice. *Proc. Natl. Acad. Sci. USA, 97,* 1247-1251.

Reddy, M.S. & Kramer, C.L. (1975). A taxonomic revision of the Protomycetales. *Mycotaxon, 3,* 1-50.

Reeve, J.N., Christner, B.C., Kvitko, B.H., Mosley-Thompson, E. & Thompson, L.G. (2002). Life in glacial ice. In: Abstracts of the 4th International Congress on Extremophiles, Naples, Italy, 22th–26th September 2002.

Rivkina, E.N., Friedmann, E.I., McKay, C.P. & Gilichinsky, D.A. (2000). Metabolic activity of permafrost bacteria below the freezing point. *Appl. Environ. Microbiol., 66,* 3230-3233.

Rogers, S.O., Ma, L.J., Zhao, Y., Theraisnathan, V., Shin, S., Zhang, G., Catranis, C.M., Starmer, W.T. & Castello, J.D. (2005). Recommendations for elimination of contaminants and authentication of isolates in ancient ice cores. In: J.D. Castello & Rogers, S.O. (Eds.), *Life in Ancient Ice* (pp 5-21). Princeton: Princeton University Press, NJ.

Rohde, R.A. & Price, P.B. (2007). Diffusion-controlled metabolism for long-term survival of single isolated microorganisms trapped within ice crystals. *PNAS, 104,* 16592-16597.

Sampaio, J.P. (1999). Utilization of low molecular weight aromatic compounds by heterobasidiomycetous yeasts: taxonomic implications. *Can. J. Microbiol., 45,* 491-512.

Sampaio, J.P. (2004). Diversity, phylogeny and classification of basidiomycetous yeasts. In: R. Agerer, P. Blanz & M. Piepenbring (Eds.), *Frontiers in Basidiomycote Mycology* (pp 49-80). Eching. IHW-Verlag.

Sampaio, J.P., Gadanho, M., Bauer, R. & Weiss, M. (2003). Taxonomic studies in the Microbotryomycetidae: *Leucosporidium golubevii* sp. nov., *Leucosporidiella* gen. nov. and the new orders Leucosporidiales and Sporidiobolales. *Mycol. Prog., 2,* 53-68.

Samson, R.A., Hoekstra, E.S. & Frisvad, J.C. (eds.) (2004). *Introduction to Food-and Airborne Fungi*. Utrecht: Centraalbureau voor Schimmelcultures.

Schadt, C.W., Martin, A.P., Lipson, D.A. & Schmidt, S.K. (2003). Seasonal dynamics of previously unknown fungal lineages in tundra soils. *Science, 301,* 1359-1361.

Schmidt-Lorenz, W. (1982). Indicator organisms in frozen foods in relation to spoilage. *Antonie van Leeuwenhoek, 48*, 625-633.

Scorzetti, G., Fell, J.W., Fonseca, A. & Statzell-Tallman, A. (2002). Systematics of basidiomycetous yeasts: a comparison of large subunit D1/D2 and internal transcribed spacer rDNA regions. *FEMS Yeast Res., 2,* 495-517.

Sibirny, A.A. (1996). *Pichia guilliermondii* In: K. Wolf (Ed.), *Nonconventional Yeasts in Biotechnology: A Handbook* (pp 255–275). Berlin: Springer.

Siegert, M.J., Ellis-Evans, J.C., Tranter, M., Mayer, C., Petit, J.R., Salamatin, A. & Priscu, J.C. (2001). Physical, chemical and biological processes in Lake Vostok and other Antarctic subglacial lakes. *Nature, 414,* 603-609.

Skidmore, M., Anderson, S.P., Sharp, M., Foght, J. & Lanoil, B.D. (2005). Comparison of microbial community compositions of two subglacial environments reveals a possible role for microbes in chemical weathering processes. *Appl. Environ. Microbiol., 71,* 6986-6997.

Skidmore, M.L., Foght, J.M. & Sharp, M.J. (2000). Microbial life beneath a high Arctic glacier. *Appl. Environ. Microbiol., 66,* 3214-3220.

Smedsgaard, J. (1997) Micro-scale extraction procedure for standardized screening of fungal metabolite production in cultures. *J. Chromatogr., 760,* 264-270.

Soinam, V.S., Vorobyova, E.A., Suzina, N.E., Dimitriev, V.V., Faizutdinova, R.N., Duda, V.I., Vishnivetskaya, T.A., Ostroumova, N.V. & Gilichinsky, D.A. (2000). Viable ancient permafrost yeast. In: J.P. Van Dijken & Scheffers, W.A. (Eds.), The Rising Power of Yeasts in Science and Industry, ISY 2000, Tenth International Symposium on Yeasts, 27 Aug–1 Sept, Papendal, Arnhem, the Netherlands. University Press, Delft, p. 140.

Sonjak, S., Frisvad, J.C. & Gunde-Cimerman, N. (2005). Comparison of secondary metabolite production by *Penicillium crustosum* strains, isolated from Arctic and other various ecological niches. *FEMS Microbiol. Ecol., 53,* 51-60.

Sonjak, S., Frisvad, J.C. & Gunde-Cimerman, N. (2006). *Penicillium* mycobiota in Arctic subglacial ice. *Microb. Ecol., 52,* 207-216.

Sonjak, S., Frisvad, J.C. & Gunde-Cimerman, N. (2007a). Genetic variation among *Penicillium crustosum* isolates from Arctic and other ecological niches. *Microb. Ecol., 54,* 298-305.

Sonjak, S., Uršič, V., Frisvad, J.C. & Gunde-Cimerman, N. (2007b). *Penicillium svalbardense*, a new species from Arctic glacial ice. *Antonie van Leeuwenhoek, 92,* 43-51.

Starmer, W.T., Fell, Y.W., Catranis, C.M., Aberdeen, V., Ma, L.J., Zhou, S. & Rogers, S.O. (2005). Yeasts in the genus *Rhodotorula* recovered from the Greenland ice sheet. In: J.D. Castello & S.O. Rogers (Eds.), *Life in ancient ice*. Princeton University Press, Princeton and Oxford.

Sterflinger, K. & Krumbein, W.E. (1997). Dematiaceous fungi as a major agent for biopitting on Mediterranean marbles and limestones. *Geomicrobiol. J., 14,* 219-230.

Sterflinger, K., de Hoog, G.S. & Haase, G. (1999). Phylogeny and ecology of meristematic ascomycetes. *Stud. Mycol., 43,* 5-22.

Takano, Y., Kobayashi, K., Marumo, K. & Ishikawa, Y. (2004). Biochemical indicators and enzymatic activity below permafrost environment, In: Extremophiles 2004, 5th International Conference on Extremophiles, September 19-23, Abstract Book. Cambridgc, Maryland, p. 84.

Takashima, M., Sugita, T., Shinoda, T. & Nakase, T. (2003). Three new combinations from the *Cryptococcus laurentii* complex: *Cryptococcus aureus*, *Cryptococcus carnescens* and *Cryptococcus peneaus*. *Int. J. Syst. Evol. Microbiol., 53,* 1187-1194.

Tosil, S., Casado, B., Gerdol, R. & Caretta, G. (2002). Fungi isolated from Antarctic mosses. *Polar Biol., 25,* 262-268.

van Uden, N. (1984). Temperature profiles of yeasts. *Adv. Microb. Physiol., 25,* 195-251.

Vishniac, H.S. (1987). Psychrophily and the systematics of yeast-like fungi. In: G.S. de Hoog, M.T. Smith & A.C.M. Weijman (Eds.), *Proceedings of an international symposium on the perspectives of taxonomy, ecology and phylogeny of yeasts and yeast-like fungi*. Delft: CBS; Amsterdam: Elsevier Science Publishers.

Vishniac, H.S. (2002). *Cryptococcus tephrensis*, sp. nov., and *Cryptococcus heimaeyensis*, sp. nov.; new anamorphic basidiomycetous yeast species from Iceland. *Can. J. Microbiol., 48,* 463-467.

Vishniac, H.S. & Onofri, S. (2002). *Cryptococcus antarcticus* var. *circumpolaris* var. nov., a basidiomycetous yeast from Antarctica. *Antonie van Leeuwenhoek, 83,* 231-233.

Vishniac, H.S. & Klinger, J. (1986). Extreme environments. In: F. Megušar & M. Gantar (Eds.), *Yeasts in the Antarctic deserts, Perspectives in Microbial Ecology* (pp. 46-51). Ljubljana: Slovene Society for Microbiology.

Vishniac, H.S. & Onofri, S. (2003). *Cryptococcus antarcticus* var. *circumpolaris* var. nov., a basidiomycetous yeast from Antarctica. *Antonie van Leeuwenhoek, 83,* 231-233.

Vishniac, H.S. (1993). The microbiology of Antarctic soils. In: E.I. Friedman (Ed.), *Antarctic Microbiology* (pp 297-341). New York: Wiley-Liss, Inc.

Wheeler, K.A. & Hocking, A.D. (1993). Ineteractions among xerophilic fungi associated with dried salted fish. *J. Appl. Bacteriol., 74,* 164-169.

Yarrow, D. (1998). Methods for the isolation, maintenance and identification of yeasts. In: C.P. Kurtzman & J.W. Fell (Eds.), *The Yeasts, A Taxonomic Study* (pp. 77-100). Amsterdam: Elsevier.

Zalar, P., de Hoog, G.S & Gunde-Cimerman, N. (2001). Fungi from the salterns: halophilic versus eurytolerant species in the genus *Cladosporium*, In: A. Ventosa and D.R. Arahal (eds.) Abstracts of the International Conference on Halophilic Organisms, September 23-27. Sevilla, Spain, p. 17.

Zalar, P., de Hoog, G.S. & Gunde-Cimerman, N. (1999). Ecology of halotolerant dothideaceous black yeasts. *Stud. Mycol., 43*, 38-48.

In: New Permafrost and Glacier Research
Editors: Max I. Krugger and Harry P. Stern

ISBN: 978-1-60692-616-1

Chapter 9

THE EXPANSION OF SUPRAGLACIAL LAKES IN THE HIMALAYAS: ITS HISTORY AND MECHANISMS

Kazuhisa A. Chikita*

Department of Natural History Sciences, Faculty of Science, Hokkaido University, Sapporo, 060-0810, Japan

ABSTRACT

Many supraglacial lakes have appeared and expanded in the Himalayas since the 1950s by glacial retreat, probably due to global warming after the Little Ice Age. Some of these lakes have produced glacial lake outburst flood (GLOF) events since the 1960s, which have occurred, on average, once every three years somewhere in the Himalayas. The three glacial lakes, Tsho Rolpa (water level 4,580m a.s.l.) and Imja (5,009m a.s.l.) in the Nepal Himalayas, and Lugge (4,539m a.s.l.) in the Bhutan Himalayas, were typical supraglacial lakes in the 1950s to 1960s, but, at present, are moraine-dammed by the subsequent horizontal expansion, enough to touch the side moraine. The lakes were investigated from the hydrological and hydrodynamic viewpoints by field observations from 1995 to 1997, and in 2001 and 2002. In particular, Tsho Rolpa has the highest potentiality of GLOF, since the lake water directly contacts the end moraine, in addition to sufficient water pressure (maximum depth, 131m). Hence, the hydrodynamics of the lake were explored in the pre-monsoon season of 1996 by mooring current meters, temperature loggers and turbidimeters under water, and by obtaining vertical profiles of water turbidity and temperature. As a result, a dynamic model of the lake basin expansion, related to calving at the glacier terminus, was proposed.

A comparison of the thermal structure between Tsho Rolpa, Imja, and Lugge shows a definite difference in the lake hydrodynamics associated with the lake expansion rate. The end-moraine and the dead-ice zone of Imja are 10–25m higher than the lake level. The topographic screening of the end moraine on valley winds, commonly blowing along the elongated lakes, tends to decrease wind velocity near the lake surface, which weakens vertical thermal circulations inducing the ice melt below the lake bottom. The characteristic thermal structure of Imja Lake was probably produced by such a screening

* Telephone and fax: +81-11-706-2764. E-mail address: chikita@mail.sci.hokudai.ac.jp

effect. A three-dimensional numerical simulation of wind around Imja definitely shows the upwind topographic screening effect. The hydrodynamics of Tsho Rolpa was simulated by creating a three-dimensional lake basin of actual size in the calculation domain. In the simulation, a return current, activated by the inflow–outflow system of the lake, was observed. Spatial distributions of the concentration and size for suspended sediment were also simulated, which was reasonable in pattern to the observation.

INTRODUCTION

Supraglacial lakes (or ponds) on the order of 100m or less in horizontal scale were located on the termini of Himalayan glaciers in the 1950s and 1960s. However, thereafter, they have been expanding by the glacial retreat after the Little Ice Age of the sixteenth to nineteenth centuries, and, at present, up to moraine-dammed lakes of 1 km scale [1]. The expansion of moraine-dammed lakes consists of horizontal expansion by calving at the glacial terminus and vertical deepening by ice-melting below the lake bottom [2]. Of all the moraine-dammed lakes, Tsho Rolpa, Rolwaling in the Nepal Himalayas, has the highest potential for GLOF (Glacial Lake Outburst Flood) [3, 4], which is produced by the collapse of the end moraine damming up the lakes. In order to prevent a GLOF at Tsho Rolpa, siphonal pumping in the monsoon season or gate construction on the end moraine by the cut-and-cover method [5] (URL: www.geologyuk.com/) was undertaken from 1995 to 2000. The contribution of the worldwide retreat of mountainous glaciers to an increase in the sea level has tended to increase in recent years [6, 7]. It should be noted that the glacial retreat in the Himalayas could not only increase the glacier-melt water input through the Ganges and Indus rivers to the Indian Ocean, but also produce serious disasters such as GLOFs.

In this chapter, the hydrodynamics of three moraine-dammed glacial lakes, Tsho Rolpa, Imja and Lugge, are compared from field observations, and the associated heat transport in the lakes is discussed by three-dimensional numerical simulations of wind and lake currents.

STUDY SITES AND OBSERVATIONS

Tsho Rolpa (water level, 4,580m a.s.l.; 27°51′N, 86°29′E) and Imja (5,009m a.s.l.; 27°54′N, 86°55′E) in the Nepal Himalayas and Lugge (4,539 m a.s.l.; 28°5′N, 90°18′E) in the Bhutan Himalayas (Fig. 1) commonly contact both side moraines and the cliff-shaped glacial terminus at uplake (Fig. 2A). Imja and Lugge abut on the dead-ice zone at downlake, which is 20 to 30m higher than the lake level, but Tsho Rolpa touches the end moraine near the lake level (Fig. 2B).

The histories of the lake expansion for Tsho Rolpa and Imja are shown as an increase of the lake surface area in Fig. 3. It should be noted that an increasing rate of the surface area is clearly linear for each lake. Tsho Rolpa and Imja laterally contacted both side moraines in the 1960s and 1980s, respectively. The width of the water surface is longitudinally roughly constant for each lake (Fig. 1). The linearity in Fig. 3 thus means that the rate of the glacial retreat is roughly constant after the 1960s or 1980s. However, it is unknown why the glacial retreat occurs at a constant rate for each lake. The increasing rate for Tsho Rolpa is consistently higher than that for Imja. Sakai et al. [11] indicated that Imja increased in surface area and water volume from 0.60 km^2 and 2.8×10^7 m^3 in 1992 to 0.86 km^2 and 3.6×10^7 m^3 in 2002, respectively. The mean water depth of Imja is thus 46.7m in 1992 and 41.9m in 2002. Hence, the expansion of the lake basin likely prevails horizontally more than vertically. Lugge has also been expanding by the glacial retreat of approximately 200 m in 1993 to 2004 [12]. However, its increasing rate is less than 0.01 km^2/yr [13], which is smaller than that of Imja. This small rate is probably due to the relatively large downstream movement of Lugge Glacier. By observing the movements of Lugge Glacier and the neighboring Thorthormi Glacier with no moraine-dammed lake, Naito et al. [14] suggests that the calving at the Lugge glacier terminus facilitates the downward glacial movement in the downstream.

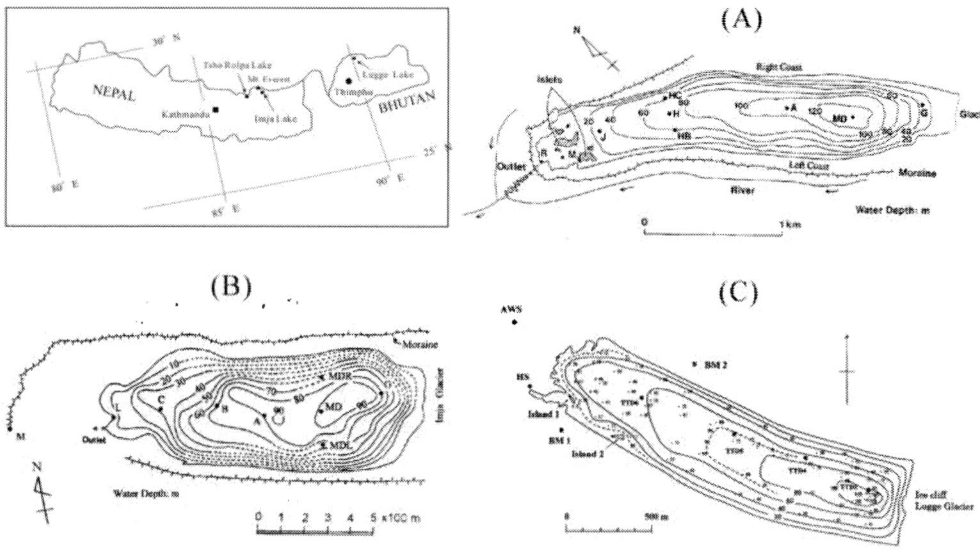

Figure 1. Location of three moraine-dammed lakes: (A) Tsho Rolpa; (B) Imja; and (C) Lugge, and observation sites on the bathymetric maps. The bathymetries of Imja, Tsho Rolpa, and Lugge were obtained in 1992 [8], 1993 [9] and 2002 [10], respectively.

Figure 2. Tsho Rolpa Glacial Lake: (A) cliff-shaped glacial terminus in contact with the lake water at uplake (27 May 1995) and (B) the end moraine only 2 to 3m higher than the water level (26 May 1995). The glacial terminus above the lake level is ca. 20m high.

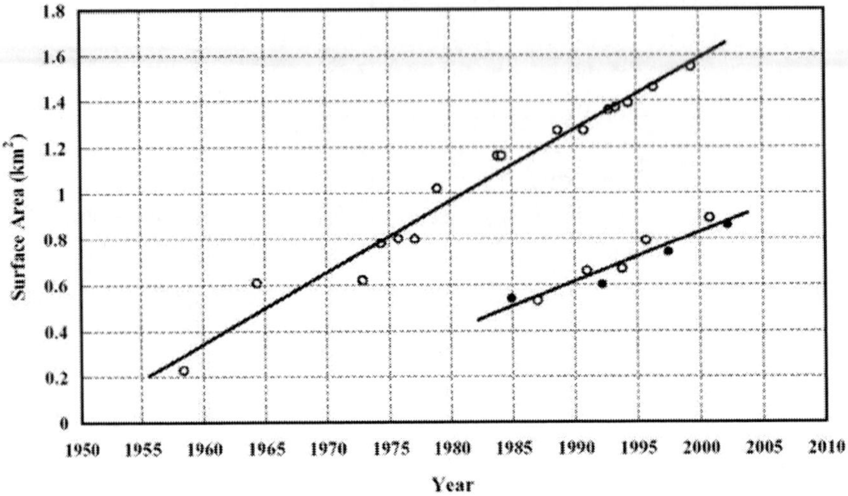

Figure 3. Transition of the surface area for Tsho Rolpa and Imja (modified after [4, 17]). The black and white circles for Imja were obtained by topographic surveys and satellite images (SPOT with resolution 10m or Landsat7 with resolution 15m), respectively.

Physical conditions of Tsho Rolpa, Imja, and Lugge were examined in the pre-monsoon seasons of 1996 (Tsho Rolpa) and 1997 (Imja) and in the post-monsoon season of 2002 (Lugge) by using a TTD (temperature-turbidity-depth) profiler (model ATU200-PK, Alec Electronics, Inc., Japan; 0.2m pitch) and sampling lake water [15, 16, 17]. The vertical measurements of water temperature and turbidity were longitudinally carried out three or four times at some points along the centerline of the elongated lakes (Fig. 1). The water turbidity (ppm) was converted into suspended sediment concentration (SSC) (g/L) by using the significant correlation (r = 0.836 to 0.996) between water turbidity and SSC. The "surplus density", σ, of lake water was numerically obtained by $\sigma = (\rho_{\theta C} - 1000) \times 10$, where $\rho_{\theta C}$ is the bulk density (kg/m^3) of lake water at SSC, C (g/L), water temperature, θ (°C), and dissolved solids, D (g/L). The bulk density $\rho_{\theta C}$ is defined by $\rho_{\theta C} = (1 - C/\rho_s)\rho_\theta + C$, where ρ_s is the particle density of suspended sediment (= 2,730 kg/m^3 for Tsho Rolpa, 2,750 kg/m^3 for Imja, and 2,760 kg/m^3 for Lugge [10, 16, 17]) and ρ_θ is the water density at θ and D. Dissolved solids in lake water were nearly uniform at ca. 0.030 g/L for Tsho Rolpa, at ca. 0.017 g/L for Imja and at ca. 0.025 g/L for Lugge [18]. Thus, the effect of dissolved solids on the water volume is negligibly small.

The monitoring of water temperature, turbidity, and flow velocity was simultaneously conducted in Tsho Rolpa by mooring turbidimeters and current meters at some depths [16]. The particle size of suspended sediment was analyzed for less than 44μm grains by the photo-extinction method and for more than 44μm grains by sieving [19].

The meteorology (wind, air temperature, relative humidity, rainfall, air pressure, and solar radiation) was measured at site AWS on the end moraine of Imja, near the end moraine of Tsho Rolpa or on the dead-ice zone of Lugge (Fig. 1) [16, 17, 20]. Using the meteorological and hydrological data, Sakai et al. [2] estimated the heat balance of Tsho Rolpa and thereby the ice-melting rate.

Observational Results and Discussion

Fig. 4 shows longitudinal distributions of water temperature, SSC, and surplus density σ in Tsho Rolpa, Imja and Lugge. It should be noted that the water of 4 °C is not located in the bottom layer, but at ca. 20 m depth of Tsho Rolpa, at ca. 6 m depth of Imja, and at the surface of Lugge. This looks very abnormal, since, if the water is clear, its density is maximum at 4°C. Most of the lake water exhibits water temperature of 1.5 to 5°C, which, if clear, gives the water density of 999.92 to 1000.00 kg/m^3 under the air pressure of 1 atm. This means that, if SSC is more than 0.1 g/L, the bulk density $\rho_{\theta C}$ of lake water is always more than 1000 kg/m^3 for the particle density of 2,730 to 2,760 kg/m^3. The SSC values in the three lakes range from 0.08 to 0.9 g/L. It is thus seen that the lake water density depends on SSC rather than water temperature. In fact, the SSC patterns in Fig. 4 are similar to the σ patterns, though it is relatively unclear in Lugge, due to the rough isoplethic intervals of SSC and σ. Hence, the density structure of the three lakes could be controlled by behaviors of suspended sediment supplied by glacier-melt or landslide from on the steep slope of side moraine. The suspended sediment in the lakes consists of more than 70 % inorganic clay particles (less than 4 μm in diameter) [16]. Such fine particles have very small settling velocity of the 10^{-4} m/s order at water temperature of 1.5 to 5°C. The sediment rich water could thus induce the long stagnation of water mass at more than 0 °C, which is effective for ice-melt below the lake bottom over the year [2].

The isotherms' pattern in the upper layer of Tsho Rolpa (top of Fig. 4A) suggests that the surface water heated by solar radiation is transported toward the glacier terminus (glacier front), and then, is moved in the opposite direction after cooling by the glacial ice. Meanwhile, the distributions of SSC and temperature in the lower layer indicate that turbid and cold water intrudes into the bottom layer. This intrusion was probably produced by turbid meltwater inflow through an englacial tunnel at the base of the glacier terminus, followed by sediment-laden underflows [21, 22]. The surface layer of Tsho Rolpa is almost isopycnal (i.e., similar σ values) at 0 to 25 m depths. This means the active vertical mixing by strong winds over the lake surface. The isotherms in Imja (top of Fig. 4B) indicate that warm water produced by solar radiation concentrates in the surface layer, and that cold water such as in Tsho Rolpa does not exist in the lower layer. This suggests weak wind mixing in the surface layer and no turbid meltwater inflow at the base of the glacier terminus. The low SSC in the whole Imja (middle of Fig. 4B) evidences no inflow of sediment rich meltwater. In Lugge, the heating of surface water is not prevalent, reflecting a thermal condition in the post-monsoon season. The patterns of isotherms and σ in the surface layer suggests the weak wind mixing as in Imja. However, some of turbid meltwater inflow from the glacier terminus appears to occur both in the surface and bottom layers, since the SSC isopleths concentrate in the layers. The σ patterns in Tsho Rolpa and Imja indicate that the lakes are pycnally stable by the sediment-laden underflow in the lower layer. However, Imja is almost neutral because of no turbid water inflow. The partly high σ or SSC values at the water surfaces of Imja and Lugge mean that the sediment supply to the lake surface by meltwater input or landslide is active.

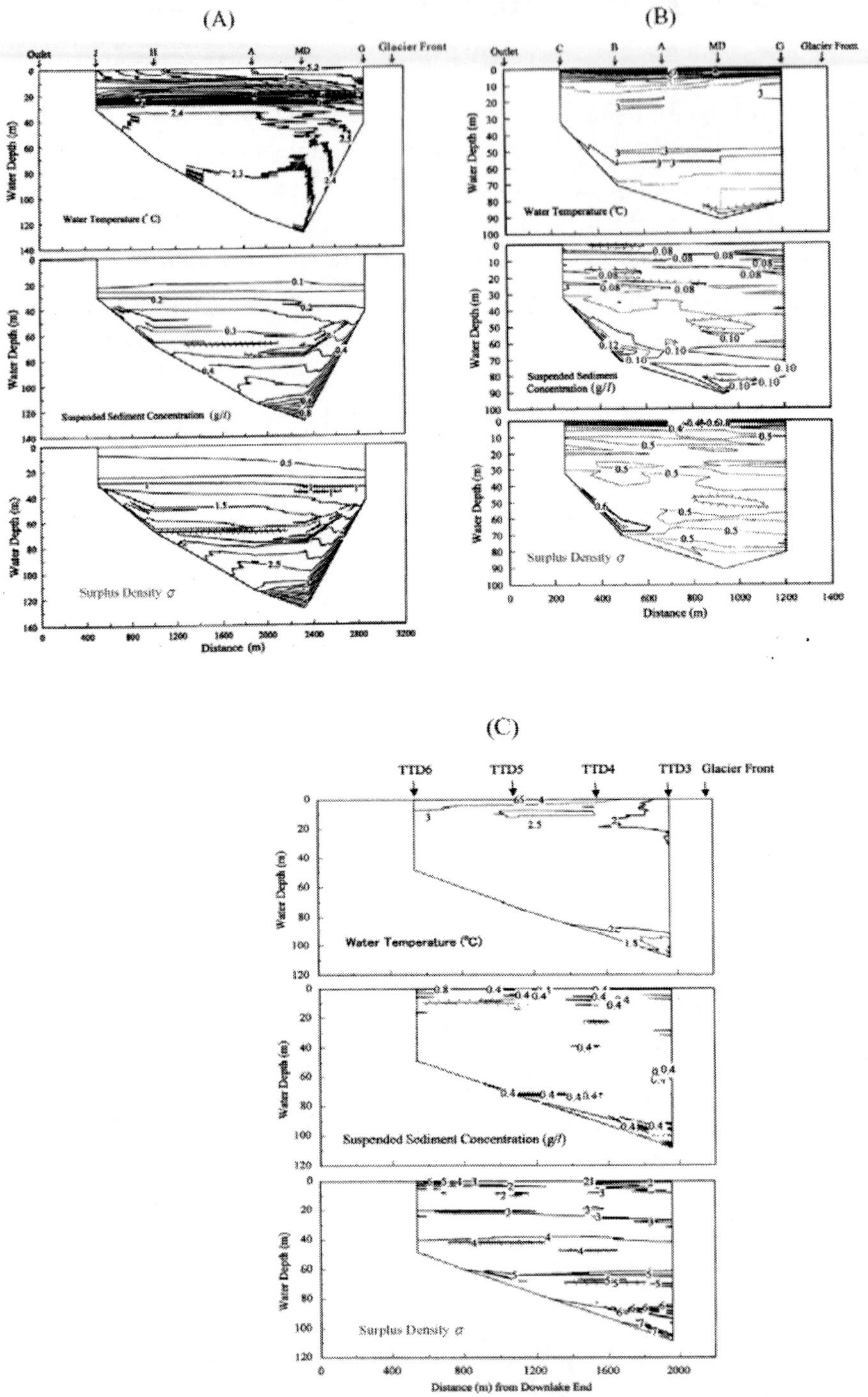

Figure 4. Longitudinal distributions of water temperature, suspended sediment concentration (SSC) and surplus density σ in (A) Tsho Rolpa (2 June 1996), (B) Imja (15 July 1997) and (C) Lugge (26 September 2002).

Fig. 5 shows time series of flow vectors and water temperature observed at 1 hr intervals at 25.4 m depth of site A in Tsho Rolpa, compared with wind vectors at site AWS (Fig. 1) [16]. As soon as valley winds prevail on daytime along the elongated lake surface, the flow velocity increases in the opposite direction with decreasing water temperature. This indicates that the countercurrents occurred near the pycnocline or the thermocline at ca. 25 m depth in order to compensate for the upper wind-driven currents flowing toward the glacier terminus (Fig. 4A). Meanwhile, the vertical water circulation built up by wind-driven currents and compensation currents is accompanied by the uplift of the lower cold water, thus decreasing water temperature [23]. In addition, the monitoring of flow velocity, water temperature and SSC at 1.5 m above the bottom at site G evidenced the generation of sediment-laden underflows from the base of the glacier terminus [16].

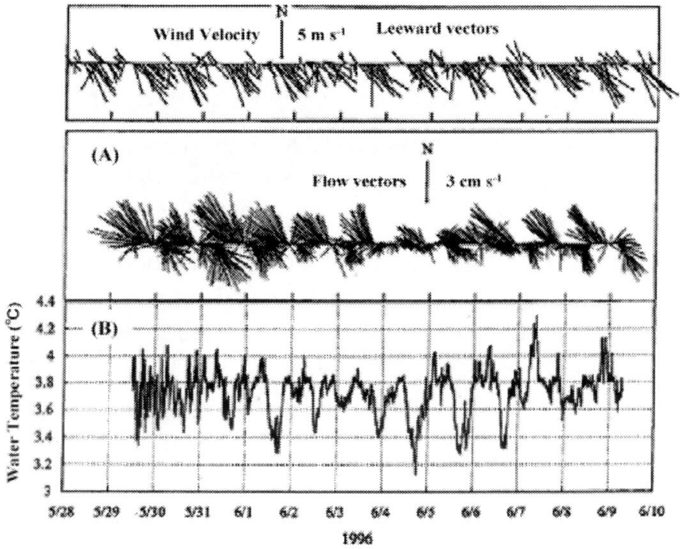

Figure 5. Hourly records of (A) flow vectors and (B) water temperature at 25.4 m depth of site A, compared with leeward wind vectors at site AWS [16].

The monitored results and the longitudinal structures in Fig .4A allowed us to introduce a conceptual model of lake currents related to heat transport (Fig. 6). The hydrodynamics of Tsho Rolpa consists of vertical water circulation in the upper layer above the thermocline, sediment-laden underflows in the bottom layer, and the intrusion in the middle layer from the interaction between the two currents near the glacier terminus. The calving at the glacier terminus could be facilitated by the active dispersal of cooled water in the downlake direction, which is produced by the interaction between sediment-laden underflows and wind-driven vertical circulation. The strong southeastern valley wind produces the setup by wind-driven currents toward the glacier terminus. During the setup, the thermocline is lifted up at the downlake of a nodal line, and lowered at the uplake of the nodal line.

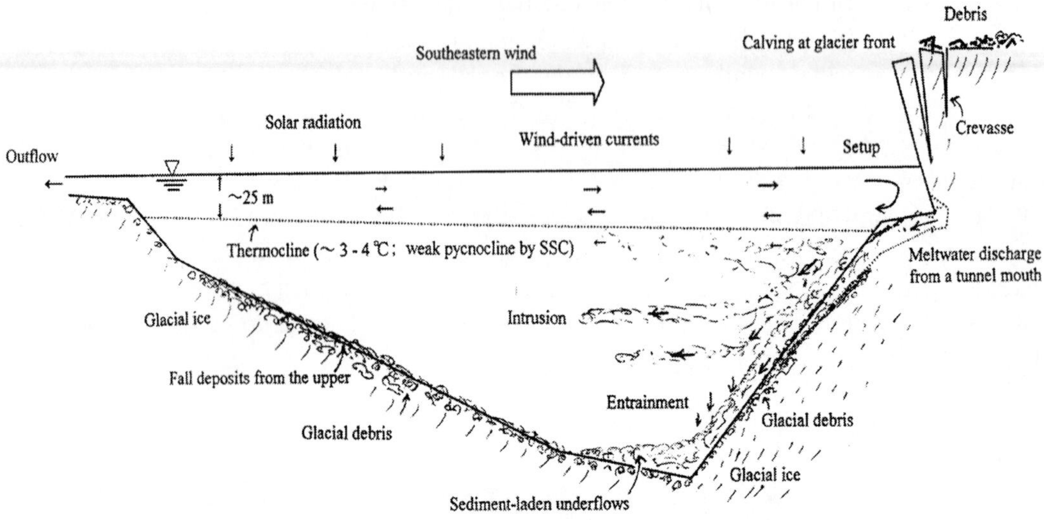

Figure 6. Conceptual model of the hydrodynamics of Tsho Rolpa related to the heat dispersal [16]. The white circle shows the location of the monitoring at site A (Fig. 5). The black circle indicates a nodal line with no oscillation of thermocline or pycnocline during the setup toward the glacier terminus.

Fig. 7 shows temporal variations of wind vectors observed at site AWS (Fig. 1). A wind system is common to the three lakes, i.e., being independent of rainfall or non-rainfall, valley winds prevail on daytime along the elongated lake surface (daily maximums of 3.6 to 8.6 m/s), but, at night, no wind or weak mountain winds occur. The wind at Lugge was observed in the post-monsoon season, and at Imja and Tsho Rolpa, in the pre-monsoon season. However, the wind system at Lugge is consistent in the pre-monsoon, monsoon and post-monsoon seasons (personal communication with Dr. N. Naito in 2004), though the wind at Lugge in Fig. 7 is relatively weak. As a feature of the terrain around the meteorological station of Tsho Rolpa, the crest of the end moraine near the station is only at 2 to 3 m above the lake level (Fig. 2B). Thus, the wind record directly exhibits wind speed and direction at near the water surface. In Imja and Lugge, the dead-ice zone is still developed at downlake at 10 to 25 m above the water level of Imja and at ca. 28 m above the water surface of Lugge (Fig. 1). The meteorological stations at the two lakes are set at near the level of the dead-ice zone. Thus, the wind records probably do not represent wind speed and direction at near the lake surface, since the relatively high dead-ice zone at downlake could obstruct the ventilation of the valley wind. The weak wind mixing in Imja and Lugge (Fig. 4B and 4C) is probably due to the attenuation of valley winds by the topographic screening effect of the dead-ice zone. The topographic effect possibly depends on the lake surface area in the downwind direction. Such a topographic effect is demonstrated by numerical simulation in the next section.

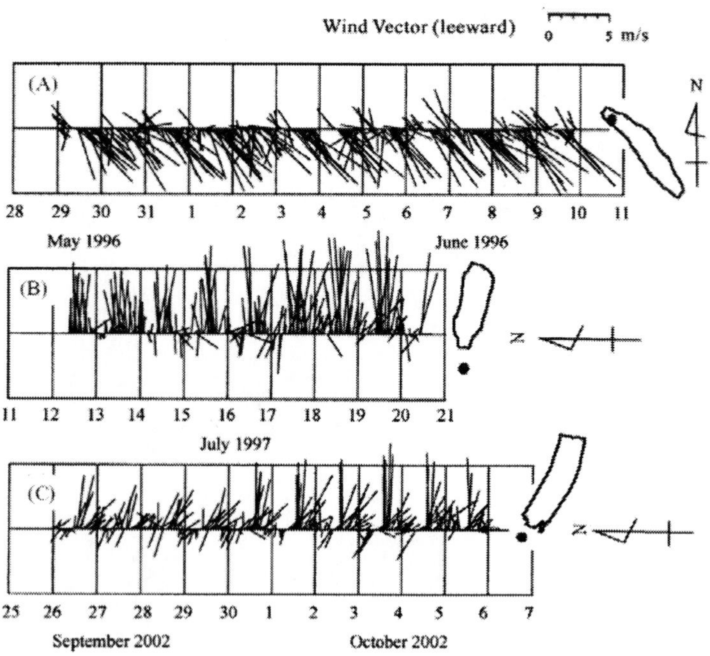

Figure 7. Leeward wind vectors recorded at site AWS of (A) Tsho Rolpa, (B) Imja and (C) Lugge (Fig. 1). Outlines of the lake surface and location (black circles) of meteorological stations are shown.

NUMERICAL SIMULATION OF WIND

Topographic Models and Numerical Conditions

In order to ascertain the topographic screening effect of the dead-ice zone on valley winds, three-dimensional numerical simulation of airflow was carried out (Fig. 8) [18]. Topographic figures of actual size for the water surfaces and surrounding moraines of Tsho Rolpa and Imja (Fig. 1A and B) were built up in the calculation domain of $x \times y \times z$ = 7,000m × 2,000m × 400m. By referring to 1:50,000 scale topographic maps of 1996 and 1997 and data of topographic surveys around Tsho Rolpa and Imja, the topographic figures were assembled by some simple geometrical figures. The roughness length of debris-covered surface and lake surface was assumed to be 0.1 m and 0 m (smooth boundary), respectively. The geometry of the lake surfaces was assumed to be rectangles 3,100m long and 400m wide for Tsho Rolpa and 1,200m long and 400m wide for Imja. The height, H_e, of end moraine and the height, H_i, of glacier terminus above the lake surface were given at 1 m and 35 m for Tsho Rolpa, respectively. For Imja, H_i = 25 m, and the height of the end moraine to the dead-ice zone was set as H_e (= 25 m). The height of the side moraine above the lake surface was commonly given at a constant of 70 m. Considering the air pressure at the high altitude [16, 17], the air density and the standard air pressure were set at 0.75 kg/m^3 and 0.6 atm, respectively. The airflow velocity at the inlet (y - z plane at x = 0) was given constant at 1.0 to

5.0 m/s for Tsho Rolpa and at 1.0 to 6.3 m/s for Imja, which correspond to the valley wind speeds observed (Fig. 7).

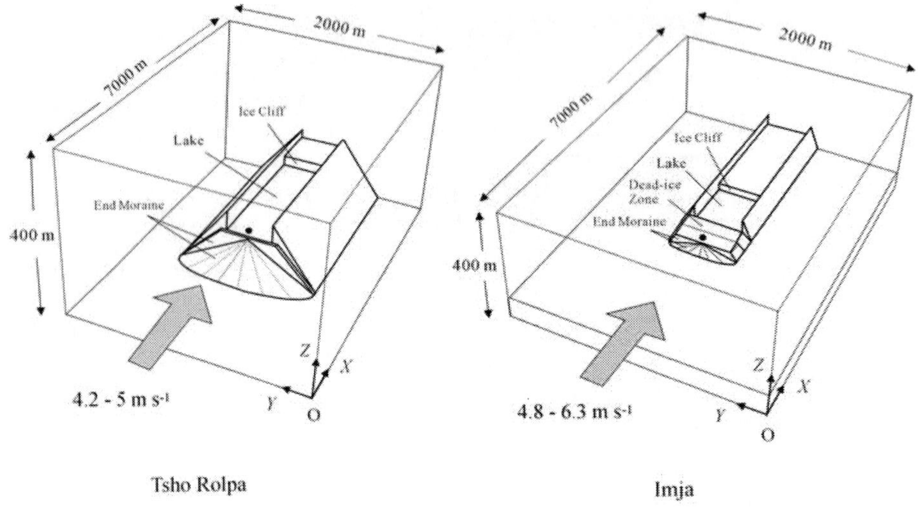

Figure 8. Topographic figures of Tsho Rolpa and Imja built up in the calculation domain of wind simulation [18].

A CFD (computational fluid dynamics) program called "PHOENICS ver. 3.5" was used for the simulation. In the program, the completely implicit and hybrid methods were used to resolve discrete types of integral equations of continuity and motion (Navier-Stokes) for incompressible fluids. The eddy viscosity in the Navier-Stokes equation was calculated by using a revised $\kappa - \varepsilon$ model (MMK model) developed by Murakami et al. [24], Mochida et al. [25] and Kondo et al. [26], where κ is the turbulence kinetic energy and ε is the turbulence dissipation rate. The basic grid number in the present simulation is $x \times y \times z = 80 \times 80 \times 80$, and the grid size near the ground and lake surface was set to be relatively small. The airflow velocity and pressure fields under steady state were repeatedly calculated until their values were converged into constants.

Simulated Results

Fig. 9 shows spatial distributions of horizontal airflow velocity, U_2, at 2 m above the lake surface for Tsho Rolpa (H_e=1 m and H_i =35 m) and Imja (H_e=H_i =25 m). In order that the same velocity of 5.06 m/s (near the maximum wind velocity observed) is obtained at the points (black circles in Fig. 9) corresponding to the sites of meteorological stations (Fig. 1), the flow velocity at the inlet was then given at 5.0 m/s for Tsho Rolpa and 4.7 m/s for Imja. The flow velocity, U_e, at 2 m above the end moraine was then 6.47 m/s for Tsho Rolpa and 5.06 m/s (black circle) for Imja. The velocity distribution in Tsho Rolpa is affected by both the end moraine and the cliff-shaped glacier terminus, especially by the glacier terminus, showing a decrease of U_2 in the downwind direction within ca. 800 m upwind of the glacier terminus. The U_2 in Imja decreases by the end moraine (and dead-ice zone) and glacier terminus within ca. 100 m and 200 m, respectively. The end moraine (and dead-ice zone) and glacier terminus thus can produce a topographic screening effect and a topographic barrier effect, respectively. The flow velocity $\overline{U_2}$ averaged over the lake area was then 5.10 m/s for Tsho Rolpa and 2.98 m/s for Imja. The flow velocity for Imja is thus 42 % smaller than that for Tsho Rolpa (depletion rate of 0.42). The average wind shear stress, τ_b, at the water surface is given by $\tau_b = \rho_a C_d \overline{U_2}^2$. The average shear stress imposed on the surface of Imja is thus ca. 66 % smaller than for Tsho Rolpa, since the drag coefficient C_d is given at 0.0012 to 0.0015

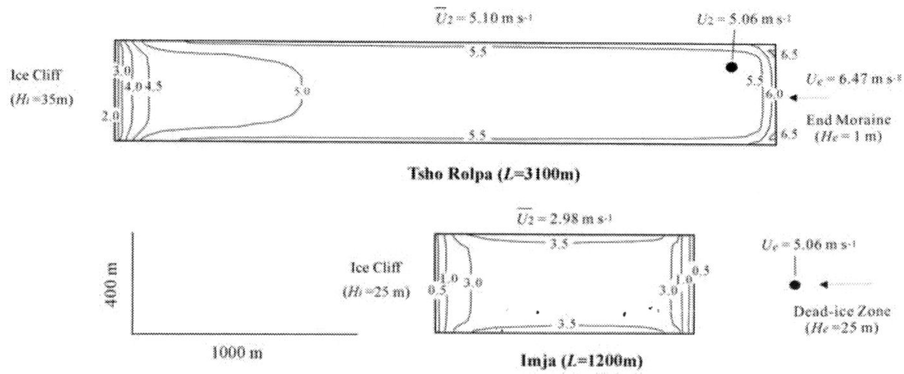

Figure 9. Spatial distributions of horizontal airflow velocity, U_2 (m/s), at 2 m above the lake surface calculated for topographic figures of Tsho Rolpa (upper) and Imja (lower) [18]. The airflow velocity, $\overline{U_2}$, averaged over the lake surface and the air flow velocity, U_e, at 2 m above the end moraine (black circles in Fig. 8) are numerically shown.

for wind velocity of 1 to 6 m/s [27, 28]. The depletion rate in flow velocity was 0.33 for 1.07 m/s calculated at the points corresponding to the weather stations. The decrease in depletion rate is thus small for the flow velocity of 1 to 5 m/s at the points corresponding to the meteorological stations. Hence, the weak mixing or vertical water circulation in the surface layer of Imja probably results from the wind shear stress being not enough to develop wind-driven currents (Fig. 4B).

Fig. 10 shows relations between the velocity ratio, α , the end moraine's height, H_e, and the glacier terminus' height, H_i, above the lake surface. For Imja (L=1,200 m), the velocity ratio α decreases linearly with increasing H_e with the correlation of $R^2 = 0.8885$, while, for Tsho Rolpa, α decreases rapidly at $H_e = 0$ to 8 m, but declines gradually at $H_e > 8$ m (Fig. 10A). Hence, it is suggested that, for glacial lakes dammed up by the relatively long dead-ice zone, the screening effect on the wind velocity over the lake surface is more effective than for glacial lakes dammed up directly by the end moraine. The screening effect of end moraine may be seasonally sensitive in Tsho Rolpa, since the lake level fluctuates at a range of ± 0.6 m in May to October with no ice cover [29], thus being $H_e < 8$ m. The three-dimensional topographic shape of the end moraine (and dead-ice zone), damming up glacial lakes, and a change in the relative height due to the fluctuation in lake level is thus important to know if or not wind-driven water (and thermal) circulation prevails in the lakes. The velocity ratio α in Imja tends to decline gradually at the lengths of 2,200 m and 3,100 m. The basin expansion of Imja may thus weaken the screening effect of the dead-ice zone and end moraine.

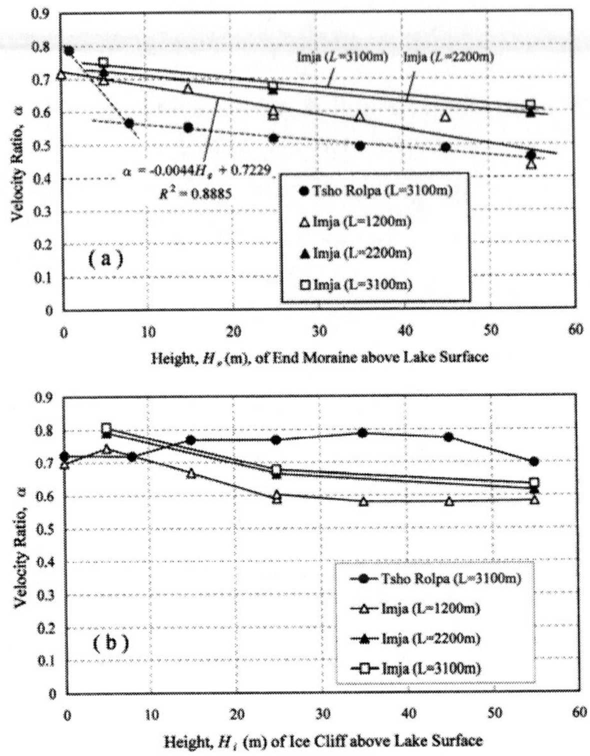

Figure 10. Relations (A) between the velocity ratio, $\alpha = \overline{U_2} / U_e$ and the height, H_e, of end moraine (and dead-ice zone) and (B) between α and the height, H_i, of glacier terminus [18].

The barrier's effect of the glacier terminus is marked for Imja 1,200m long at H_i = 5 to 25 m, but nearly constant at H_i = 25 to 55 m (Fig. 10B). This tendency seems to be unchangeable for the lake expansion up to the lengths of 2200 m and 3100m, though the velocity ratio α wholly increases with increasing lake length. The glacier terminus of Tsho Rolpa tends to give nearly constant α at 0.72 to 0.79 for H_{ic} = 0 to 45 m (Fig. 10B). The three-dimensional geometry of the end moraine (and dead-ice zone) could thus affect the efficiency in the barrier's effect of glacier terminus. Any changes in the height of the side moraine and/or the width of the end moraine did not change significantly spatial distributions of the horizontal airflow velocity at 2 m above the lake surface. The existence of the dead-ice zone upwind of glacial lakes and the increase in the height above the lake surface lead to efficiently decrease the wind velocity near the lake surface and control the thermal structure within the lakes.

Numerical Simulation of Lake Currents

Preparation

In order to quantitatively certify the hydrodynamic system in Tsho Rolpa (Fig. 6), lake currents were three-dimensionally simulated by making a three-dimensional figure of the Tsho Rolpa lake basin in the calculation domain (Fig. 11). The lake basin of actual size was made up by referring to the bathymetric map of 1992 (Fig. 1) [9]. The new version of the CFD program for the airflow simulation, "PHOENICS 2006", was used for the lake current simulation. Considering spatial distributions of ice mass below the lake bottom [29], the calculation domain of $x \times y \times z$ = 3,249m × 760m × 130m was set to be occupied by a solid block (0°C) of $\Delta x \times \Delta y \times \Delta z$ = 200m × 760m × 130m at x = 0 to 200m and a solid block (adiabatic) of $\Delta x \times \Delta y \times \Delta z$ = 3,049m × 760m × 130m at x = 200 to 3,249m, which uniformly contain stone-chippings inside. Corresponding to the outflow observed in Tsho Rolpa [2], the glacier-melt water inflow of 3 to 15m^3/s was given constant at the inlet of $\Delta y \times \Delta z$ =20m × 5m at x = 0. The outlet was set at the place corresponding to its actual location on the lake surface (Fig. 1A). Considering the grain size distributions of suspended sediment [19], the meltwater (0°C) was mixed with sediment (particle density, ρ_s =2,730kg/m^3) consisting of particles of 250μm (2%), 62.5μm (3%), 16.25μm (15%), 3.91μm (40%) and 0.977μm (40%) in diameter. The mixture was given the bulk density, $\rho_{\theta C}$ = 1,003kg/m^3 (C = 5.0g/L). Flow velocity (m/s) at the lake surface was changed from (u, v) = (-0.05, -0.01) to (0, 0) as a wind-driven current, where u and v are the x and y components of flow velocity.

As initial conditions, the whole lake water was set to be uniformly at 3.5°C, and the blocks around the lake basin were given at 0°C. Considering the net radiation observed at site AWS [2], the heat flux at the lake surface was given constant at 130 W/m^2. The Coriolis effect on the lake currents was taken into account by giving the Coriolis parameter $f = 2\Omega\sin\theta$ =6.813 × 10^{-5} s^{-1}, where Ω is the angular speed of the earth (=7.29 x 10^{-5} rad/s) and θ is the latitude (=27°51′ for Tsho Rolpa). The islets near the end moraine were set as ice mass of 0°C. As in the airflow simulation, the completely implicit and hybrid methods were used to resolve discrete types of integral equations of continuity and motion (Navier-Stokes) for incompressible fluids. The eddy viscosity in the Navier-Stokes equation was calculated by using the standard $\kappa - \varepsilon$ model. The following advective diffusion equation, Eq. (1), of suspended sediment concentration, C, was resolved by using the flow velocity field calculated and assuming the eddy diffusivity to be equal to the turbulent diffusivity of suspended sediment.

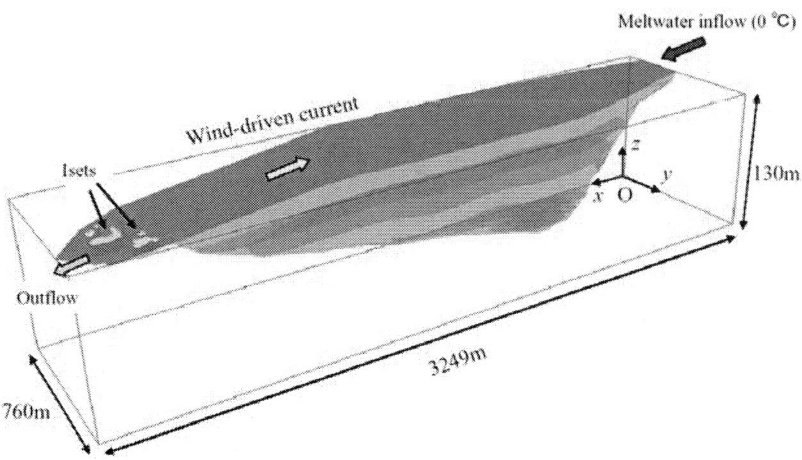

Figure 11. Tsho Rolpa lake basin of actual size built up in the calculation domain. The basin shape is made by referring to the bathymetric map in Fig. 1.

$$\frac{\partial C}{\partial t}+u\left(\frac{\partial C}{\partial x}\right)+v\left(\frac{\partial C}{\partial y}\right)+\left(w-w_s\right)\left(\frac{\partial C}{\partial z}\right)=\frac{\partial}{\partial x}\left(K_x\frac{\partial C}{\partial x}\right)+\frac{\partial}{\partial y}\left(K_y\frac{\partial C}{\partial y}\right)+\frac{\partial}{\partial z}\left(K_z\frac{\partial C}{\partial z}\right) \quad (1)$$

where w is the z component of flow velocity, w_s is the settling velocity of suspended sediment in still water, and K_x, K_y and K_z are the turbulent diffusivity of sediment. The suspended sediment concentration, C, was given as the volume fraction in water, where numerical values of near unit mean sediment deposition. The molecular diffusion is neglected, since the turbulent diffusion is relatively very large.

SIMULATED RESULTS

Fig. 12 shows simulated longitudinal distributions of (A) - (C) flow velocity and (D) water temperature at y = 290m. In Fig. 12B, a wind driven current of u = -0.05m/s and v = -0,01m/s was given at the surface. It is noted that the cold water from the inlet (glacier terminus) is dispersed by sediment-laden underflow, and that the strong return flow toward the inlet occurs in the middle layer. These flows are not intensified by increasing inflow from 3 to 10 m^3/s (Fig. 12A and C), but by the wind-driven current at the surface toward the inlet (Fig. 12B). Fig. 12A indicates that, in a glacial lake, the coupling of outflow at an outlet and subaqueous inflow at the glacier terminus tends to strengthen the inner lake currents, especially on the basin slope probably underlain by dead-ice mass [29]. The inflow - outflow system could thus play the role on the lake deepening by ice-melt, since the relatively large heat transport occurs on the slope. The water temperature distribution (Fig. 12D) calculated in the velocity field of Fig. 12B exhibits the stratification of 5°C at the surface to 2°C at the bottom except 1.0 – 1.5°C at near the inlet, which is similar in pattern to the observation (Fig. 4A). However, the setup toward the glacier terminus by the valley wind is not simulated, since the free oscillation of water surface is not taken into account in the model (i.e., rigid lid model). Hence, the uplift of the thermocline (Fig. 5 and Fig. 6) is not also simulated. The unsteady calculation by the free surface model is needed for more realistic simulation of water temperature.

Fig. 13 shows simulated, longitudinal distributions of suspended sediment concentration in volume fraction for the five particle sizes of 0.977 to 250μm at y = 290m, where the velocity field is given in Fig. 12B. Most of the 250μm particles concentrate or deposit at or around the deepest point, where the volume fraction is more than 0.6. On the whole, the suspended particles are mostly converged in the layer of more than 90m in depth. Considering the total volume fraction of particles in the vertical water column at the deepest point, suspended particles at depths of more than 90m become coarser toward the lake bottom with increasing sediment concentration. This tendency is reasonable to the observation at site MD of Fig. 1 (Fig. 5 in [19]). The sediment-laden underflow from the inlet appears to selectively carry silt and clay particles (less than 62.5μm). Of the particles, the clay particles of 0.977μm and 3.91μm behave similarly, showing the dispersal larger than for the coarser particles.

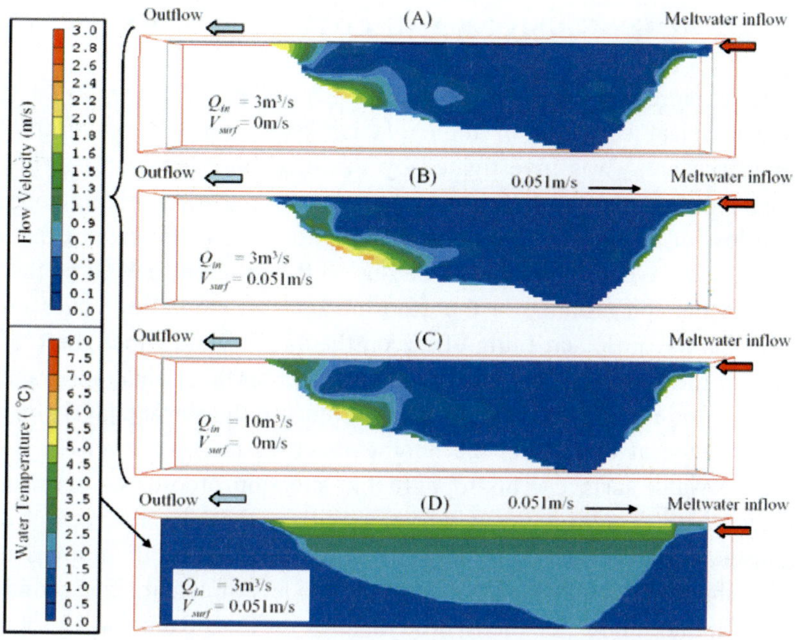

Figure 12. Simulated results of (A)-(C) flow speed and (D) water temperature. The meltwater inflow, Q_{in}, and flow speed, V_{surf}, of wind-driven current at the surface were given constant (or zero) at (A) $3m^3/s$ and 0m/s, (B) $10m^3/s$ and 0.051m/s, and (C) $3m^3/s$ and 0m/s, respectively.

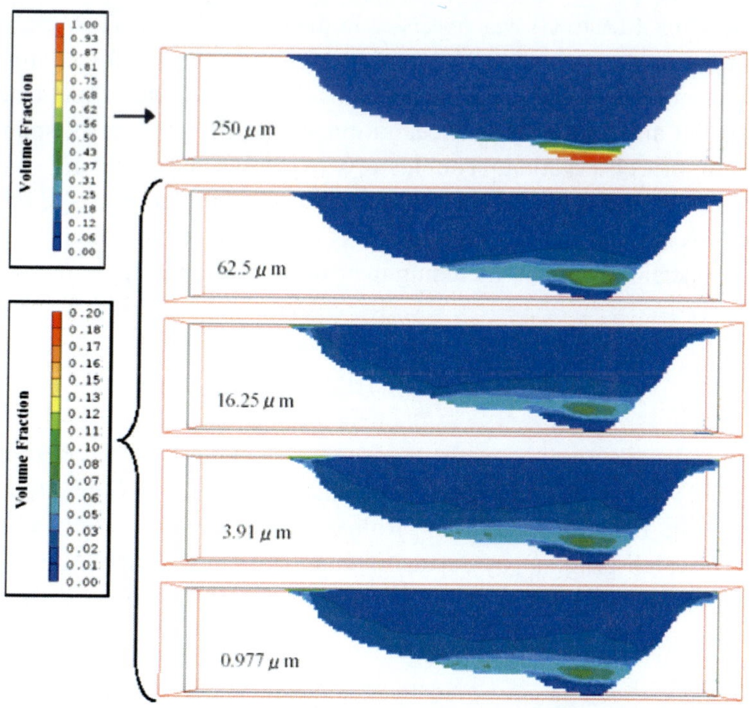

Figure 13. Simulated results of suspended sediment concentration in volume fraction in the velocity field of Fig. 12B. The unit volume fraction for 250μm particles means complete deposition of suspended sediment.

Conclusions and Future Works

In order to explore the mechanism of the expansion of supraglacial lakes in the Himalayas, the physical structure of three glacial lakes, Tsho Rolpa, Imja, and Lugge, increasing in surface area, was investigated by the filed observations. The lakes exhibited different thermal and sedimentary structures, depending on the wind-mixing condition and the turbidity and inflow of meltwater at the glacier terminus. The difference of the wind-mixing condition in the upper layer was judged to be due to the different degree of the topographic screening effect of the end moraine or the dead-ice zone on the valley winds near the water surface. Tsho Rolpa has little screening effect on the valley wind, since the crest of the end moraine upwind of the lake is almost level with the water surface. This topographic condition can induce strong wind mixing and wind-driven currents in the surface layer. Imja and Lugge have the dead-ice zone at downlake, the height of which is 10 to 25m (Imja) or ca. 28m (Lugge) above the water surface. This topographic situation produces a screening effect on the valley wind, which decreases the wind velocity near the lake surface and weakens the wind-mixing in the surface layer. This suggestion was certified by the numerical simulation of airflow, using the three-dimensional topographic models of actual size around Tsho Rolpa and Imja. A conceptual model for the hydrodynamics of Tsho Rolpa was proposed from in situ measurements of water temperature, turbidity, and flow velocity. The lake hydrodynamics were simulated by making up a three-dimensional lake basin in the calculation domain. The fields of flow velocity, pressure, water temperature and suspended sediment concentration in the lake were calculated in steady state. In the simulation, the strong return flow (Fig. 12A to C) was observed in the middle layer, where flow velocity was not monitored in the field observation. As the next step for observations in Himalayan lakes, referring to such simulated results, the location of observation sites should be determined. The simulated spatial distributions of concentration and particle size for suspended sediment were agreeable to the observation. The history of the lake expansion or the temporal development of the lake basin by the melting of debris-covered ice needs to be simulated by considering the seasonal and interannual variations of thermal environments around the lake and the upstream glacial movement, including the calving at the glacier terminus.

Acknowledgements

I am grateful to Dr. Tomomi Yamada, Representative, Snow and Ice Network, Nonprofit Organization (NPO) Corp. for consistent encouragement and invaluable advice to my Himalayan study. The field survey in Tsho Rolpa was supported as a GLOF (Glacier Lake Outburst Flood) research program by Japan International Cooperation Agency (JICA). The support team of CHAM-Japan, Inc. gave me advice to improve the numerical simulations.

REFERENCES

[1] Yamada, T. (2000): Glacier lake outburst floods in Nepal. *Seppyo*, 62, 137-147.

[2] Sakai, A., Chikita, K. and Yamada, T. (2000): Expansion of a moraine-dammed glacial lake ,Tsho Rolpa, in Rolwaling Himal, Nepal Himalaya. *Limnology and Oceanography*, 45, 1401-1408.

[3] Yamada, T. (1993): Glacier lakes and their outburst floods in the Nepal Himalaya. *Report of Water and Energy Commission Secretariat (WECS), Kathmandu, Nepal,* 37 pp.

[4] Yamada, T. (2001): Meteorological-hydrological environment and thermal regime of a moraine-dammed glacier lake, and their contribution to the lake expansion. *Seppyo*, 53, 223-243.

[5] Rana, B., Shrestha, A. B., Reynolds, J. M., Aryal, R., Pokhrel, A. P. and Budhathoki, K. P. (2000): Hazard assessment of the Tsho Rolpa Glacier Lake and ongoing remediation measures. *Journal of Nepal Geological Society*, 22, 563-570.

[6] Braithwaite, R. J. and Raper, S. C. B. (2002): Glaciers and their contribution to sea level change. *Physics and Chemistry of the Earth*, 27, 1445-1454.

[7] Dyurgerov, M. (2003): Mountain and subpolar glaciers show an increase in sensitivity to climate warming and intensification of the water cycle. *Journal of Hydrology*, 282, 164-176.

[8] Yamada, T. and Sharma, C. R. (1993): Glacier lakes and outburst floods in the Nepal Himalaya. *Snow and Glacier Hydrology (Proceedings of the Kathmandu Symposium, November 1992), IAHS Publication*, no. 218, 319-330.

[9] Kadota, T. (1994): Report for the field investigation on the Tsho Rolpa glacier lake, Rolwaling Valley, February-June 1994. *Report of Water and Energy Commission Secretariat (WECS), N551.489 KAD.*

[10] Yamada, T., Naito, N., Kohshima, S., Fushimi, H., Nakazawa, F., Segawa, T., Uetake, J., Suzuki, R., Sato, N., Karma, Chhetri, I. K., Gyeden, L., Yabuki, H. and Chikita, K. (2004): Outline of 2002-research activities on glaciers and glacier lakes in Lunana region, Bhutan Himalayas. *Bulletin of Glaciological Research*, 21, 79-90.

[11] Sakai, A., Yamada, T. and Fujita, K. (2003): Volume change of Imja Glacial Lake in the Nepal Himalayas. *International Symposium on Disaster Mitigation & Basin Wide Water Management*, 556-561.

[12] Fujita, K., Suzuki, R., Nuimura, T. and Sakai, A. (2008): Performance of ASTER and SRTM DEMs, and their potential for assessing glacial lakes in the Lunana region, Bhutan Himalaya. *Journal of Glaciology*, 34, 220-228.

[13] Bajracharya, S. R., Mool, P. K. and Shrestha, B. R. (2007): *Impact of climate change on Himalayan glaciers and glacial lakes*. International Centre for Integrated Mountain Development (ICIMOD), Kathmandu, Nepal, 119pp.

[14] Naito, N., R. Suzuki, Y. Matsuda and Y. Ageta (2004): Glaciological observation. Report of Japan-Bhutan Joint Research 2003 on the assessment of Glacier Lake Outburst Flood (GLOF) in Bhutan, Graduate School of Environmental Studies, Nagoya University; Graduate School of Biosciences and Biotechnology, Tokyo Institute of Technology; and Geological Survey of Bhutan, 19-33.

[15] Chikita, K., Jha, S. P. and Hasegawa, H. (2000): The expansion mechanism of supraglacial lakes in the Nepal Himalaya. *Verh. Internat. Verein. Limnol.*, 27, 2145-2148.

[16] Chikita, K. A., Jha, J. and Yamada, T. (1999): Hydrodynamics of a supraglacial lake and its effect on the basin expansion: Tsho Rolpa, Rolwaling Valley, Nepal Himalaya. *Arctic, Antarctic and Alpine Research*, 31, 58-70.

[17] Chikita, K. A., Joshi, S. P., Jha, J. and Hasegawa, H. (2000): Hydrological and thermal regimes in a supraglacial lake: Imja, Khumbu, Nepal Himalaya. *Hydrological Sciences Journal*, 45, 507-521.

[18] Chikita, K. A. (2007): Topographic effects on the thermal structure of Himalayan glacial lakes: Observations and numerical simulation of wind. *Journal of Asian Earth Sciences*, 30, 344-352.

[19] Chikita, K., Jha, J. and Yamada, T. (2000): Sedimentary effects on the expansion of a Himalayan supraglacial lake. *Global and Planetary Change*, 28, 23-34.

[20] Suzuki, R., Fujita, K., Ageta, Y. Naito, N., Matsuda, Y. and Karma (2007): Meteorological observations during 2002-2004 in Lunana region, Bhutan Himalayas. *Bulletin of Glaciological Research*, 24, 71-78.

[21] Chikita, K. (1989): A field study on turbidity currents initiated from spring runoffs. *Water Resources Research*, 25, 257-271.

[22] Chikita, K. A., Smith, N. D., Yonemitsu, N. and Perez-Arlucea, M. (1996): Dynamics of sediment-laden underflows passing over a subaqueous sill: glacier-fed Peyto Lake, Alberta, Canada. *Sedimentology*, 43, 865-875.

[23] Mortimer, C. H. (1952): Water movements in lakes during summer stratification; evidence from the distribution of temperature in Windermere. *Philosophical Transactions of Royal Society*, ser. B, 236, 355-404.

[24] Murakami, S., Mochida, A. and Ooka, R. (1993): Numerical simulation of flow field over surface-mounted cube with various second-moment closure models. *Proceedings of 9th Symposium on Turbulent Shear Flow*, 13-5-1-13-5-6.

[25] Mochida, A. and Ishida, Y. and Murakami, S. (1993): Numerical study on flow field around structures with oblique wind angle based on composite grid system. *Proceedings of 7th U. S. National Conference on Wind Engineering*, 15-16.

[26] Kondo, K., Murakami, S., Mochida, A. and Ishida, Y. (1995): Numerical prediction of flow filed around 2D square rib using revised $\kappa - \varepsilon$ model. *Journal of Wind Engineering*, 63, 35-39.

[27] Charnock, H. (1955): Wind stress on a water surface. *Quarterly Journal of the Royal Meteorological Society*, 81, 639-640.

[28] Hsu, S.A. (1986): A mechanism for the increase of wind stress (drag) coefficient with wind speed over water surfaces: a parametric model. *Journal of Physical Oceanography*, 16, 144-150.

[29] Yamada, T. (1996): Report on the investigations of Tsho Rolpa Glacier Lake, Rolwaling Valley. Report to Water and Energy Commission Secretariat (WECS), Nepal and Japan International Cooperation Agency (JICA), 129 pp.

In: New Permafrost and Glacier Research
Editors: Max I. Krugger and Harry P. Stern
ISBN: 978-1-60692-616-1

Chapter 10

SURGE-TYPE GLACIERS ON DISKO ISLAND, GREENLAND

Jacob C. Yde [1,2] ***and N. Tvis Knudsen*** [1]

[1] Department of Earth Sciences, Ny Munkegade, Bygning 1520, University of Aarhus, DK-8000 Århus C, Denmark

[2] Center for Geomicrobiology, Department of Biological Sciences, Ny Munkegade, Bygning 1540, University of Aarhus, DK-8000 Århus C, Denmark

ABSTRACT

This study identifies 75 surge-type glaciers on Disko Island (Qeqertarsuaq), central West Greenland, using old maps, aerial photographs, satellite imageries and field observations. The surge-type glaciers comprise about 30% of the glaciers larger than 1 km^2 and 59% of the glacierised area. The duration of the surge cycle is at least 100 years or more, which is relatively long compared to surge clusters in other parts of the world. The surge events are very dramatic and include some of the longest frontal advances ever recorded such as 10.5 km advance of Kuannersuit Glacier in 1995-98, where the glacier advance velocity reached at least 70 m d^{-1}. The surge-type glaciers are responsible for valley floor geomorphology such as complex moraine systems and extensive dead ice areas. This implies that glacier surging has to be considered when glacier recession is related to climatic fluctuations.

INTRODUCTION

Glacier surging is one of the most fascinating and enigmatic topics in glaciological research. When glacier surging suddenly triggers, the glacier becomes heavily crevassed and in most cases the glacier terminus rapidly advances. The short dramatic active phase is followed by a long quiescent phase, where the glacier terminus undergoes thinning and recession while the accumulation area regains in mass before the next surge event occurs. This cyclic behaviour superimposes any climatic variations. Hence, it is very important to identify surge-type glaciers in order to exclude them in attempts to relate glacier terminus fluctuations to climate change.

Although research on glacier surging has been in progress for more than 50 years, it is still unclear why some glaciers surge while adjacent glaciers do not and what exactly triggers and terminates surge events. There seems to be consensus in the glaciological community in that several mechanics may lead glacier surging (e.g. Murray et al., 2003).

Also, surge-type glaciers appear as a regional phenomenon in many glacierized areas, while they are almost absent in other regions. In recent years, a number of papers have identified surge-type glaciers in regions such as Iceland (Thorarinsson, 1969; Björnsson et al. 2003), Svalbard (Hagen, 1988; Hagen et al., 1993; Dowdeswell et al., 1995; Hamilton and Dowdeswell, 1996; Jiskoot et al., 1998, 2000), Russian High Arctic (Dowdeswell and Williams, 1997), Caucasus (Dolgoushin and Osipova, 1975), Pamirs (Dolgoushin and Osipova, 1975), Tien-Shan (Dolgoushin and Osipova, 1975), Karakoram Himalaya (Barrand and Murray, 2006), Kamchatka (Dolgoushin and Osipova, 1975), Alaska-Yukon (Post, 1969; Clarke et al., 1986), Canadian High Arctic (Copland et al., 2003), and East Greenland (Jiskoot et al., 2003).

This paper follows this line of research and presents the results of a systematic examination of the incidence of surge-type glaciers on Disko Island, West Greenland. These surge-type glaciers constitute together with surge-type glaciers on Nuussuaq Peninsula, situated north of Disko Island, the Disko-Nuussuaq surge cluster. At present, only two glaciers (glacier 1IA02034 and glacier 1IB26003) on Nuussuaq Peninsula are categorized as surge-type glaciers (Weidick, 1988; Weidick et al., 1992), but a thorough examination is likely to reveal more. The current literature categorizes 24 glaciers on Disko Island as surge-type glaciers, and only four of these have been observed during their active surge phase (Weidick, 1988; Weidick et al., 1992; Gilbert et al., 2002).

In Greenland two surge clusters have been recognized; the East Greenland surge cluster and the Disko-Nuussuaq surge cluster. Both surge clusters are mainly situated in basaltic provinces. The East Greenland surge cluster comprises 71 surge-type glaciers and is the best described (Rucklidge, 1966; Henriksen and Watt, 1968; Olesen and Reeh, 1969; Rutishauser, 1971; Colvill, 1984; Weidick, 1988; Jiskoot et al., 2001, 2003; Murray et al., 2002; Woodward et al., 2002; Pritchard et al., 2003). In addition, surge events have been reported on outlet glaciers in South Greenland (Weidick, 1984a,b), Northwest Greenland (Mock, 1966; Rignot and Kanagaratnam, 2006), North Greenland (Higgins and Weidick, 1988) and Northeast Greenland (Reeh et al., 1994).

As only a few glaciers on Disko Island have official names, reference to specific glaciers is given to their hydrological code in the regional glacier inventory (Weidick et al., 1992).

STUDY AREA

Disko Island (70° N, 54° W) is situated in central West Greenland (Figure 1). The island is the largest in Greenland and covers 8575 km^2 of which about 19% is glacierized (Humlum, 1987). According to the regional glacier inventory (Weidick *et al.*, 1992), the island contains 1070 glaciers, ice and snowfields of which 350 are larger than 1 km^2. Of these, 247 are glaciers, while the remaining comprises ice and snowfields. From the two major ice caps,

Sermersuaq (Storbræen) and Bræpasset, in the central part of the island, outlet glaciers extend into the characteristic U-shaped valleys that drain meltwater to the large fjords and the Disko Bay. All glaciers on Disko Island terminate on land, and there are no indications of glaciers reaching the sea during the Little Ice Age (LIA), except for some rock glaciers.

The climate is polar maritime with discontinuous permafrost along the southern and western coastal areas, whereas the central and northeastern parts of the island have colder continental climate with continuous permafrost (Humlum, 1999; Yde and Knudsen, 2007). The 1961–1990 mean annual air temperature in the coastal town of Qeqertarsuaq (Godhavn) was –4.0°C (Humlum, 1999), and it is about 3°C colder at sea-level in the central part of the island (Humlum *et al.*, 1995).

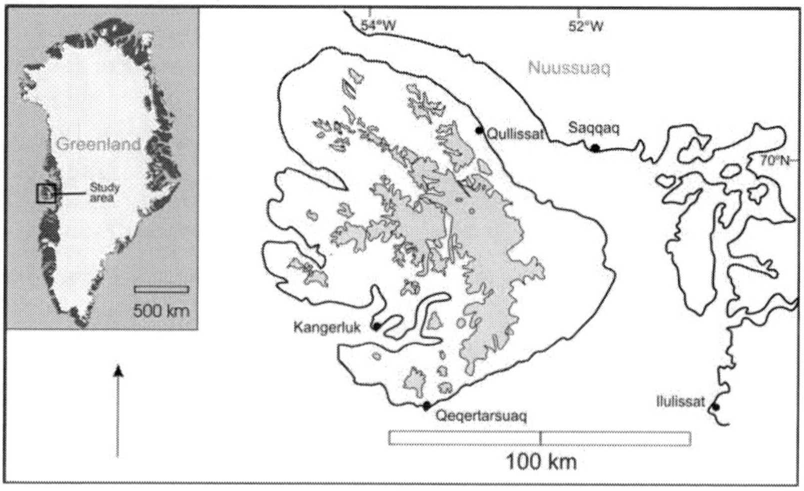

Figure 1. Location map of Disko Island, West Greenland

The geology is characterised by Early Tertiary basalt plateaux at 600–1200 m a.s.l. with numerous glacier-carved cirques (Pedersen, 1977; Piasecki et al., 1992). The lowest layers of basalt lava have intercalated layers of contemporary shale. In the eastern part of Disko Island, the Tertiary basalt overlays Paleocene sand- and mudstones of the Atanikerluk Formation and Cretaceous sandstones of the Atane Formation (Pedersen et al., 2001). Both of these formations contain coal seams. The basement comprises the Precambrian north-south striking Disko Gneiss Ridge, which only is exposed at Qeqertarsuaq and in some of the central fjords. Many homothermal (3-12°C) springs are situated at the interface between gneiss and the overlying rocks.

Method

Information on 19th-century and early 20th-century glacier fluctuations on Disko Island was compiled from expedition accounts, drawings, maps and photographs (Yde and Knudsen, 2007). Remote sensing data were obtained from a 1931-33 topographic map (GID, 1941) covering the south and central part of the island, aerial photographs from 1953-54, 1964 and 1985, and satellite images from 1973 (Landsat 1 Multispectral Scanner (MSS)), 1976

(Landsat 2 MSS), 1985 (Landsat 5 Thermal Mapper (TM)), 1991 (Système Probatoire pour l'Observation de la Terre 2 (SPOT 2)), and 1999, 2002 and 2005 (Landsat 7 Enhanced TM Plus (ETM+)). The 1964 aerial photographs and the 1991 SPOT image did not cover the northern part of the island. Fieldwork in 2001, 2002, 2003 and 2005 contributed with supplementary information on glaciological features and proglacial landforms.

It has become standard practice to classify surge-type glaciers according to the strength of evidence for surge behaviour, often by using a surge index (e.g., Clarke et al., 1986). Hence, glaciers with observed surge events or one or more distinct surge-related characteristics are assigned as surge-type glaciers. As many characteristics are not unique to glacier surging, this quantitative approach was combined with a qualitative assessment of individual features. However, there will always be uncertainties in the identification process, and many surge-type glaciers may remain unidentified because no surge features appear on aerial photographs and satellite imagery during the quiescent phase. For instance, the small glacier 1HD06009 surged in 1953, but no glaciomorphological surge features appeared on later images. Even fieldwork may not reveal unequivocal evidence of past glacier surging.

Surge-type glaciers were diagnosed using glaciomorphological features (elements) such as surge fronts (Kamb et al., 1985), heavily crevassed surface (Post, 1969), looped medial moraines (Post, 1969; Jiskoot et al., 2000; Post and LaChapelle, 2000), and circular depressions termed potholes or lacunas (Sturm, 1987; Sturm and Cosgrove, 1990; Post and LaChapelle, 2000). The first two features occur during the active surge phase, whereas the latter two features may remain for one or more surge cycles. Other features, such as the occurrence of glacier-derived naled (Yde and Knudsen, 2005b), thrust-block naled (Yde et al., 2005a), stranded ice along valley walls in the ice reservoir area (Post and LaChapelle, 2000; Copland et al., 2003), and large circular depressions termed chasms (Yde and Knudsen, 2005a) are also related to glacier surging, but may also occur on normal glaciers. It can be difficult to distinguish between potholes and chasms as they may have the same geometric dimensions. Potholes are formed as remnants of supraglacial lakes between transverse crevasse ridges immediately after surge events and therefore appear as assemblages in the central parts of surge-type glaciers, whereas chasms are supraglacial subsidence depressions formed as a consequence of roof collapse in subglacial drainage cavities. Chasms generally occur along lateral margins during periods of enhanced melt rates and high subglacial water discharge such as during the initial quiescent phase.

Proglacial landforms such as thrust-block moraines, eskers, flutes, drumlins and crevasse-squeezed ridges are associated with fast glacier flow and, thus, may indicate previous surge activity (Evans and England, 1991; Evans and Rea, 1999, 2003). Many glaciers on Disko Island have extensive proglacial areas with debris-covered dead ice and complex moraines, which indicate that these glaciers have experienced significant advances and recession in historical time, which may relate to past surge events. Rapid recession rates some years after the termination of a surge event is common as surge-type glaciers may advance down-valley where relatively high temperatures prevail.

Table 1. List of surge-type glaciers on Disko Island.

No.	Glacier code	Diagnostic surge features and comments
1	1HB07004	Potholes, undulated surface
2	1HB10008	Potholes
3	1HB10013	Active surge in 1953, crevassed surface
4	1HB10027	Active surge between 1985-1999, looped medial moraine, potholes
5	1HB10035	Potholes
6	1HB10036	Active surge in 1964. Potholes, looped medial moraine, stranded ice, extensive dead ice area
7	1HB11009	Potholes
8	1HB11013	Potholes, complex proglacial moraines
9	1HB11014	Potholes, complex proglacial moraines
10	1HB11015	Potholes, complex proglacial moraines
11	1HB11029	Potholes, looped medial moraines
12	1HB11033	Extensive dead ice area
13	1HB15005	Potholes, undulated surface, extensive dead ice area
14	1HB15008	Same glacier as 1HB11034. Active surge between 1931-1953 (?), which erased the end moraine of glacier 1HB11033. Proglacial surge landforms
15	1HB15009	Potholes, undulated surface
16	1HB15012	Active surge in 1953, crevassed surface, potholes, proglacial naled
17	1HB15013a	Potholes, chasms, looped medial moraine
18	1HB15013c	Potholes, looped medial moraine, extensive dead ice area
19	1HB15016	Potholes
20	1HB15017	Active surge in 1995-1998, crevassed surface, potholes, looped medial moraine, chasms, undulated surface, proglacial naled, thrust-block naled, hydrofractures
21	1HB15029	Extensive dead ice area, looped medial moraine, chasms

Table 1. (Continued)

No.	Glacier code	Diagnostic surge features and comments
22	1HB15031	Extensive dead ice area, chasms
23	1HB16010	Undulated surface, extensive dead ice area
24	1HB16017	Potholes, extensive dead ice area, undulated and crevassed surface
25	1HB16020	Extensive dead ice area, undulated surface, proglacial naled, potholes, looped medial moraine
26	1HB16047	Undulated and crevassed surface
27	1HC03010	Crevassed terminus in 1985, but no frontal advance
28	1HC03011	Looped medial moraine
29	1HC03013	Looped medial moraine
30	1HC04019	Potholes
31	1HC05004	Potholes
32	1HC05006	Potholes
33	1HC05007	Extensive dead ice area
34	1HD02004	Very steep terminus in 2002 and 2005
35	1HD02005	Looped medial moraine, extensive dead ice area
36	1HD02011	Looped medial moraine, extensive dead ice area
37	1HD04007	Tributary glacier cut off, proglacial naled, chasms, looped medial moraine?
38	1HD06005	Crevassed surface, potholes, looped medial moraine, extensive dead ice area
39	1HD06009	Active surge in 1953, crevassed surface
40	1HD06012	Potholes
41	1HD06020	Potholes
42	1HD06030	Active surge between 1953-1964, steep terminus in 1953, potholes
43	1HD06035	Chasms, looped medial moraine, extensive dead ice area
44	1HD06041	Looped medial moraine, extensive dead ice area, potholes?

No.	Glacier code	Diagnostic surge features and comments
45	1HD06042	Potholes, tributary glacier cut off, looped medial moraine, complex moraine
46	1HD06046	Potholes, proglacial naled
47	1HD06048	Potholes, looped medial moraine
48	1HD06051	Potholes, looped medial moraine
49	1HD06053	Extensive dead ice area, looped medial moraine?
50	1HD06055	Potholes, looped medial moraine
51	1HD07013	Potholes, crevassed surface
52	1HD07018	Active surge between 1931-1953? Potholes, looped medial moraine
53	1HD07028	Looped medial moraine, tributary glacier cut off, proglacial naled, extensive dead ice area, two proglacial moraines, crevassed surface
54	1HD07040	Potholes, undulated surface, looped medial moraines, complex of three glaciers where at least one is categorized as surge-type
55	1HE01024	Looped medial moraine, extensive dead ice area, chasms
56	1HE01036	Active surge between 1931-1953? Extensive dead ice area, looped medial moraine, proglacial naled
57	1HE01038	Active surge between 1953-1964. Potholes, chasms
58	1HE06001	Tributary glaciers cut off, extensive dead ice area
59	1HE08010	Looped medial moraine
60	1HE09002	Looped medial moraine, extensive dead ice area, proglacial naled
61	1HE09014	Chasms, extensive dead ice area
62	1HE09042	Extensive dead ice area superimposed upon a lateral moraine, which is formed during a presumed surge event of glacier 1HE09044
63	1HE09044	Potholes, extensive dead ice area, tributary glaciers cut off, proglacial naled
64	1HE09050	Crevassed surface, looped medial moraine?
65	1HE09051	Potholes, chasms, tributary glacier cut off, rapid recession

Table 1.(Continued)

No.	Glacier code	Diagnostic surge features and comments
66	1HE09062	Looped medial moraine, proglacial naled, extensive dead ice area
67	1HE09083	Looped medial moraine, potholes, proglacial naled, chasms, looped end moraine
68	1HE09089a	Looped medial moraine by glacier 1HE09090, crevassed surface, proglacial naled, potholes?
69	1HE09090	Potholes, looped medial moraine, chasms, proglacial naled
70	1HE09094	Potholes, looped medial moraine by glacier 1HE09095, surge could be triggered by glacier 1HE09095?
71	1HE09095	Potholes, proglacial naled, extensive dead ice area, looped medial moraine
72	1HE09105	Extensive dead ice area
73	1HE10005	Looped medial moraine
74	1HE10007	Looped medial moraine by glacier 1HE10005, surge could be triggered by glacier 1HE10005?
75	1HE11013	Potholes, chasms, proglacial naled

RESULTS AND DISCUSSION

Table 1 shows a list of the surge-type glaciers on Disko Island classified in this study. Eight glaciers have experienced surge events since 1953, three glaciers advanced between 1931 and 1953, which are likely to have been surge events, and 64 glaciers are inferred as surge-type glaciers due to one or more significant surge characteristics. Out of the 247 glaciers larger than 1 km^2, 30% are classified as surge-type glaciers. Of the previous 24 glaciers classified as surge-type, six (glaciers 1HB15018, 1HB15039, 1HD06026, 1HE01011, 1HE01015, 1HE01010) are now reclassified as normal flow glaciers. There is no evidence of glaciers, which have experienced two or more surge events since 1931. This indicates that the surge cycle is relatively long compared to other surge clusters in the world. The quiescent phase is likely to be 100 years or longer, whereas the active surge phase is about 2-3 years. The number of surge events per decade seems to have declined through the second half of the 20th-century. This is in concordance with observations from Svalbard (Dowdeswell et al., 1995), and may be a consequence of climatic amelioration resulting in decrease in glacier extent, higher equilibrium line altitude, and small glaciers may change their thermal regime from warm-based or polythermal to entirely cold-based due to decreased ice thickness.

Yde and Knudsen (2007) compared physical parameters of surge-type and normal flow glaciers larger than 1 km^2 on Disko Island. The results showed that surge-type glaciers are larger (12.0 km^2) than normal flow glaciers (3.6 km^2), indicating that surge-type glaciers comprise 59% of glacierized area. Their mean terminus altitude is nearly similar (620 m a.s.l.) as for normal flow glaciers (670 m a.s.l.). Between 1953 and 2005, the mean recession rate of surge-type glaciers was about 16 m yr^{-1}, and as much as 20 m yr^{-1} if only quiescent glaciers are included. In comparison, normal flow glaciers experienced a mean recession rate of about 8 m yr^{-1} during the second half of the 20th-century. This has implications when recession rates are related to climate change. The current positions of the terminus of surge-type glaciers reflect the magnitude of their last surge events and the time elapsed since the last surge events rather than climatic responses.

All surge-type glaciers on Disko Island are located on basaltic rocks. Although basaltic rocks constitute the dominant lithology, it is likely that basalts increase the potential for glacier surging as both Nuussuaq peninsula and the East Greenland surge cluster are situated on basalts. More information on enhanced geothermal heat flux or subglacial groundwater drainage within cracks in the basalt beds or in the intercalated layers of laterite soil residue will elucidate the significance of the presence of a basalt substratum on glacier surging.

Most surge-type glaciers are found in the large central valleys. Complex moraine systems, old moraine ridges and dead ice areas are often found on the valley floors, and in most cases they can be associated with one or more surge-type glaciers. In Kuganguaq (catchment 1HE09), 13 glaciers show signs of past surge activity, but none have surged since 1953. Twelve surge-type glaciers are identified in Stordalen/Kussuaq (catchment 1HD06) of which two have surged since 1953. Ten glaciers are classified as surge-type in Kuannersuit Kuussuat (catchment 1HB15). There seems to be no surge-type glaciers among the small maritime glaciers on the western- and southern-most parts of the island. Below is a description of four of the largest surge-type glaciers on Disko Island, and their informal names are applied.

Kuannersuit Glacier (1HB15017)

Between 1995 and 1998 this outlet glacier, which descends from Sermersuaq ice cap into Kuannersuit Kuussuat, experienced a 10.5 km surge advance with maximum velocities of more than 70 m d^{-1} (Figure 2). This is one of the longest terrestrial frontal advances ever recorded, and several aspects of the consequences of the surge event have been described (Gilbert et al., 2002; Yde et al., 2003, 2005a,b; Yde and Knudsen 2005a,b; Knudsen et al., 2007; Roberts et al., 2009; Yde, Chapter 5). About 3 km^3 of ice was moved from the reservoir area to the newly formed glacier tongue, significantly increasing the extent of the ablation area to constitute about 80-90% of the total area. The glacier surface became heavily crevassed and undulated, and a looped medial moraine was moved several kilometres down-glacier, but survived the surge event. The glacier overrode a pingo and proglacial naled assemblages, and subglacial sediment was entrained within several shear planes. During the initial quiescent phase rapid thinning lowered the glacier surface behind the debris-covered glacier terminus. Lateral recession commenced immediately after the termination of the surge events with about 10 m yr^{-1}, and chasms formed along the margins above the main subglacial channels. In 2005 when the glacier was last visited, a series of evolving chasms caused subsidence across the glacier tongue up-glacier of the uppermost shear plane. This will eventually lead to detachment of the entire debris-covered glacier terminus.

Figure 2. The central part of Kuannersuit Glacier (1HB15017) in July 2002, looking north towards the surge reservoir area. The glacier surface is pitted with numerous potholes, and the lateral margins are crevassed. The glacier is moving towards the right.

Jost (1940) visited Kuannersuit Glacier on an expedition in May 1913. At that time the glacier was located at about 1 km up-valley from its 1998 post-surge position. Jost (1940) was very intrigued by the glacier recession of more than 5 km between his observations in 1913 and surveys for the first local topographic map in 1931-33 (GID, 1941), suggesting a recession rate of at least 250 m yr^{-1}. He described an at least 1.5 km long ice-gorge crossing the glacier, which could indicate that a detachment of the glacier terminus had occurred. He also observed transverse crevasse ridges, looped medial moraines and supraglacial "craters" (i.e. potholes or chasms), and compared his observations with descriptions of glaciers in Karakoram Himalaya, making his observations some of the first scientific observations of surge-type glacier glaciomorphology.

Based on the observations of Jost (1940) and what is known about the evolution of glacier surface features after the 1995-98 surge event, it is suggested that the previous surge event occurred in year 1900 ± 5.

Sorte Hak Glaciers (1HB15029 and 1HB15031)

These glaciers are most likely identical to the one observed by Steenstrup in 1898 (Steenstrup, 1901) and described by Jost in 1913 (Jost, 1940) occupying the central part of Kuannersuit Valley. Based on field observations of end-moraines and the topographic map (GID, 1941) the glaciers maintained their largest extent until at least 1931. Since then, the glaciers have receded about 10.5 km from their outermost moraine through a steep, narrow gorge called Sorte Hak. The recession was very rapid between 1931 and 1953 with rates of about 200 m yr^{-1} and resulted in detachment of the debris-covered glacier snout, forming an extensive dead ice area with many circular dead ice lakes. Parts of the debris-cover constitute of glaciofluvial outwash sediments rather than supraglacial meltout till, indicating that bulk meltwater emanating from a glacier portal located in the Sorte Hak Gorge covered the lower part of the glaciers. In the 1970s, the confluent glacier tongue split into a northern (1HB15029) and southern branch (1HB15031), which means that the next surge event of either branch will not affect the other branch. The last surge event must have triggered while both branches were together, or both branches must have surged separately within a short period. Hence, it is suggested that the last surge event occurred during the second half of the 19th-century, when the Little Ice Age peaked.

Avdlaugissat Glacier (1HB11029)

This outlet glacier, which flows southward from Bræpasset ice cap into Avdlaugissat Valley, has the longest tongue of all glaciers on Disko Island. Its surface is covered by numerous potholes (Figure 3), which have existed for more than 50 years. One of the most interesting characteristics of Avdlaugissat Glacier is that the glacier has only receded about 2.2 km since 1931, where the terminus was at the outermost moraine. Most of this recession has occurred in recent decades, where the glacier tongue has thinned significantly. This slow recession from the maximum position suggests that the last surge event of Avdlaugissat Glacier must have been very dramatic, moving many km^3 of ice down-valley in order to make the glacier tongue thick enough to resist the enhanced ablation at low altitudes. The current recession is witnessed in the field by a marked trimline, stranded ice on the valley walls hundred meters above the current glacier surface, and mature chasms along the lateral margins. It is very likely that the recession rate will accelerate in the near future.

Figure 3. The central part of Avdlaugissat Glacier (1HB11029) in July 2003, looking west. Several potholes are found on the glacier surface. The glacier is moving from right to left.

Conclusion

This analysis of maps, aerial photographs and satellite images has revealed that more surge-type glaciers exist on Disko Island than previously estimated. A total of 75 glaciers are classified as surge-type, constituting 30% of the glaciers larger than 1 km^2 and 59% of the glacierised area. The duration of the surge cycle is 100 years or more. The quiescent phase can be divided into three periods: (1) an initial period, where the glacier terminus remains at the maximum position, (2) an intermediate period, where the debris-covered terminus is detached and the recession rate is as high as 250 m yr^{-1}, and (3) a late period, where the recession rate is slow and the mass balance is positive. Due to the relatively long duration of the surge period, climate change may affect the frequency of surge events and the evolution of the quiescent phase. Especially, large glaciers with ablation areas facing in the arc south to northwest are sensitive to climatic variations (Yde and Knudsen, 2007). In this context, it must be emphasised that the potential of glacier surging has to be considered when paleoclimatic conditions (e.g. temperature) are interpreted from glacier fluctuations and moraine positions.

Acknowledgements

The Commission for Scientific Research in Greenland (KVUG) funded this project. Arctic Station in Qeqertarsuaq/Godhavn provided local logistic support for fieldwork on Disko Island. We thank Ruth Nielsen for map design.

References

Barrand, N.E. and Murray, T. (2006). Multivariate controls on the incidence of glacier surging in the Karakoram Himalaya. *Arctic, Antarctic, and Alpine Research*, 38, 489-498.

Björnsson, H., Pálsson, F., Sigurdsson, O. and Flowers, G.E. (2003). Surges of glaciers in Iceland. *Annals of Glaciology*, 36, 82-90.

Clarke, G.K.C., Schmok, J.P., Ommanney, C.S.L. and Collins, S.G. (1986). Characteristics of surge-type glaciers. *Journal of Geophysical Research*, 91(B7), 7165-7180.

Colvill, A.J. (1984). Some observations on glacier surges, with notes on the Roslin glacier, East Greenland. In: Miller, K.J. (editor). *The International Karakoram Project, volume 1.* Cambridge: Cambridge University Press. 64-75.

Copland, L., Sharp, M.J. and Dowdeswell, J.A. (2003). The distribution and flow characteristics of surge-type glaciers in the Canadian High Arctic. *Annals of Glaciology*, 36, 73-81.

Dolgoushin, L.D. and Osipova, G.B. (1975). Glacier surges and the problem of their forecasting. *IAHS Publications*, 104, 292-304.

Dowdeswell, J.A., Hodgkins, R., Nuttall, A.-M., Hagen, J.O. and Hamilton, G.S. (1995). Mass balance change as a control on the frequency and occurrence of glacier surges in Svalbard, Norwegian High Arctic. *Geophysical Research Letters*, 22, 2909-2912.

Dowdeswell, J.A. and Williams, M. (1997). Surge-type glaciers in the Russian High Arctic identified from digital satellite imagery. *Journal of Glaciology*, 43, 489-494.

Evans, D.J.A. and England, J. (1991). High Arctic thrust block moraines. *The Canadian Geographer*, 35, 93-97.

Evans, D.J.A. and Rea, B.R. (1999). Geomorphology and sedimentology of surging glaciers: a land-systems approach. *Annals of Glaciology*, 28, 75-82.

Evans, D.J.A. and Rea, B.R. (2003). Surging glacier landsystem. In: Evans, D.J.A. (editor). *Glacial landsystems*. London: Arnold. 259-288.

Geodetic Institute of Denmark (GID). (1941). *Topographic map 1:250.000. Map no. 69 V.1 Godhavn*. Copenhagen, Geodetic Institute of Denmark.

Gilbert, R., Nielsen, N., Möller, H., Desloges, J.R. and Rasch, M. (2002). Glacimarine sedimentation in Kangerdluk (Disko Fjord), West Greenland, in response to a surging glacier. *Marine Geology*, 191, 1-18.

Hagen, J.O. (1988). Glacier surge in Svalbard with examples from Usherbreen. *Norsk Geografisk Tidsskrift*, 42, 203-213.

Hagen, J.O., Liestøl, O., Roland, E. and Jørgensen, T. (1993). Glacier atlas of Svalbard and Jan Mayen. *Norsk Polarinstitutt*, 129, 141 pp.

Hamilton, G.S. and Dowdeswell, J.A. (1996). Controls on glacier surging in Svalbard. *Journal of Glaciology*, 42, 157-168.

Henriksen, N. and Watt, W.S. (1968). Geological reconnaissance of the Scoresby Sund Fjord complex. *Grønlands Geologiske Undersøgelse rapport*, 15, 72-77.

Higgins, A.K. and Weidick, A. (1988). The world's northernmost surging glacier? *Zeitschrift für Gletscherkunde und Glazialgeologie*, 24, 111-123.

Humlum, O. (1987). Glacier behaviour and the influence of upper-air conditions during the Little Ice Age in Disko, central West Greenland. *Danish Journal of Geography*, 87, 1-12.

Humlum, O. (1999). Late-Holocene climate in central West Greenland: meteorological data and rock-glacier isotope evidence. *Holocene*, 9, 581-594.

Humlum, O., Christiansen, H.H., Hansen, B.U., Hasholt, B., Jakobsen, B.H., Nielsen, N. and Rasch, M. (1995). Holocene landscape evolution in the Mellemfjord area, Disko Island, central west Greenland: area presentation and preliminary results. *Danish Journal of Geography*, 95, 28-41.

Jiskoot, H., Boyle, P. and Murray, T. (1998). The incidence of glacier surging in Svalbard: evidence from multivariate statistics. *Computer Geosciences*, 24, 387-399.

Jiskoot, H., Murray, T. and Boyle, P. (2000). Controls on the distribution of surge-type glaciers in Svalbard. *Journal of Glaciology*, 46, 412-422.

Jiskoot, H., Pedersen, A.K. and Murray, T. (2001). Multi-model photogrammetric analysis of the 1990s surge of Sortebræ, East Greenland. *Journal of Glaciology*, 47, 677-687.

Jiskoot, H., Murray, T. and Luckman, A. (2003). Surge potential and drainage-basin characteristics in East Greenland. *Annals of Glaciology*, 36, 142-148.

Jost, W. (1940). Gletscherschwankungen auf der Insel Disco in Westgrönland. *Zeitschrift für Gletscherkunde, für Eiszeitforschung und Geschichte des Klimas*, 27, 20-28.

Kamb, B., Raymond, C.F., Harrison, W.D., Engelhardt, H., Echelmeyer, K.A., Humphrey, N., Brugman, M.M. and Pfeffer, T. (1985). Glacier surge mechanism: 1982-1983 surge of Variegated Glacier, Alaska. *Science*, 227, 469-479.

Knudsen, N.T., Yde, J.C. and Gasser, G. (2007). Suspended sediment transport in glacial meltwater during the initial quiescent phase after a major surge event at Kuannersuit Glacier, Greenland. *Danish Journal of Geography*, 107, 1-7.

Mock, S.J. (1966). Fluctuations of the terminus of the Harald Moltke Bræ, Greenland. *Journal of Glaciology*, 6, 369-373.

Murray, T., Strozzi, T., Luckman, A., Pritchard, H. and Jiskoot, H. (2002). Ice dynamics during a surge of Sortebræ, East Greenland. *Annals of Glaciology*, 34, 323-329.

Murray, T., Strozzi, T., Luckman, A., Jiskoot, H. and Christakos, P. (2003). Is there a single surge mechanism? Contrasts in dynamics between glacier surges in Svalbard and other regions. *Journal of Geophysical Research*, 108(B5), 2237, doi:10.1029/2002JB001906.

Olesen, O.B. and Reeh, N. (1969). Preliminary report on glacier observations in Nordvestfjord, East Greenland. *Grønlands Geologiske Undersøgelse rapport*, 21, 41-53.

Pedersen, A.K. (1977). Tertiary volcanic geology of the Mellemfjord area, south-west Disko. *Grønlands Geologiske Undersøgelse rapport*, 81, 35-51.

Pedersen, A.K., Larsen, L.M., Ulff-Møller, F., Pedersen, G.K. and Dueholm, K.S. (2001). 1:100.000 69 V.2 N Pingu. Copenhagen: Geological Survey of Denmark and Greenland, [Geological map].

Piasecki, S., Larsen, L.M., Pedersen, A.K. and Pedersen, G.K. (1992). Palynostratigraphy of the lower tertiary volcanics and marine clastic sediments in the southern part of the West Greenland Basin: implications for the timing and duration of the volcanism. *Grønlands Geologiske Undersøgelse rapport*, 154, 13-31.

Post, A. (1969). Distribution of surging glaciers in western North America. *Journal of Glaciology*, 8, 229-240.

Post, A. and LaChapelle, E.R. (2000). *Glacier ice*. Washington: IGS. 145 pp.

Pritchard, H., Murray, T., Strozzi, T. Barr, S. and Luckman, A. (2003). Surge-related topographic change of the glacier Sortebræ, East Greenland, derived from synthetic aperture radar interferometry. *Journal of Glaciology*, 49, 381-390.

Reeh, N. Bøggild, C.E. and Oerter, H. (1994). Surge of Storstrømmen, a large outlet glacier from the Inland Ice of North-East Greenland. *Grønlands Geologiske Undersøgelse rapport*, 162, 201-209.

Rignot, E. and Kanageratnam, P. (2006). Changes in the velocity structure of the Greenland Ice Sheet. *Science*, 311, 986-990.

Roberts, D.H., Yde, J.C., Knudsen, N.T., Long, A.J. and Lloyd, J.M. (2009). Ice marginal dynamics and sediment delivery mechanisms during surge activity, Kuannersuit Glacier, Disko Island, West Greenland. *Quaternary Science Reviews*, 28, 209-222.

Rucklidge, J. (1966). Observations of hollows in the snow surface of Torv Gletscher, East Greenland. *Journal of Glaciology*, 6, 446-449.

Rutishauser, H. (1971). Observations on a surging glacier in East Greenland. *Journal of Glaciology*, 10, 227-236.

Steenstrup, K.J.V. (1901). Beretning om en undersøgelsesrejse til øen Disko i sommeren 1898. *Meddelelser om Grønland*, 24, 249-306.

Sturm, M. (1987). Observations on the distribution and characteristics of potholes on surging glaciers. *Journal of Geophysical Research*, 92, 9015-9022.

Sturm, M. and Cosgrove, D.M. (1990). An unusual jökulhlaup involving potholes on Black Rapids Glacier, Alaska Range, Alaska, U.S.A. *Journal of Glaciology*, 36, 125-126.

Thorarinsson, S. (1969). Glacier surges in Iceland, with special reference to the surges of Brúarjökull. *Canadian Journal of Earth Sciences*, 6, 875-882.

Weidick, A. (1984a). Location of two glacier surges in West Greenland. *Grønlands Geologiske Undersøgelse*, 120, 100-104.

Weidick, A. (1984b). Studies of glacier behaviour and glacier mass balance in Greenland – a review. *Geografiska Annaler*, 66A, 183-195.

Weidick, A. (1988). Surging glaciers in Greenland – a status. *Rapport Grønlands Geologiske Undersøgelse*, 140, 106-110.

Weidick, A., Bøggild, C.E. and Knudsen, N.T. (1992). Glacier inventory and atlas of West Greenland. *Grønlands Geologiske Undersøgelse*, 158, 194 pp.

Woodward, J., Carver, S., Kunzendorf, H. and Bennike, O. (2002). Observations of surge periodicity in East Greenland using molybdenum records from marine sediment cores. In: Dowdeswell, J.A. and Ó Cofaigh, C. (editors). *Glacier-influenced sedimentation on high-latitude continental margins*. London: Geological Society, Special Publications, 203, 367-373.

Yde, J.C. Periglacial landforms and processes on Disko Island, Greenland. In: Krugger, M.I. and Stern, H.P. (editors). *New Permafrost and Glacier Research*. Nova Science Publishers, 131-145.

Yde, J.C. and Knudsen, N.T. (2005a). Glaciological features in the initial quiescent phase of Kuannersuit Glacier, Greenland. *Geografiska Annaler*, 87A, 473-485.

Yde, J.C. and Knudsen, N.T. (2005b). Observations of debris-rich naled associated with a major glacier surge event, Disko Island, West Greenland. *Permafrost and Periglacial Processes*, 16, 319-325.

Yde, J.C. and Knudsen, N.T. (2007). 20th-century glacier fluctuations on Disko Island (Qeqertarsuaq), Greenland. *Annals of Glaciology*, 46, 209-214.

Yde, J.C., Knudsen, N.T. and Gasser, G. (2003). Oxygen isotope studies of bulk meltwaters at a surge-type glacier. *Proceedings of the Northern Research Basins 14th International Symposium and Workshop, Kangerlussuaq/Sdr. Strømfjord, Greenland, August 25-29*. 209-216.

Yde, J.C. and Knudsen, N.T., Larsen, N.K., Kronborg, C., Nielsen, O.B. Heinemeier, J. and Olsen, J. (2005a). The presence of thrust-block naled after a major surge event: Kuannersuit Glacier, West Greenland. *Annals of Glaciology*, 42, 145-150.

Yde, J.C., Knudsen, N.T. and Nielsen, O.B. (2005b). Glacier hydrochemistry, solute provenance, and chemical denudation at a surge-type glacier in Kuannersuit Kuussuat, Disko Island, West Greenland. *Journal of Hydrology*, 300, 172-187.

INDEX

A

B

C

D

E

F

G

H

M

N

O

T

U

V

W

Y